The Politics of Energy Forecasting

The Politics of Energy Forecasting

A Comparative Study of Energy Forecasting in Western Europe and North America

Edited by

THOMAS BAUMGARTNER AND ATLE MIDTTUN

CLARENDON PRESS · OXFORD
1987

Oxford University Press, Walton Street, Oxford OX2 6DP
Oxford New York Toronto
Delhi Bombay Calcutta Madras Karachi
Petaling Jaya Singapore Hong Kong Tokyo
Nairobi Dar es Salaam Cape Town
Melbourne Auckland
and associated companies in
Beirut Berlin Ibadan Nicosia

Oxford is a trade mark of Oxford University Press

Published in the United States
by Oxford University Press, New York

British Library Cataloguing in Publication Data
The Politics of energy forecasting: a comparative study of energy forecasting in Western Europe and North America.
1. Power resources—Forecasting—Social aspects
I. Baumgartner, Thomas II. Midttun, Atle
333.79'12 HD9502.A2
ISBN 0–19–828547–7

Library of Congress Cataloging in Publication Data
The Politics of energy forecasting.
Includes index.
1. Energy industries—Forecasting—Case studies.
2. Power resources—Forecasting—Case studies.
3. Energy policy—Case studies. I. Baumgartner, Thomas (Thomas Martin) II. Midttun, Atle, 1952–
HD9502.A2P66 1987 333.79 86–23885
ISBN 0–19–828547–7

Set by Promenade Graphics, Cheltenham, Glos.
Printed in Great Britain by
Butler and Tanner Ltd, Frome, Somerset

Preface

The plan for this book was conceived during a lunch discussion at the end of the Second Energy & Society Conference in September 1982 in Dubrovnik. On this occasion, we presented a preliminary version of the Norwegian case. We discovered that a number of conference participants shared our perspective and had similar experiences to relate. No doubt the combined effects of the Dubrovnik sun, shaded terraces, spicy food, and heavy Dalmatian wine helped to make us aim for a book that would be sober and a little subversive at the same time. The book would counterbalance the conventional technical approach to energy modelling and forecasting, and it would concentrate on the modelling, forecasting, and decision-making processes instead of on the final products, the models, and the forecasts. It would provide the basis for a broader societal appraisal of the pros and cons of model-supported policy-making. David Crossley, Reinier de Man, and Arne Kaijser had joined us in that afternoon's discussion.

We set out to find authors able and willing to cover the topic in a number of important OECD countries by using the common perspective proposed by us that has now evolved into chapter 2. Our sample was to be socially and politically diverse, and had to face a variety of energy problems. We were looking for critical insiders to the energy modelling and forecasting scene, who were also fluent in social science theorizing. This search took its time, the more so as we did not have much to offer in the way of financial rewards. But we were successful beyond our initial hopes, as proven by this book.

It was natural that Tormod Lunde and Atle Midttun should complete their Norwegian case. Reinier de Man used the insights into Dutch energy policy that he had gained while at Leiden University to put together a Dutch study. He luckily also agreed to research the British case by drawing on some very preliminary work done by Steve Thomas from SPRU at Sussex University. Steve unfortunately could not continue with the project, due to other demands on his time. Reinier's British analysis is the only one based on a look from the outside in, and not to its detriment, we are happy to report.

David Crossley, unfortunately, had to drop a planned Australian case. His appointment as Co-ordinator to the Victoria Energy Plan precluded engagement in such an extra-curricular activity. But David put us in contact with Cliff Hooker from Newcastle University, who had been working on Canadian energy policy issues from a 'Lovinian' perspective. Cliff, in turn, drew in John B. Robinson from the University of Waterloo. John had extensive experience with energy modelling focusing on scenario and energy service analyses, an experience that has now led him to attempt an ambitious backcasting exercise.

John also introduced us to Bill Keepin, then already at the Beijer Institute in Stockholm. He had just completed writing down his inside analysis of the IIASA world energy model. He had reached conclusions quite similar to our own. Luckily, we managed to persuade Bill and Brian Wynne from the University of Lancaster, both then still associated with IIASA, to summarize and integrate their two long, detailed, and complementary studies of the IIASA mega-modelling adventure.

In France, we contacted Louis Puiseux, who, with his long service in the Department of Economic Research of Electricité de France and his subsequent work on French energy policy, was just the right inside-outsider to describe to us the events in a country that does not follow exactly the same energy path as the other countries in our sample.

We knew our German contributors, Hans Diefenbacher and Jeffrey P. Johnsen, from previous comparative research on energy problems. We were pleased that they agreed to write a comprehensive overview of the West German developments. Both had been involved in the energy work of IFEU, the Protestant Research Institute in Heidelberg, and the Institute of Applied Ecology in Freiburg—all three institutes major actors in the nuclear energy debate and the development of alternative energy scenarios. Our bet on committed objectivity has fully paid off.

We had originally planned a mini-comparison between Scandinavian countries. Arne Kaijser from Linköping University was unfortunately prevented from exploiting his inside knowledge, and we could not convince any of our other contacts in Sweden to engage in a similar task. But we were lucky to come across Knud Lindholm Lau, working on political issues for Danish Radio, who had just finished his thesis on Danish energy forecasting. However, it required some patience, leavened with bouts of social pressure on our part, to drag this busy journalist away from current events and have him sit down at the typewriter in order to condense his findings for us.

The United States was a real challenge to us. The volume and variety of energy modelling and forecasting activity made surveying this scene a daunting task for anyone. We finally found our way to Martin Greenberger at the University of California (Los Angeles), who had been a pioneer in comparative energy model assessment. We were pleased to have him join William Hogan from Harvard University, founding director of the Energy Modelling Forum and a former head of analysis with the Federal Energy Administration. They had just published a highly original study of the energy (modelling) decade in the United States.

We have tried to achieve a certain unity while maintaining the originality of each case. A discussion of preliminary versions of some of the cases took place in London in June 1984. Then followed personal visits by the editors to the European contributors, extensive correspondence, and sometimes even heavy editing on our part. We therefore accept the blame for any remaining weaknesses and errors, while originality remains the responsibility of each one of the authors. We would like to thank them very much for their enthusiastic support for our project and their patient response to our often heavy demands.

A number of persons have contributed substantially to making this a better book. Besides those contacted individually by our co-authors, we would especially like to thank Maja Arnestad, Jørgen Randers, and Charles Stabell from the Norwegian School of Management, Tom R. Burns from Uppsala University and Reinhard Ueberhorst, former chairman of the German parliamentary commission on energy forecasting, for their valuable comments and suggestions. We gratefully acknowledge the financial support from the Norwegian Research Council for Science and the Humanities which covered some of our co-ordination expenses. The Norwegian School of Management has several times provided a congenial environment for us editors to meet. Its Text Centre, with its very competent staff, has been of great help in getting this book to the printing stage. Finally, but not least, we would like to thank our families, who have gracefully accepted our often lengthy absences and periods of intense preoccupation with this book.

Contents

Notes on the Contributors xi

Part I Theory and Praxis of Energy Modelling

1 Energy Forecasting: Science, Art, and Politics
Thomas Baumgartner and Atle Midttun 3

2 The Socio-Political Context of Energy Forecasting
Thomas Baumgartner and Atle Midttun 11

Part II Pitfalls of Mega-Modelling

3 The Roles of Models—What Can We Expect from Science? A Study of the IIASA World Energy Model
Bill Keepin and Brian Wynne 33

Part III Negotiated Futures

4 Energy Forecasting in West Germany: Confrontation and Convergence
Hans Diefenbacher and Jeffrey Johnson 61

5 The Dutch Energy Scenario Game: Corporatist Search for Consensus
Reinier de Man 85

6 United Kingdom Energy Policy and Forecasting: Technocratic Conflict Resolution
Reinier de Man 110

Part IV Élites and Professionals among Themselves

7 Electricity Forecasting in Norway: Administrative Centralism
Tormod Lunde and Atle Midttun 137

8 Electricity Forecasting in Denmark: Conflicts between Ministries and Utilities
Knud Lindholm Lau 155

9 The Ups and Downs of Electricity Forecasting in France: Technocratic Élitism
Louis Puiseux 180

Part V **Market Competition**

10 Future Imperfect: Energy Policy and Modelling in Canada: Institutional Mandates and Constitutional Conflict
John B. Robinson and Clifford A. Hooker 211

11 Energy-Policy Modelling in the US: Competing Societal Alternatives
Martin Greenberger and William W. Hogan 241

Part VI **Comparative Notes on Politics and Methodology**

12 Energy Forecasting and Political Structure: Some Comparative Notes
Thomas Baumgartner and Atle Midttun 267

13 Modelling and Forecasting in Self-Reactive Policy Contexts: Some Meta-Methodological Comments
Thomas Baumgartner and Atle Midttun 290

Index 309

Notes on the Contributors

Thomas M. Baumgartner is a research consultant based in Zurich. He has economics degrees from the Universities of Zurich and New Hampshire. His current research interests are in energy and the environment, inflation, technology, social accounting, and the concept of actor-system dynamics.

Hans Diefenbacher is working at the Protestant Institute for Interdisciplinary Research (FEST) in Heidelberg, doing research in the areas of economy, development, and industrial democracy. He has studied economics and holds degrees from the Universities of Heidelberg and Kassel. In the past, he has done energy research at the Institute of Applied Ecology in Freiburg (FRG).

Martin Greenberger is Professor of Public Policy and Analysis at the UCLA Graduate School of Management, where he holds the IBM Chair in Computers and Information Systems. While associated with the Electric Power Research Institute he initiated a number of research projects leading to the establishment of the Energy Modelling Forum at Stanford University, the Energy Model Assessment Program at MIT, and the Utility Modelling Forum for electric companies.

William W. Hogan is Professor of Political Economy and Director of the Energy and Environmental Policy Center at Harvard University's John F. Kennedy School of Government. He holds the IBM Chair of Technology and Science. Previously, he was Deputy Assistant Administrator for Data and Analysis at the Federal Energy Administration. While professor at Stanford University, he was founding director of the Energy Modelling Forum.

Clifford A. Hooker is Professor of Philosophy at the University of Newcastle, Australia. Originally a research physicist, he has published extensively in philosophy, philosophy of science, and foundations of physics. He taught for a decade at the University of Western Ontario, Canada, where he began research on energy and environmental questions, co-authoring *Energy and the Quality of Life* (1980, University of Toronto Press). He is now developing retrospective energy methodologies into a general approach to public policy-making.

Jeffrey P. Johnson is presently working on energy questions for the Gesellschaft für Organisation, Planung und Ausbildung (GOPA) in Bad Homburg, West Germany. He has studied at California State University (Hayward) and the University of California (Berkeley). He has a diploma in economics from Heidelberg University. He has done research on soft energy paths at the Institute for Energy and Environmental Research

(IFEU) in Heidelberg and the Institute for Applied Ecology in Freiburg.

Bill Keepin is presently a Research Fellow at the Centre for Energy and Environment at Princeton University. He has degrees in physics and mathematics from the Universities of New Mexico, Colorado, and Arizona. After research in quantum physics, mathematical sociology, differential equations, and nonlinear dynamics, he joined the Energy Systems Program at the International Institute for Applied Systems Analysis (IIASA), switching later to IIASA's Institutional Settings and Environmental Policies project. He then worked for a short period at the Beijer Institute in Stockholm.

Knud Lindholm Lau is a journalist with the Department of Cultural Affairs at Danish Radio in Copenhagen. He is there working on questions of energy and technology, and is also engaged in experimental radio journalism. He has a master's degree in the social sciences from Aarhus University.

Tormod Lunde is Assistant Professor at Columbia University and holds degrees in sociology from Oslo University and Stanford University. He participated in the Energy and Society Project at Oslo University, where he began his research on energy. His other interests include sociological theory and technology.

Reinier de Man is currently a researcher at the Rotterdam School of Management. He studied chemistry at the State Universities of Utrecht and Groningen, then started to get involved in energy policy research at the Graduate Management Institute in Delft and the Working Group on Energy and Environment at Leiden State University. He also worked as Research Co-ordinator at the Institute for Systems Analysis in Amsterdam.

Atle Midttun is currently a researcher at the Department of Energy and Industrial Policy of the Norwegian School of Management in Oslo. Energy policy, organization theory, and institutional economics are his main fields of interest. He has a degree in sociology from the University of Oslo, where he also worked as a researcher with the Energy and Society Project. Subsequently, he did energy research with the Resource Policy Group in Oslo.

Louis Puiseux is presently Study Director at the École des Hautes Études en Sciences Sociales in Paris. He was for many years head of the forecasting, and then the futurology division within the Economic Research Service of Electricité de France. He has been writing critically on French nuclear energy policy since 1973.

John B. Robinson is Associate Professor, Co-ordinator of Energy Theme in the Department of Man–Environment Studies, and director of the Waterloo Simulation Research Facility at the University of Waterloo

(Canada). He has been working in the area of methodological and philosophical dimensions of energy modelling and forecasting. Currently, he is engaged in research on backcasting approaches to societal modelling and in co-ordinating an international collaborative research project that has developed out of the work done for this book.

Bryan Wynne is Reader in Science Studies and Director of the Centre for Science Studies and Science Policy at the University of Lancaster (UK). Previously, he was a Research Leader of the Institutional Settings and Environmental Policies Project at IIASA in Laxenburg (Austria). He has published a book on nuclear decision-making, *Rationality and Ritual: The Windscale Inquiry and Nuclear Decisions in the UK* (British Society for the History of Science, 1982).

PART I

Theory and Praxis of Energy Modelling

1

Energy Forecasting: Science, Art, and Politics

Thomas Baumgartner and Atle Midttun

MODELLING and forecasting of energy development has become a large 'industry' in the wake of the oil crisis of 1973–74. This development is part of a larger trend towards increased use of modelling and forecasting for planning and decision-making. The steady increase in computer size and the rapid fall in computing costs have revolutionized this business. Econometric models have become large and complex constructions.

The IIASA study of the global energy system (described in chapter 3 of this book), which began in 1973 and reached publication stage in 1981, is a case in point. It represented 225 person-years of effort, involved more than 140 scientists, and cost approximately $10 million. One run of the core model required the specification of some 1,600 constraint variables and 2,600 activity variables.

It should come as no surprise that these 'mega-models' have become so complex that nobody really understands them. We have all heard of the back-of-an-envelope corrections that even reputed forecasting organizations have been forced to make to the output of their models when they turn out impossible or simply implausible results. Even modellers and forecasters themselves are becoming outsiders. Thompson (1984) points out that the average stay of energy modellers at IIASA was less than seven months, and that in the end the memory of even the most senior researcher did not extend back for more than four or five years of this ten year project.

No wonder that modellers now in many instances require computer programs to extract the structure of these complex models (Gilli, 1979; Ritschard, 1980). Only in this way can one obtain an idea of the degree of endogeneity of variables, of implicit feedback linkages and loops, and of sub-model integration. Such analyses can reveal quite surprising structural model features. They also show how small specification changes may have significant consequences for model structure.[1]

The complexities involved in building and running energy models, and the nature of the scientific process underlying them, seem to justify fully the question raised by Sir James Ball in his opening address to the Fourth International Symposium on Forecasting:[2] where do the forecasting artists get their inspirations from? This is a rather important question. After all, Sir James suggested that up to fifty per cent of forecasting work is art and not science. Unfortunately, he did not attempt to begin to answer his own question. A search of the literature does not lead much further. Model builders and forecasters, when asked about it, all readily agree that their trade is not fully scientific and that they revert to *post hoc* adjustments, use intuition when choosing data, modify variables, and bend equations so that reasonable results do emerge from their travail. But in official presentations, they fall back on a discussion of discounted least squares, adaptive growth curves, Kalman Filters, ARIMA Models, the Poisson Smoothing Process, and so on.[3]

The artistic element in modelling and forecasting is, however, only part of the problem. Exactly because there is art in this business, there is also an opening for societal influences: what we have called, in shorthand, the politics of modelling and forecasting. Artists are somehow inspired for their work. Often they themselves are less than clear about their Muses, and in most cases they have a hard time giving a verbal account of those Muses' influence on their work. We know that scientists similarly have problems explaining the factors that lead them to choose one approach over another, that lead them to the triumphant finding of a solution. The explanations given are often *post hoc* rationalizations of a process that remains obscure.

These societal influences are many, and they become more important the greater the part of art in this scientific endeavour of modelling and forecasting. The lengthening of the time horizon of forecasting—up to fifty years in the case of the IIASA study or the scenario analysis undertaken by the West German Parliament—and the consideration of societal changes increase the weight of such influences. Those societal influences can range from the individual—a scientist's values, interests, and ideology—to the sociological—professional orientation, group dynamics, the need of forecast users to justify their hunches and gambles—and to the political—the acceptability of assumptions and policies that are implied in the modelling.

Furthermore, the art of forecasting must also find its beneficiaries. Modellers have to secure finance for their increasingly expensive efforts. As in all such situations, the supply of models and forecasts normally adjusts to the demand for them. That is, those that have the money can choose among competing models and modellers and influence the application of their artistic talents. In many areas big business is the buyer of models and their products. In the energy field, important buyers are the government and a range of commercial or semi-commercial organizations that try to influence government policy in the energy area.

Partly because of their attachment to political and commercial interests, and partly because of their alleged attachment to scientific ideals of objectivity, forecasters have tended to underscore the institutional and ecological constraints of their *modus*. By leaving important goals and political values implicit in forecasting models, many forecasts have therefore projected short-term growth rates into the long-term future without critical examination. By neglecting also the wider ramifications of the energy sector development in other societal sectors, revolutionary societal change has in fact been projected into the future as business as usual. The seemingly trivial assumption of constant growth rates masks an exponential development with inconceivable consequences in the long run (Meadows *et al.*, 1972). The discussion of such revolutionary energy developments without considering their economic and societal ramifications is at best politically naïve, and at worst leads nations into societal crises.

Most energy forecasts around 1970 made predictions of this exponential type and still continued to do so after the energy crisis, to accommodate the expansive production plans of different energy sectors. The challenge from environmentalists, and later on the stagnation of western economies, led to considerable downward revisions throughout the decade. However, this revision has not been universal, as illustrated by the IIASA world energy study (Häfele, 1981) and the Odell/Rosing oil forecasts (1983). In spite of the fact that these forecasts are prepared with quite different resources and methodologies, they are astonishingly similar: both predict, or assume the possibility of, a radiant energy future of linear growth, even accelerating after 2000. They also abstract from ecological and institutional constraints.

The conflicting energy forecasts both at the national and global levels are—besides their scientific content—also expressions of a political negotiation process: a process that has been transferred to a meta-level, where political, administrative, and industrial interests compete for cognitive and methodological hegemony over the definition of the energy future through modelling.

The political influences—and we stress that we mean here 'politics' in the large sense given it above—do not operate on the cognitive level alone. Our case studies document many instances when people forced or tried to force modellers and forecasters to adopt specific models, methods, and values when preparing their forecasts.

There are, of course, real difficulties involved in choosing the appropriate values for certain key variables in forecasting models. In part, knowledge about an area such as energy was simply not very substantive when it suddenly became an issue of great political importance—as happened with energy after 1973. Furthermore, the dramatic change in energy development, the very reason for energy's prominence as a political issue, also implied changes in the values of key variables that had remained stable for a long time in the past—and which, therefore, had been assumed to be constant. People can honestly differ in the evaluation of the probability of change and what this change is likely to be, qualitatively and quantitatively. Yet small changes in the values of crucial variables can have significant policy effects. To use the IIASA case again as an example: only a minor adjustment of the uranium price and the coal extraction limit changes the future energy scenario dramatically from a nuclear to a coal future (see chapter 3 for further details).

Choices obviously have to be made in the face of uncertainty. Informed people can obviously and honestly differ in their opinions about the likelihood of this or that event occurring, at what moment, with what strength, and in connection with what other events. The difficulty arises from the persistent attempt—through neglect, professional indoctrination, or Machiavellian design—of the modelling and forecasting profession and organizations that administer the forecasts to hide this reality by not acknowledging it. This is often done by failing to document the decisions that lead up to the model and its product, and by letting decision-makers and buyers of forecasts use them over and over again to push through decisions with the argument that scientific (that is, objective) evidence suggests that this is an appropriate, if not the only, course of action.

In this way forecasting contains a policy bias that is not explicitly accounted for. Leaving this policy bias implicit enables the modeller and forecaster to avoid considering the probability that:

(1) the economic and social structures, institutions and behaviours underlying the past development will break down, evolve and possibly conflict in the future;
(2) the forecast energy development itself will set in motion economic, social, and political processes that will feed back into policy changes, thus invalidating the basis on which the forecast was made in the first place.[4]

Thus most forecasting has an inherently conservative bias, embedded in even the most 'policy-distant' way of modelling and forecasting, a bias that for several years made the forecasting establishment incapable of dealing adequately with the dramatic changes in energy development throughout the 1970s.

All our cases exhibit falling energy forecasts throughout the 1970s; sometimes after considerable mobilization of political opposition and counter expertise; sometimes after conflicts within government administration. These conflicts indicate that the falling economic growth rates are only part of the explanation of the falling energy forecasts. Furthermore, the correspondence between the national energy forecast adjustments and their respective growth rate development is only partial. In any case, economic development has to be interpreted before it is included in modelling. And such interpretation is particularly controversial and uncertain in cases of major change. The events surrounding the 1973–4 oil crisis illustrate the problem of scientifically evaluating unexpected change and the likelihood of political prejudices being involved in models in such situations.

For a long time the meaning and the consequences of the oil crisis of 1973–4 were far from clear. There were those who argued that OPEC's cohesion could not be maintained and that oil prices were shortly going to fall back to the levels of the early 1970s. Others were less certain of the soundness of this argument from the economic theory of cartels and resolved that it would be best if the industrialized countries helped speed up this return to 'normal' oil price levels by sending the Marines into the Gulf countries (Tucker, 1975). The sharp recession that some countries experienced in 1975–6 was therefore seen as a short-term event without any significance for future economic and energy-consumption growth. Other countries tried to use Keynesian demand-management tools to overcome the economic recessionary effects of the oil price increase, and were, like France, only forced into recessionary policies to combat inflation after the occurrence of the second oil price shock in 1979–80.

Looking back, it is striking how modellers and forecasters of traditional 'establishment' forecasts received and put down as unscientific the first zero-growth or low-energy growth forecasts and scenarios when they appeared, for example in the UK, Germany, and the Netherlands. These forecasts had all assumed lower economic growth rates, in part because their proponents were more likely than the establishment to believe in the possibility of prolonged economic stagnation in the industrialized countries, in part because they desired zero economic growth. Similarly, they did not have a political commitment to energy-intensive industries and therefore, for example, did not have to accept unquestioned the official view of the glorious future for the British steel industry that the establishment forecasters had to adopt. Furthermore, the alternative forecasters did not have to discover the economic and energy implications of alternative life styles because they were beginning to live them or were positively welcoming their widening appeal. In sum, their political values and cognitive models allowed them to recognize economic and socio-cultural developments that only much later were to become established official truths.

By entitling our book *The Politics of Forecasting* we intend to suggest that although past establishment modelling and forecasting have been curiously apolitical, they cannot escape the political consequences of this neglect. Because they failed to consider political and institutional facts concerning the production and use of energy, these forecasts did not recognize that the continuous unbroken expansion of the energy system would ultimately generate its own political seeds of destruction: the revolt of the oil producing countries once the concentration of reserves in their countries had shifted power in their favour; and the emergence of alternative and ecological aspirations and life styles in response to material saturation and the creation of life-inimical conditions.

The energy turbulence of the 1970s indicates the necessity for a supplementary institutional/political approach to forecasting and modelling. This could help modellers make better guesses and be more aware of their assumptions in situations where:

(1) many relations can no longer be inferred from past events. New relations emerge, old relations have changed in magnitude and quality.
(2) Exogenous variables become open to public choice, and therefore start to become endogenous variables, endogenous to the modelling forecasting/decision-making system itself.
(3) New variables and relationships are beginning to emerge, or seem to acquire an importance for mastering the energy problem. But it is not quite clear what they are, whether they are important, and how they are going to link up to the already existing system.
(4) New knowledge and data are required, but do not exist or are only slowly being acquired
(5) These new problems and contexts may require new methods. But their scientific value is still contested or questioned exactly because it is difficult to foresee what the methods' implications for energy forecasts are going to be.

By modelling the modelling process and its political and institutional environment, we are taking a first step towards a programme of including institutional factors in the production and presentation of societal modelling; although, as we point out in the final chapter, we do not believe that this can be easily accommodated within the existing forecasting paradigm.

Moreover, we believe that the increased role of forecasting and long-term planning has important consequences for political democracy. There is today a tendency for the development of key societal sectors to be left in the hands of technocratic/administrative élites to decide on the basis of long-term forecasts. If the political values and biases embedded in these forecasts are never fully exposed, and never become the subject of political discussions, future societal development becomes closed to political choice.

The analysis contained in this book is not intended to be exhaustive. The goal is rather to start by examining a relatively neglected area of forecasting. In doing so we believe we demonstrate the need for further analysis and a more detailed treatment for the broader issues we raise.

The following discussion of these issues is divided into six parts. Part I, including this and the next chapter, introduces the issues we are concerned with and some of our main concepts and analytical tools used in dealing with them.

Parts II to V present a number of empirical studies of energy forecasting. The study of the IIASA World Energy Model in Part II presents various methodological problems related to large-scale energy modelling along with a discussion of some institutional and political implications.

Parts III to V present eight cases of national developments in energy modelling in major western industrial countries. These cases are grouped according to characteristic features of the institutionalization of energy forecasting and political decision-making on energy futures, features that range from élitism via political negotiation to market competition.

The final section, part VI, gives a comparative analysis of the variety of political and institutional conflicts and solutions around energy forecasting found in our case studies. In this section, we also draw some institutional and methodological implications from our analysis.

Notes

1. DeVillé (1982) illustrates with a simple model the concepts of 'autonomy' and 'dependency' using such an approach. DeVillé (1984) and Boutillier (1984a and 1984b) look at econometric models.
2. Organized by the International Institute of Forecasters at the London Business School, 8–11 July 1984. Sir James Ball is a professor at the London Business School and one of Britain's major contributors to the development of macroeconomic forecasting.
3. These terms are taken from the titles and abstracts of papers presented at the first session of the above mentioned forecasting symposium.
4. This has, among other things, led to the inability to foresee the emergence of OPEC and the shift of power in its favour brought about by redistribution of known oil reserves, and the recent change in oil markets brought about by Saudi Arabian strategies to discipline its OPEC colleagues and to prevent non-OPEC countries from gaining new market shares. Similarly, such forecasting led the nuclear power and electricity industries in many countries to believe that the production of electricity could be expanded several-fold without encountering socio-political resistance. Correcting this forecasting error would require the development of multi-level modelling (Burns *et al.* 1985: ch. 9; Baumgartner *et al.* 1985: ch. 10).

Bibliography

Barr, H. (1984), 'Electricity Forecasting for New Zealand'. Paper presented at the 4th International Symposium on Forecasting, London, 8–11 July.

Baumgartner, T., T. R. Burns, and P. DeVillé (1986), *The Shaping of Socio-economic Systems*. London: Gordon and Breach.

Boutillier, M. (1984a), 'Reading Macroeconomic Models and Building Casual Structures', in J. P. Ancot (ed.), *Analysing the Structure of Economic Models*. The Hague: Martinus Nijhoff.

—— (1984b), 'The Concept of Reading as Analysis of Macroeconomic Models', in T. Basar and L. F. Pau (eds.), *Dynamic Modelling and Control of National Economics*. Oxford: Pergamon.

Brunner, G. D., R. D. Brunner, and R. F. McNown (1984), 'Editors' Introduction: The IIASA Energy Study'. *Policy Sciences*, 17: 197.

Burns, T. R., T. Baumgartner, and P. DeVillé (1985), *Man, Decisions, Society*. London: Gordon and Breach.

Chateau, B. (1985), 'La prévision énergétique en mutation?' *Revue de l'énergie*, 370 (January), 1–11.

DeVillé P. (1982), 'Dependent Development: A Multilevel System Approach with

Reference to Mexico's Development', in R. F. Geyer and J. van der Zouwen (eds.), *Dependence and Inequality*. Oxford: Pergamon, 165–84.
—— (1984), *Marché du travail et modèles macro-économique*. Ch. 3, 'Quéstions et méthodes'; ch 4, 'Les modèles belges'. Université Catholique de Louvain: Institut des Sciences Économiques, Rapport CIPS.
Gilli, M. (1979), *Étude et analyse des structures causales dans les modèles économiques*. Bern: Peter Lang.
Häfele, W. (1981), *Energy in a Finite World: A Global Systems Analysis*, vols. i and ii. Cambridge, Mass: Ballinger.
—— and H. H. Rogner (1984), 'A Technical Appraisal of the IIASA Energy Scenarios? A Rebuttal'. *Policy Sciences*, 17: 341–65.
Keepin, B. (1984), 'A Technical Appraisal of the IIASA Energy Scenarios'. *Policy Sciences*, 17: 199–276.
Meadows, D. (1981), 'A Critique of the IIASA Energy Model'. *The Energy Journal*, 2: 17–28.
—— *et al.* (1972), *Limits to Growth*. Washington DC: Potomac Associates.
Odell, P. R. and K. E. Rosing (1983), *The Future of Oil*. Erasmus University, Rotterdam: Centre for International Energy Studies, Eurices Paper 83–1a.
Ritschard, G. (1980), *Contribution à l'analyse des structures qualitatives des modèles économiques*. Bern: Peter Lang.
Robinson, J. B. (1982), 'Apples and Horned Toads: On the Framework and Determined Nature of the Energy Debate'. *Policy Sciences*, 15: 23–45.
Thompson, M. (1984), 'Among the Energy Tribes: A Cultural Framework for the Analysis and Design of Energy Policy'. *Policy Sciences*, 17: 321–39.
Tucker, R. (1975), 'Oil: The Issue of America Intervention'. *Commentary*, 59: 21–31.
Wynne, B. (1984), 'The Institutional Context of Science, Models and Policy: The IIASA Energy Study'. *Policy Sciences*, 17: 277–320.

2

The Socio-Political Context of Energy Forecasting

Thomas Baumgartner and Atle Midttun

INTRODUCTION

INCLUDING institutional factors in the production and presentation of societal modelling essentially involves engaging in a self-reflective process on behalf of the modelling community. We must turn our attention away from the energy scenarios or the direct modelling output to the system that generates the assumptions which determine these scenarios. As Thompson (1984) points out, we simply cannot give a scientific answer to the question of which output is correct. Most of the energy scenarios presented in this book are within the limits of what could be called physically and technologically feasible futures. In this sense they could all be correct.

Social and institutional linkages and political cultures influence assumptions about nature and society: assumptions that in turn underlie the cognitive and normative framework which produces the variety of competing models. In this perspective, models are systematic codifications of cognitive and social structures, developed by actors to promote their interests and/or world views.

Understanding the modelling process therefore also fundamentally implies understanding the structure and process of political systems that surround it.

By contrasting the simplistic representations of the relationship between model and society found in, for example, the social-engineering perspectives with the more complex social-systems perspective proposed here, we wish to draw attention to some key elements of our analysis.

Forecasting models in the social-engineering perspective are, to express it in a simplified manner, part of an almost mechanistic regulatory system (see Figure 2.1). The decision-making element of a system is provided with information from a model representation of the system (or part of it). An adequate data base is presumed to exist. And the regulator itself is somehow unaffected by the system. In this way the social-engineering perspective draws a picture of reality as if:

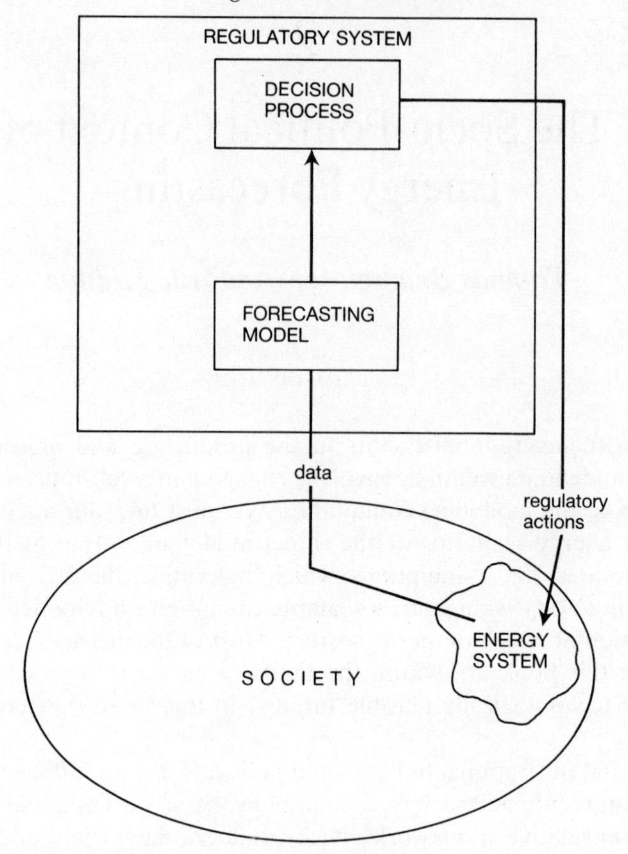

Fig. 2.1 Forecasting model in a simple regulatory system

(1) there is only one 'correct' model of society possible;
(2) knowledge of social processes is sufficient to formulate objective laws that could serve as a basis for exact prediction of the future;
(3) data are available to map such processes in detail;
(4) political and administrative organs are operating under consistent and stable goal-structures, and are able to pursue long-term policies without fear of losing legitimacy;
(5) modelling and forecasting are purely technical activities, applying well-specified norms and rules, relying on perfect data, and free from commercial and political considerations.

Research in the social sciences as well as past experience with modelling in several countries contradict these assumptions. By neglecting serious political analysis, this perspective underestimates the complexity and the conflicts between interest groups, institutions, and ideologies that characterize modern societies, and it overestimates the authority and effectiveness of administrative and political organs. A more realistic model of forecasting activities and its societal context is, by necessity, more complex.

First of all, as we have already argued, there does not exist a 'correct' model of society. Mental and formal models are always products of social interaction. They reflect the interests of actors and their experiences (see Berger and Luckmann, 1976; Bråten, 1973). The controversies of energy modelling and forecasting in most western countries have pointed this out and have suggested the need to include other perspectives in the planning of our energy futures.

Secondly, the political and administrative systems of modern societies are highly diversified and often pursue contradictory policies (Bacharach and Baratz, 1970; Nadel and Rourke, 1975). In fact, in modern political science and sociology, public authorities are seen more as representatives of sector-interests than as a unified regulatory system (Schmitter and Lembruch, 1979; Olsen, 1980; Hernes, 1978). The interplay between sector-interests and energy forecasting and decision-making thus becomes a central theme for research.

Thirdly, knowledge of social processes is limited and to a large extent defies exact quantification and, indeed, often registration too. This is partly due to the creative and problem-solving character of human actors, which implies self-transforming potentialities and tendencies (Burns *et al.*, 1982). Basic premisses for reaching rational decisions in an objective sense as well as objective modelling are thus lacking.

Determining energy futures is partly a process of democratic decision-making where modelling and forecasting may play a number of roles in the democratic process. Depending on how modellers present future possibilities and constraints, they may in fact shape the future. By stimulating development in certain directions and excluding other development paths, forecasts may be self-fulfilling, or may lead to reactions which are counter to its assumptions.

The critique of the social-engineering perspective, mentioned above, calls for an alternative analytical framework that is able to include the political and institutional implications of forecasting and to discuss methodological questions with reference to their societal implications. By addressing three core issues concerning forecasting and its societal context we hope to outline the contours of such an alternative analytical framework.

1. The first issue is how and why data selection and technical/methodological questions concerning modelling and forecasting have political aspects. The short history of energy modelling has given numerous examples of political debate linked to controversial methodological choices. The gradual inclusion of price elasticities in energy demand forecasting in the 1970s, for instance, opened up new political options for

energy policy. The determination of these elasticities has ever since been subject to considerable dispute in many countries.

2. The second issue concerns the interplay between modelling and its institutional and professional environment. Why are professions and institutions crucial in determining forecasting output? How do professionalization and organizational patterns provide information-selecting filters and channels of influence that affect forecasting output? Some of our cases show that forecasting may in fact be linked to an institutional system and a dominant professional paradigm in such a way that forecasts and decisions mutually support each other, turning the forecast into what is virtually a self-fulfilling prophecy.

3. The third issue is the interplay between forecasting and political decision-making in general. What is the political function of forecasts? How do forecasts influence the maintenance of political power and the legitimation of political decisions? In several countries the opponents of current energy policy have in fact accused the established élites of using forecasting mainly as a tool for political legitimation, with less regard to the informative and explorative aspects of forecasting. The energy establishment, in turn, has accused the ecological movement of irresponsible and unrealistic forecasts.

In the following, we shall discuss each of these issues in some detail, referring to ways in which they affect the social-engineering perspective of modelling. Finally, we shall consider the aggregated implications of our analysis and its relation to the general questions of political authority and democratic participation.

POLITICAL AND INSTITUTIONAL ASPECTS OF METHODOLOGY

Social systems are so complex that models of them can only represent a few parts or aspects. Information has to be selected, if not created. Model builders have to select key features for modelling; they have to use aggregated and averaged data; and they have to make assumptions about the development of key inputs into the system, unless, of course, somebody else provides them with this information. It is especially clear, in the latter case, that implicit and explicit political considerations may determine the values assumed for key variables such as the economic growth rate and price and exchange rate developments.

Methodological choices tend to be formulated in technical language and discussed within a technical frame of reference. But this should not divert our attention from seeing the 'invisible hand' of political interests at work.

The causal relation between politics and methodology is reciprocal. On the one hand methodological choices have political and institutional consequences through their determination of forecasting results and through the authority of forecasts in societal planning and decision-making. On the other hand, political and institutional conditions also influence the choice of methodology, for instance through the selection of forecasting organizations and professions, and also through the power to control methodological choices by political and institutional actors.

The interaction between politics and methodology can be seen at two levels: (1) in the choice of data and estimation of parameters; (2) in the choice of modelling methodologies.

The choice of data bases and parameters

At the level of choice of data and estimation of parameters apparently technical questions, such as the determination of the time period used for estimating functions, and the choice of values for supposedly stable parameters, are often highly politically loaded. This is especially true when forecasts are based on weak theoretical knowledge and there are conflicting interests attached to them.

Many methodological choices are, in fact, weakly founded in social theory. The determination of energy price elasticities and energy coefficients of GNP is a case in point. For a long time the elasticities and coefficients of the 1960s were thought to hold even in the changed circumstances of the 1970s and 1980s. In the case of price elasticities, this choice even clearly went against theoretical arguments which suggest that the elasticities for small price changes at low price levels cannot be used to calculate the consequences of large price changes at high price levels.

This preference for the assumption of continuity probably reflects the cognitive limitations of a generation of modellers as well as the wishes of dominant institutional interests. A presentation in the form of forecasts, however, easily has the effect of undermining arguments about changing systems, system crises, the need to push for alternatives, and so on. Because forecasts also influence behaviour, such assumptions may have self-fulfilling consequences.

Economists, today's dominant modelling profession, have achieved an (apparent) precision in their theoretical and analytical statements through the use of the *ceteris paribus* clause and ideal types such as the free trade market. This is often done without due consideration of the realism and consequences of this practice. Thus behavioural and institutional conditions that implicitly determine model structures and functional estimates are seldom investigated and even less often made public—a fact that leads to an unrealistic reification of forecasts and a neglect of their political assumptions.

Institutional conditions, for example, have a direct bearing on the choice of the period used for estimating the functions of the model. Statistical reasons suggest that the longest possible time series are best. Yet this means that important structure-level changes are averaged out of the estimates. It also means that changes towards the end of the period are likely to be swamped by the stability of the earlier parts of the period (Burns *et al.*, 1985). All our case studies of energy forecasting in fact show that the change from the stability of the 1960s to the turbulence of the 1970s was nowhere recognized by establishment forecasters. Therefore, it took a number of years of learning before new developments were assimilated. In fact the persistent opposition of environmental groups was a major factor in this assimilation.

This shows that even the interpretation of the past calls for a careful institutional analysis before estimates are undertaken. As suggested by Armstrong (1978), in situations that seem to change, models have to be estimated with the help of short time series, whereas stable situations can be analysed with a longer time perspective.

Choice of modelling methodologies

Political considerations are more obviously involved in the choice of models than in the collection of data and choice of parameters. Models necessarily have to simplify reality, but in the simplification process certain assumptions have to be made, assumptions which later on limit the themes that can be raised and questions that can be asked about reality. Choice of forecasting methodology and models therefore usually restricts the variety of perspectives possible.

Different schools in modelling may disclose very different repertoires of data collection, of basic assumptions, theoretical structure, and estimation techniques, which give great variety in analytical perspectives. This variety may be used strategically by institutional actors to further their sector interests.

Our Danish case (chapter 8), for instance, shows how the choice of a macro-economic versus a sector methodology (that is a top down versus a bottom up perspective) was the source of considerable dispute between the utilities and the Ministry of Energy before the publication of energy forecasts in 1981.

The present focus and methodology of the dominant econometric models with their neglect of organizational factors probably prevent understanding of the political/institutional dynamics which, for instance, follow a certain energy development. These also lead to lack of insight into political solutions that presuppose reorganization or organizational innovation (see Koreisha and Stobaugh, 1979).

Against econometric models it may furthermore be argued that forecasting models which are heavily anchored in the past and the present, neglect our freedom of political choice. If one instead went about energy modelling the other way round, that is by depicting one or more desired future

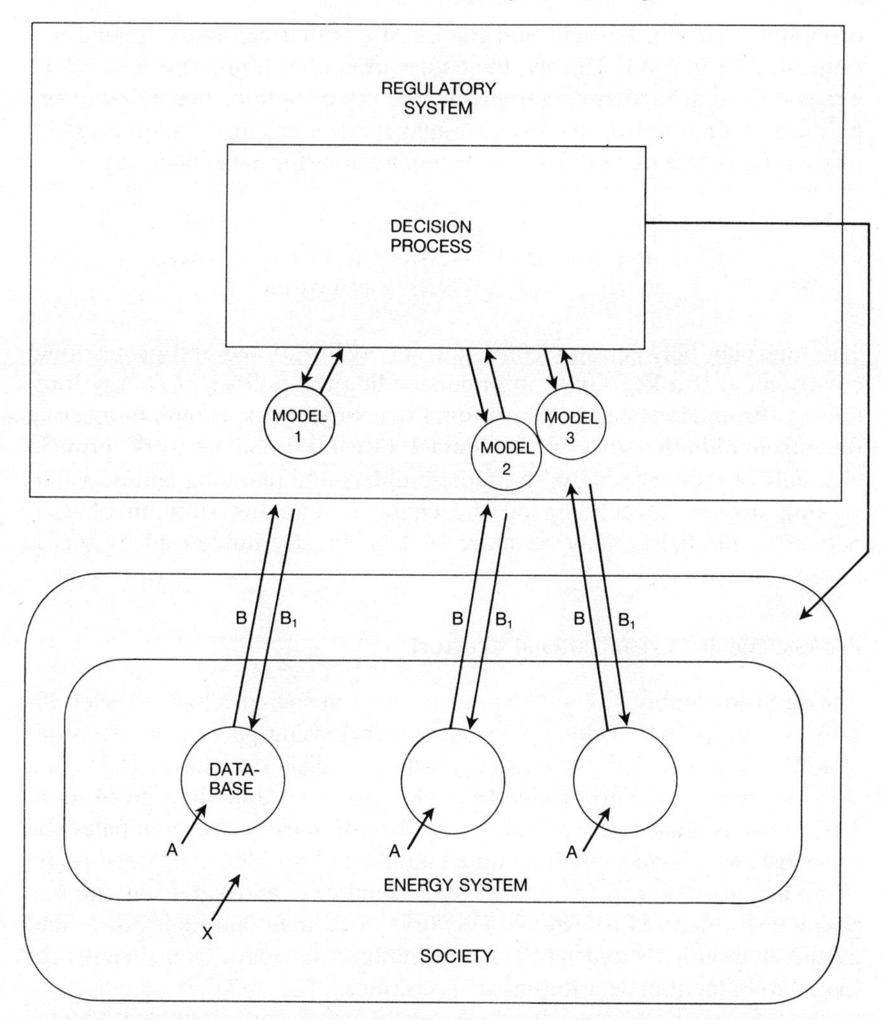

Fig. 2.2 Forecasting and political aspects of methodological choices

states and then sketching out how they could be realized—using impact analysis—one would convey a very different and open future. It is not surprising that many 'alternative' forecasters who went for a more ecological energy future wished to take this approach to energy modelling.

The above discussion of the politics of forecasting methodology implies that we have to complicate the sketch of the forecasting model in a simple regulatory system presented in Figure 2.1 (see Figure 2.2). First, the definition of the energy system to be modelled is highly dependent on the modeller's judgement (as indicated by link X). Secondly, the selection of data-bases is also open to judgement; this is a severe constraint on the type of models that can be built and the questions that can later be analysed (indicated by link A). Thirdly, the choice of models represents a choice of perspective that further constrains the variety of options presented to the political and administrative decision-makers (see link B). As indicated by link B_1 the choice of models also has implications for data collection.

FORECASTING, PROFESSIONAL STRUCTURE, AND ORGANIZATIONAL NETWORKS

The interplay between modelling and its organizational and professional environment is a key factor in understanding the politics of energy forecasting. Professions and organizational structures work as filters, selecting the information that goes into the model. Organizational networks provide channels of influence between model-builders and planning and decision-making organs. Depending on the character and the strength of such networks, modelling may be more or less directly influenced by sector interests in society.

Professional and organizational structure

The need for highly qualified labour in modern societies has led scientifically trained people to take over an ever-increasing part of the decision-making apparatus. This tendency towards specialization also brings about a situation where similarly trained people come to dominate a given area. The result is that more and more sectors of society are dominated by specialists who have a common understanding of problems, proceed in the same way to define them and develop solutions, and agree on suitable policy instruments. Professional networks often maintain their power and cognitive monopoly efficiently by preventing outsiders from being promoted to influential decision-making positions.

The long period of specialized training, which ensures that its graduates approach the problems they are faced with according to the tenets of their specialized science, is after all a heavy investment made by the professional, an investment in knowledge acquisition which creates in him a great concern to avoid fundamental changes that would threaten his cognitive monopoly.

The linkage between professional cognitive monopolies and organizational resources creates networks of great societal influence. The past influence of the legal profession in government administration and the present influence of macro-economists in the planned economies of many European countries are good examples. In the latter case the integration of data collection through statistics bureaux, with economic training and theorizing in economic institutes and planning and implementation through finance ministries, provided an iron triangle that in many cases exercised decisive influence on political decisions.

It is not surprising, given this situation, that macro-economic professionals and models have also come to dominate the modelling and planning of the energy area. Nor is it surprising that ecological perspectives have had a hard time challenging this entrenched and influential modelling tradition.

Organizational networks and self-fulfilling prophecies

Modelling societal systems presents different problems from modelling technical or natural systems (Schulze, 1978). In the former case, only the action of individuals, social groups, and organizations can bring about the outcome forecast by the models. Actors and their behaviour are therefore important input factors in systems development; but they are seldom well represented in the models. This also goes for the model-builders and forecasters, and the decision-makers and their relationship to the executing actors.

As a consequence, institutional linkages and feedback loops between forecasting, planning, and policy implementation are systematically underestimated. This becomes pertinent as energy forecasting in many countries is undertaken by organizations that also are energy producers or have strong links to energy producers. Through tariffs, pricing policies, infrastructure investments, and so on, forecasting organizations that are also energy suppliers can work to fulfil their prophecies.

The current crisis of energy forecasting can thus be seen just as much as a breakdown on the implementation side as a failure of correct prediction. The systematic expansion of energy consumption in the 1960s and early 1970s may be seen in this light. The link between growth-orientated organizations and forecasting milieux both in terms of cognitive and organizational structure secured the implementation of expansive energy plans in response to corresponding forecasts. The resultant growth in energy consumption confirmed the forecasts and validated the modelling techniques and models used.

Due to economic development and political challenge, this system broke down on the implementation side in most western industrialized countries in the late 1970s and early 1980s. The problem was that this was not realized by forecasting and planning organizations until several years later. Their commitment to the high growth paradigm and their links to expansively orientated energy supply systems were strong.

The preceding discussion suggests that we have to complicate our analytical model of forecasting even further. We shall now include the modelling organization, the modelling professions, and the links to other organizations as important contexts (see Figure 2.3).

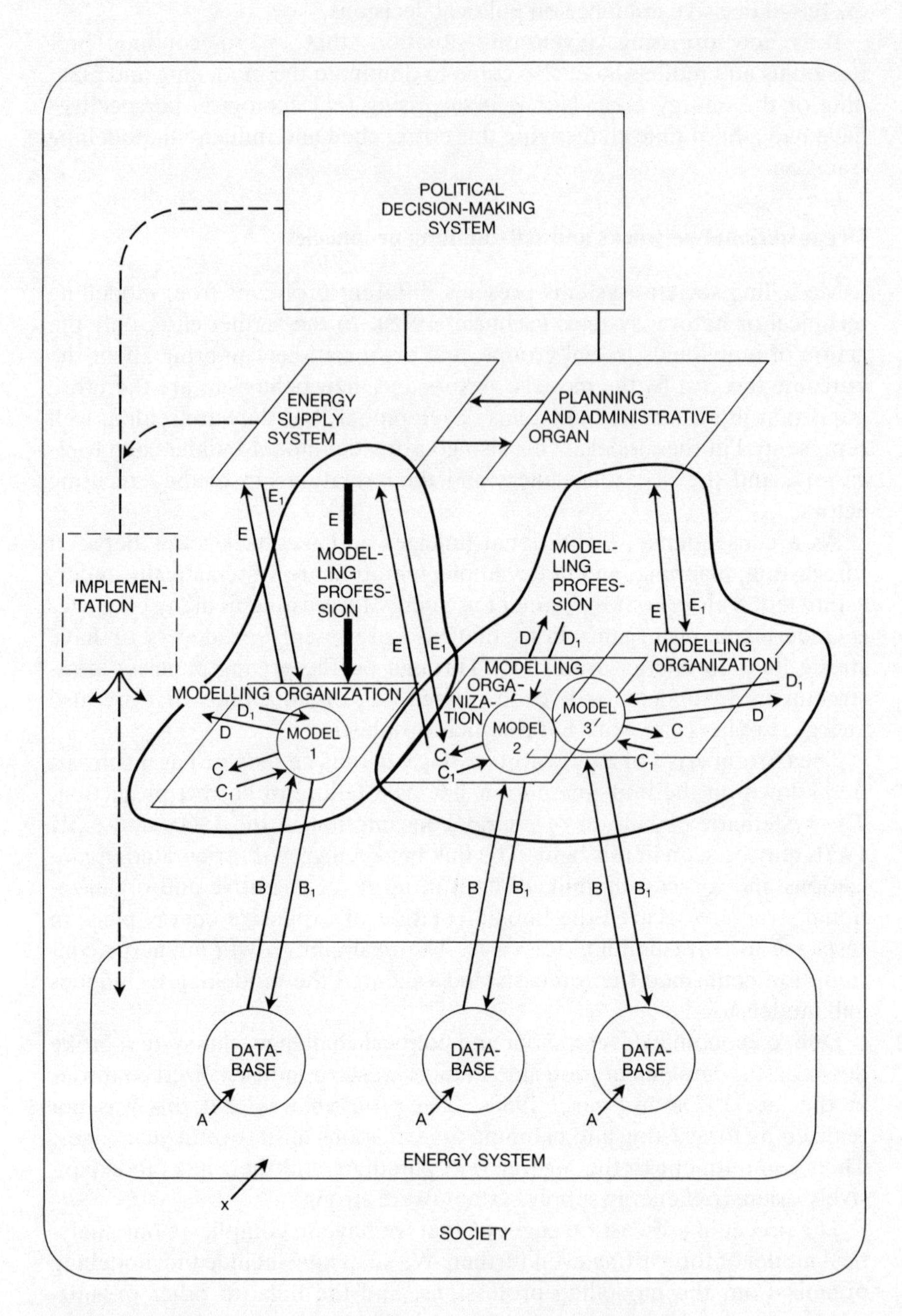

Fig. 2.3 Forecasting, professional structure, and organizational networks

C and C_1 symbolize the relations between the model and its organizational context. D and D_1 symbolize the relations between the modelling organization and its professional basis. Finally, E and E_1 symbolize the links between the modelling system and the administrative and energy supply systems that are involved in implementation.

The overlapping between the professional system and other systems indicates that this is often an important informal link between organizations. Furthermore the figure indicates the possibility of integration of models as well as of modelling organizations.

FORECASTING-MODELS IN POLITICAL DECISION-MAKING

The significance of forecasting models in societal planning finally rests on their acceptance in political decision-making. The interplay between modelling institutions and political decision-making, however, is complex. Political decision-making involves not only organizations with explicit political mandates like the national assembly, but also the upper levels of public administration and more or less organized interest groups inside and outside the public sector.

Besides their alleged purpose to contribute to rational decision-making, modelling and forecasting also have at least two important side effects. Firstly, modelling is a way of defining reality and shaping political debates. Secondly, it is also a powerful tool in legitimating political decisions.

Modelling as a way of defining reality

Modelling may be a way of monopolizing the definition of problems in a fragmented political environment. The energy issue may, for instance, concern energy pricing and competition on the world market as far as energy-intensive industry is concerned. The trade unions, however, are more concerned with employment consequences. To consumers, on the other hand, access to cheap fuel for transport and heating purposes may be of chief importance. The environmental movement may, finally, see the energy issue as a core element in the question of maintaining the ecological balance of our natural environment. Not only do different segments of society and different interest groups have conflicting and contradictory goals, but the public at large also exhibit inconsistent attitudes towards the energy issue. A recent survey in Norway indicates that people wish to limit the further expansion of energy production because they dislike the attendant environmental consequences. Yet the same people are also in favour of an expansion of energy-intensive industries and of maintaining current energy consumption levels in households (Albrechtsen and Moum, 1984).

By introducing complex modelling methodology which only very few are able to challenge, one legitimates certain problem-definitions and solution-strategies. The perspectives of dominant models are usually also the perspectives of powerful social actors.

Dominant modelling paradigms in public decision-making usually rely on support from administrative bodies as well as dominant political interests. Administrative élites often have the same professional background as the modellers. They are subject to the same professional norms and influences which, among other things, prescribe a positive attitude towards planning and the use of modern planning techniques, such as econometric modelling. These activities and the use of their products also provide challenging professional problems and interesting career opportunities.

In controlling expertise, knowledge, and information, the administration can influence the political decision-making process by both defining the problems to be dealt with and the solutions to deal with them. The integration of energy modelling into the public administrative apparatus may, however, vary according to national political and administrative structures. The case studies in the following chapters indicate that a general political characteristic like the political party structure, for instance, clearly has some bearing on the management of energy forecasting. Countries with long and stable political majorities, such as Norway and France, have escaped extra-governmental opposition. The Netherlands and partly Germany, on the other hand, with more unstable political alliances in government, have had considerable extra-governmental energy forecasting opposition.

The US policy on energy issues, *laissez-faire* as it is in many policy areas when compared to western Europe, fits in well with the general lack of political continuity due to shifting political majorities without any stable political party machine and top-level administration behind them.

On the administrative level, the degree of centralized planning traditions seems to have a bearing on the institutionalization of forecasting. Countries with strong centralized planning traditions, like Norway and France, seem to have escaped extra-governmental opposition. Countries with less developed central planning and a less uniform corps of planning professionals, like Britain and Germany, have had considerable extra-governmental opposition. The German institution of court hearings and the British public inquiry system on siting of major controversial projects, proved to be important factors in mobilizing extra-governmental political opposition and generating the need for alternative energy scenarios.

Forecasting: analysis or legitimation?

In relation to the political sphere, models and forecasting generally serve a dual function. They may serve as analytical tools to explore premisses for and consequences of policy options. Or they may serve as instruments for political legitimation. In many cases these two functions are closely interlinked. Since there are a number of model types and forecasting techniques available and the political system is complex, forecasting may have different functions in various parts of the political system.

The mixture of analytical and legitimating functions may, however, differ at various stages of the policy process. At an early stage, political actors with access to forecasting models may be in a position to try out several alternatives and to adjust assumptions and results in a continuing interactive process until an acceptable outcome is reached. At later stages the process becomes more closed, important exogenous parameters are fixed, the number and character of alternatives to be calculated are narrowed down. At this stage, the legitimating function of forecasting becomes more prominent.

The legitimating function arises from the position attributed to models in scientifically orientated societies. Reference to advanced mathematical and technical properties is an important political argument used by groups possessing a model monopoly to silence critical questions and challenges from groups without access to models. In fact, the diffusion of forecasting results from the core group that produces them to the wider public that is supposed to decide on the basis of them very often implies a transformation of uncertainty to certainty. On the way from the producers to the consumers, the traces of ambiguous data, elasticities that are difficult to estimate, and so on, are lost, and the results stand out as more and more authoritative.

Another effect of this use of models is that the conflicts and arguments can be shifted from the political to the technical field. This is one method to reduce the circle of 'legitimate' challengers by raising technical 'barriers to entry', a reduction which limits the challengers to the group of professionalized and socialized people.

A subtle effect of this shift in the debate from the political to the technical is that the discussion of problems is necessarily limited to those who are fluent in this technical discourse. Concepts from the world of modelling become necessary to understand reality. Reality is soon defined as the reality represented in the models, and no longer the reality experienced by those directly concerned. This is possibly one of the reasons why we have experienced a split in the way politics is lived and made between the professional ways of technocrats and politicians and the emotional reaction of protest and civil disobedience.

By adding a political level to Figure 2.3 we can now, in Figure 2.4, present the complete analytical model of forecasting and the interplay between modellers and their social and political contexts.

Link E indicates that modelling results are used as inputs into planning and decision processes. We are here primarily interested in the use of forecasting results in decision-making in the public sphere. The institutional and professional bias built into the models and the forecasts on the second

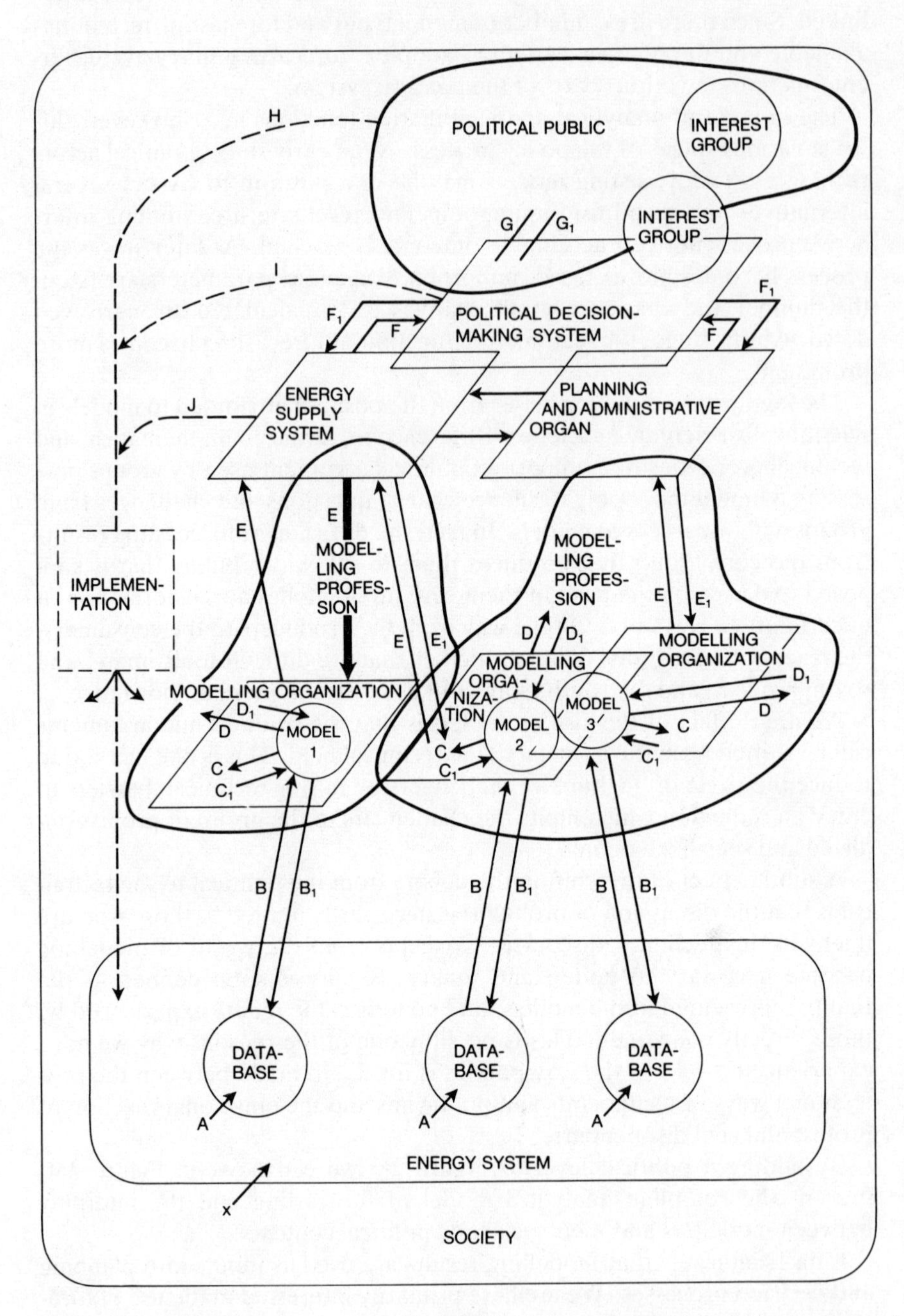

Fig. 2.4 Forecasting and the interplay between modellers and their social and political contexts

level now implies a constraint on planning possibilities on the political/administrative level. Of course, this bias is passed on by the administrative decision-makers to the political ones (link F). The bias, however, may well work the other way round (link E_1). Economists dominating relevant ministries are, for example, likely to fund, draw upon, and use modelling efforts and forecasts which are inspired by economic thinking.

The importance of forecasts in political decision-making varies with several factors that have already been discussed. In some cases, the forecast may have decisive influence over policy-making (indicated by F). In other cases policies are already spelt out and plans and forecasts are ordered and used for purposes of legitimation.

As we have already noted, political processes do not take place only within political organs. In a number of cases politically interested sections of the general public and organized interest groups have an important say. Although the political authorities are mostly able to lay down the premisses (as indicated by G), in certain cases interest groups and the public (through debate) may be able to oppose official policies so as to have a dominating influence on public policies (as indicated by G_1).

Feedback loops

H, I, and J represent feedback loops back to the social system. Depending on the level at which decisions are made, they are fed into implementation systems and have more or less impact on social reality. The link between forecasting and implementation may be weak or strong, depending on institutional characteristics. As the figure indicates, there may be overlaps between elements at several levels. Forecasting models may have common elements and utilize common data, and forecasting and planning organizations as well as professions may overlap. This overlap can indicate co-operation as well as competition.

META-POLITICAL CONSEQUENCES OF MODELLING AND FORECASTING

Forecasting not only has effects on planning and political decision-making processes, but may also have important consequences for societal development in general. Many energy technologies and many developments in energy consumption are only compatible with certain overall principles of social organization, such as small decentralized systems or large-scale centralization.

The choice of development paths for strategic resources like energy therefore often also implies choosing social structure (see Ryan, 1980; Christakis, 1973; Schurr, 1980). Choosing a growth-orientated nuclear energy future obviously has very different implications from choosing a low-growth solar energy path. How this issue is treated in the forecasting and planning systems obviously has a strong bearing on the development of the general political debate. If political attention were to be given to the linkages between different social sub-systems, energy models would have to include social variables that could more clearly depict the options available or political choice (see Greenberger *et al.*, 1976). This would be difficult to incorporate into existing models, and would probably need a development of new approaches.

This would probably also imply giving greater emphasis to normative elements in forecasting. Jantsch (1970), for example, introduces three levels in forecasting: normative, strategic, and operational. Forecasting has different functions on each of these levels. On the first two, the function is just as much to generate ideas as to give predictions. Here models are used to clarify the relationship between developments in different parts of society, given, for instance, different developments in energy demand. At this stage openness to different goals and norms is a necessary model characteristic.

However, the majority of today's forecasting models clearly belong to the operational level and are often based on administrative sector perspectives or sectorial interests. By narrowing the focus they effectively promote their sectoral perspectives and reject alternative perspectives and ideas as weakly grounded, lacking substantial support. The result is often that strong sector interests with modelling resources acquire a key role in that they succeed in laying down the premisses for political debate (Bråten, 1973). Forecasting models may thus contribute to cementing traditional development patterns and traditional problem-solving, thereby freezing our concepts of the future, and limiting the range of political options open to us (Armstrong, 1978; Greenberger *et al.*, 1976; Gershuny, 1979; Wenk, 1979).

A strong and integrated modelling, planning, and implementation system can therefore have democratic disadvantages. Such a system may raise the threshold for political participation by shutting out political movements that do not have access to models. This is particularly a problem in the case of grass-roots movements that challenge established policies and political institutions from below. If such movements represent perspectives that more adequately express the essence of new developments, there may, in fact, be a close connection between societal adaptation to new situations and the issue of democratic participation.

Our cases in fact reveal that so-called alternative energy forecasts were more able than the establishment, at an early point, to incorporate the new energy development of the late 1970s into their models. Those early low-growth scenarios in West Germany, Britain, and the Netherlands that were received as outrageous and unrealistic proved to anticipate the later development far better than the high-growth establishment predictions. On the other hand a possible future return to exponential growth in energy consumption would probably be recognized quickly by forecasters with links to industry.

An unintended, but nevertheless politically important outcome of the crisis of energy forecasting in the 1970s seems to be the delegitimation of the forecasting activity as a whole. In many of our cases, political oppositions have demonstrated that they were able to play the forecasting game just as well, if not better, than technocratic élites in more industrial and governmental positions. This outcome of the energy forecasting struggle may also, in fact, have long-term political consequences for other policy areas. The spread of computers and forecasting software furthermore promises better access to modelling technology for more people at a lower price in the future. This may make it difficult for élite groups to re-establish forecasting monopolies.

NOTES ON THE FOLLOWING CHAPTERS

The general questions raised in this and the previous chapter are discussed with reference to concrete case studies for energy forecasting in the following chapters. The following parts II to V of this book present a number of empirical studies of energy forecasting. Eight of the nine discussions are case studies of national developments in energy modelling in western industrialized countries. One is a study of the IIASA World Energy Model.

Although these studies to some extent follow a common perspective, they highlight different aspects of the politics of energy forecasting. This is partly due to the variety of energy forecasting developments in individual countries. But it is also due to the outlook and positions of the authors involved.

While some of the chapters are written by insiders who have actually participated in forecasting, such as our French author, other chapters are written by outsiders, who were studying forecasting as social scientists, such as our Norwegian writers. This range of personal experiences, of course, gives access to different types of data and inevitably affects the analytical focus. In one way this is a problem because it complicates the comparative analysis. In another way it is an asset, because it provides a variety of aspects that need to be highlighted in our analysis.

The IIASA study stands out as a special case, both because of its focus and its perspective. Both its methodological complexity and global scope make it a unique undertaking. Because of the good access of its authors to detailed documentation, it is perhaps our best documented case when it comes to the discussion of methodology and its political implications.

The three following cases of forecasting in West Germany, Holland, and Britain all highlight the political games around energy forecasts. In all these countries the official energy forecasts were challenged by outsiders, who provoked heated public debates. And the forecasting development throughout the 1970s and early 1980s was heavily influenced by political considerations.

The subsequent cases of Norway, France, and to some extent Denmark are characterized by less overt political struggles around energy forecasting. The forecasting disputes were to a large extent confined to professionals and decision-making élites within the state apparatus.

The US and Canadian cases represent countries with weaker state roles in energy forecasting and energy policy; in the case of Canada because of the unclear division of authority between the federal and the regional governments, and in the case of US because of the commitment to liberal ideology and *laissez-faire* government. Although for different reasons, the development of energy forecasting in these two countries has a certain likeness with a market process.

In the final two chapters we try to sum up the main results of our analysis. Chapter 12 concentrates on comparative political and institutional aspects. We here discuss how national political and administrative traditions have influenced energy-forecasting developments. Chapter 13 draws some methodological and institutional conclusions from our study. We here point out how the turbulence of the past decade has led to important methodological and institutional developments in energy forecasting. We discuss how this has affected the status and function of forecasting and raise some normative questions about future methods and roles of forecasting in political decision-making.

Bibliography

Albrechtsen, E. H. and S. Moum (1984), 'Vekst og vern', in K. Haagensen, and A. Midttun (eds.): *Kraftutbygging, Konflikt og aksjoner*. Oslo, Norwegian University Press.

Armstrong, J. S. (1978), *Long-Range Forecasting, From Crystal Ball to Computer*. New York, Wiley.

Bacharach, P. and M. Baratz (1970), *Power and Poverty*. New York, Oxford University Press.

Berger, P. L. and T. Luckman (1976), *The Social Construction of Reality*. Harmondsworth, Penguin.

Blackaby, F. (1979), 'Economic Forecasting', in T. Whiston (ed.), *The Uses and Abuses of Forecasting*. London, Sussex University, Science Policy Research Unit.

Bråten, S. (1973), 'Model Monopoly and Communication'. *Acta Sociologica*, 16 (2): 98–707.

Burns, T. R. *et al* (1985), *Man, Decisions, Society*. London, Gordon & Breach.

Christakis, A. N. (1973), 'A New Policy-Science Paradigm'. *Futures*, 5 (6).
Dahl, R. and C. E. Lindblom (1953), *Politics, Economics and Welfare*. New York, Chicago University Press.
Gershuny, J. (1979), 'Transport Forecasting: Fixing the Future', in T. Whiston (ed.), *The Uses and Abuses of Forecasting*. Chicago University Press.
Greenberger, M. *et al.* (1976), *Models in the Policy Process*. New York: Russell Sage Foundation.
Hernes, G. (1978), *Forhandlingsøkonomi og blandingsadministrasjon*. Bergen, Norwegian University Press.
Jantsch, E. (1970), 'Toward a Methodology for Systemic Forecasting'. *Technological Forecasting*, 1.
Koreisha, S. and R. Stobaugh (1979), 'Limits to Models', in R. Stobaugh and D. Yergin (eds.), *Energy Future*. New York, Random House.
Lindblom, C. E. (1973), 'The Science of Muddling Through', in A. Falludi (ed.), *A Reader in Planning Theory*. Oxford, Pergamon.
Nadel, M. V. and F. E. Rourke (1975), 'Bureaucracies', in F. I. Greenstein and N. Polsby (ed.), *Handbook of Political Science*, vol. v, Massachusetts, Addison-Wesley.
Nordhaus, W. D. (1977), 'The Demand of Energy: An International Perspective', in W. D. Nordhaus and R. Goldstein (eds.), *International Studies of the Demand for Energy*. Amsterdam: North Holland Elsevier.
Olsen, J. P. (1980), *Politisk organisering*. Bergen, Norwegian University Press.
Ryan, C. J. (1980), 'The Choices in the Next Energy and Social Revolution'. *Technological Forecasting and Social Change*, 16.
Schmitter, P. C. and G. Lembruch (eds.) (1979), *Trends Toward Corporatist Intermediation*. London: Sage Publications.
Schulze, D. (1978), 'On the Logical-Methodological Foundations of Forecasting'. *Technological Forecasting and Social Change*, 12.
Schurr, S. H. (1980), 'Energy in America's Future'. *Technological Forecasting and Social Change*, 18.
Tank-Nielsen, C. (1983), Paper Presented at a Seminar on Energy Models the 14th of June, arranged by the Norwegian Council for Scientific and Industrial Research.
Taylor, L. D. (1977), 'Decreasing Block Pricing and the Residential Demand for Electricity', in W. D. Nordhaus and R. Goldstein (eds.), *International Studies of the Demand for Energy*. Amsterdam: North Holland Elsevier.
Thompson, M. (1984), 'Among the Energy Tribes: A Cultural Framework for the Analysis and Design of Energy Policy'. *Policy Sciences*, 17: 321–39.
Wenk, E. jun. (1979), 'The Political Limits to Forecasting', in T. Whiston (ed.), *The Uses and Abuses of Forecasting*. Chicago University Press.

PART II

Pitfalls of Mega-Modelling

3

The Roles of Models—What Can We Expect from Science? A Study of the IIASA World Energy Model

Bill Keepin and Brian Wynne

MODELS, ANALYSIS, AND INTUITION

POLICY analysis and policy-making (or at least its justification) have come to rely to a colossal extent upon complex mathematical models. This is despite the paradox occasioned by their broadened role in policy argument, that they are used more and more, but believed less and less.[1] At first sight, the whole point about models is their formalism, which should allow mathematically rigorous consistency, discrimination, and testability to be achieved, to the benefit of policy. One large symposium on energy modelling was introduced by reference to such models as the policy response to the judicial call for greater accountability and explication of decision and inference rules in science for public policy.[2] A common (idealized) justification is that

> formal models are first, testable, and second, documented, so that assumptions are clear and you can examine the data being used. Too often in energy policy matters the assumptions being made in a judgemental statement are neither obvious nor testable, also the data cannot be accessed . . . judgemental models are models that are not open to scrutiny, their prejudices are obscured.[3]

Unfortunately, however, formal models may also lead to the opposite effect of obscuring prejudices even from their authors in a labyrinth of apparently pure, technical language.

The most rigorous attempts at formalism still require the exercise of judgement, and it is widely accepted that not all of this can be made explicit and precise. Informal, tacit knowledge is part of science and modelling. Ironically, the more elaborate models become in the attempt to fulfil expectations of formal accountability, the more their incorporated informal judgements may actually multiply, rather than diminish.

This chapter examines the scientific content of the IIASA World Energy Study and compares it with the content put forward or implied in public presentations of the study. In so doing, it points also to the gap between modelling aspirations and modelling results. This case study, therefore, sheds light on the broader phenomenon of the transformation of technical formalism into the materialization of unknown prejudices.

The IIASA energy study

The IIASA study of the global energy system began in 1973 and took seven to eight years, approximately $10 million, and 225 person-years of effort to complete. Its very scale dwarfs other efforts at energy policy analysis. It has been widely taken as the most comprehensive such analysis ever, and has apparently achieved considerable impact. It has been widely described as the most impressive, even unprecedented study of the global energy problem, 'an unprecedented, detailed analysis . . . analysing options in a quantitative, mathematical form'.[4] A US Congressman observed that it had discovered the 'objective' structure of the global energy problem, separating these from 'organizational' problems in 'an elegant and coherent system solution to a global problem', which had 'changed our image of the world and man's place in it'.[5] And several recent, major analyses of global energy and climate issues have adopted it as a definitive frame of reference,[6] thus tending prematurely to leave behind any questions over its origins and validity.

The very scale of the project has effectively become a measure of its apparent objectivity. 'More than 140 scientists participated in the study, including economists, physicists, engineers, geologists, mathematicians, psychologists, a psychiatrist, and an ethnologist. Thus it is impossible for us to hold an extreme one-sided view.'[7] An impressive network of international bodies collaborated in the project,[8] and according to its Director, Wolf Häfele, it has shaped energy policy discussions within several national and international government bodies.[9] The major publication of the project was the two-volume *Energy in a Finite World* (Häfele, 1981a), hereafter cited as *EiFW*.

The IIASA analysis combined mathematical modelling with scenario construction and informal processes of judgement to analyse over a fifty-year period the possible transition to what is taken to be a sustainable world energy system. The elements of sustainability were resource supply, excluding environmental, price, technological, or social factors. The study involved 'the design of a set of energy models that were subsequently used for developing two scenarios—the principal tool of our quantitative analysis' (*EiFW*, p. xiii). The scenarios were thus constructed with the aid of the models and were the heart of the study, from which certain key policy conclusions were drawn.

The main conclusions of the study, as reported in *EiFW* and in several summary articles, were that a transition to fast breeder nuclear reactors, centralized solar, and coal synfuels must be made, and could be achieved beyond the year 2030 if the world acted decisively now to accelerate the installation of the necessary plants. These 'robust conclusions' have been forcefully publicized by Häfele.[10]

There has been strong criticism of the substance of the conclusions and some central premisses of the study,[11] such as its lack of recognition of diverse, decentralized approaches to energy supply or of ways to reduce energy demand. There is also confusion and dispute as to the status of the models and their role in generating the scenarios and policy insights. This chapter is based on the recent methodological and technical criticisms by Keepin[12] that the models involved are analytically empty, have had no real iteration or sensitivity analysis (despite claims to the contrary), and when so tested are extremely brittle to minor changes in important variables, contradicting the claims for 'robust conclusions'.

THE MODEL SET

The original aim of the IIASA modelling effort was to link at least four main models—MEDEE, MESSAGE, IMPACT, MACRO—in a full feedback loop (see Figure 3.1).[13] This was never achieved, but some papers are, to say the least, ambiguous about the real state of implementation of this full set.

Only a brief description of the IIASA energy models will be given here, necessary for the understanding of this argument. For further detail, the reader is referred to *EiFW* and Energy Systems Program (1982). Logically, the description of a loop of consistent subscenarios could set out with any of its parts. We begin with the energy consumption model, MEDEE-2. This is a static accounting model which combines basic assumptions about population and economic growth with a large array of asssumptions about lifestyles, requirements for energy services, technical efficiencies of energy-using devices, and so on, to produce profiles of final energy demand from 1980 to 2030. In all, several thousand coefficients and parameters are required for the full specification of the fourteen regional scenarios which make up the model. The major output is a time-series projection of final energy demand by sector and fuel type. Note that this demand is not the standard 'demand curve' from economics, but rather a projection of future requirements for energy as a function of time. This is then converted to a demand for secondary energy (also a time series), the principal components of which are requirements for electricity and liquid fuels. This secondary energy demand is then furnished as an input to MESSAGE, the energy supply model.

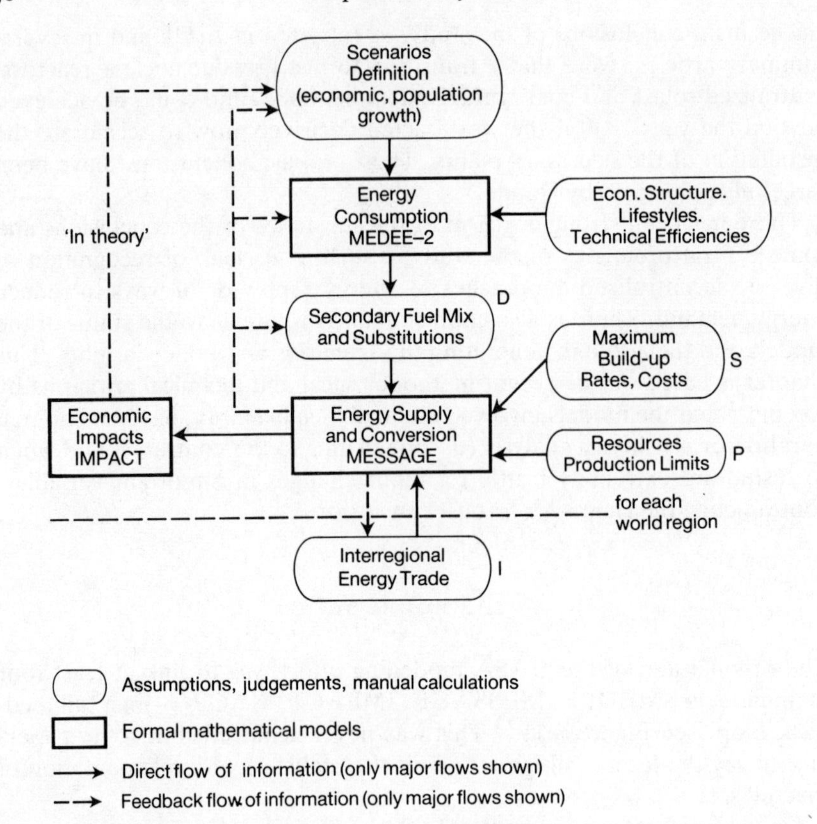

Fig. 3.1 The IIASA energy models
(Reproduced from EiFW, Figure 13.1.) The capital letters (D, demand; S, secondary; P, primary; I, imports) and the words 'In Theory' have been added as discussed in the text. The MACRO model was claimed to have been implemented in some publications (see Wynne, 1984) but was dropped from EiFW.

MESSAGE is a dynamic linear programming model that minimizes the total discounted cost of fulfilling a given secondary demand, subject to a variety of constraints on resources and technologies. Thus, under several exogenous assumptions about availability of resources, costs, and build-up of technologies, and so on, MESSAGE computes the optimal (for example, least-cost) energy supply strategy for the next fifty years that fulfils the energy demand specified by MEDEE-2. Notice that this is not an economic equilibrium model; in the IIASA study, the terms demand and supply refer to the consumption and production of energy, respectively, as functions of time. Each run of MESSAGE requires the specification of some 1600 constraint variables and 2600 activity variables (Meadows, 1981), although many of these are simply zero or constant across different regions and scenarios (Basile, 1981). The outputs from MESSAGE include the marginal costs (shadow prices) of supplying secondary energy, which are fed back to MEDEE-2, resulting in a sub-loop iteration that adjusts supply and demand. The major outputs from MESSAGE are then fed into IMPACT, the economic model.

IMPACT is a dynamic input–output model which assesses the overall economic consequences of the energy strategy spelt out by MESSAGE. Specifically, the model calculates the direct and indirect requirements for capital investment, land, water, materials, manpower, equipment, and additional energy. These variables are then fed back to modify the original assumptions about the overall development of the economy: 'after a first round of model runs, the built-in feedback mechanism changed the original assumptions so there is no real "beginning" of the model loop' (Schrattenholzer, 1981). Thus, 'the main model loop is closed with IMPACT' (ESP, 1982: ix), and the resulting updated economic growth assumptions are supplied to MEDEE-2, leading to corrected estimates of final energy demand (Kononov and Por, 1979, Figure 1).

The flow of information has now returned to the original starting point, completing the description of one full iteration of the main model loop. The entire process is now repeated several times until an internally consistent scenario is obtained. Since there are three models in the loop, each of which addresses a different facet of the energy system, a balanced scenario is expected from this process, as the outputs from each model are adjusted and corrected by the other two models.

As explained in *EiFW*, this procedure is not yet fully streamlined and computerized—most of the feedbacks are manual and the interfaces between the models are not completely formalized, leaving room for 'judgemental interventions' at various stages. But according to the director, this does not weaken the formalized, iterative process itself (Häfele, 1980b). As stated in *EiFW*, (400), 'the flow of information is mechanized' and the streamlining is currently in the process of being developed (Häfele, 1981a).

FORMALISM AND SUBJECTIVISM IN MODEL OPERATIONS

The initial aspirations of the modelling group were to link together a multiple-model set of at least four models. Even as late a 1978, through sincere wishful thinking and faith that it was going to work, this was being described as *accomplished reality*. Consequently, the role of subjective, judgemental intervention was understated and probably under-recognized. When the originally central models of MACRO and IMPACT did not deliver, the role of subjective judgement and 'interactive' iteration as opposed to formally controlled iteration was naturally amplified, though its full extent was never clearly and forthrightly stated. Indeed, the Executive Summary was directly misleading on the point, as was a later technical document. The *Executive Summary*, for example, stated with respect to the three-model figure of the model set (see Figure 3.1):

In reality, as is usually the case with such sets of models, they were used in parallel and iteratively. The object was internal consistency within each scenario, which in turn required several iterations of the model set. The major consistency checks are suggested by the dotted lines of Figure [3.1].

The lack of any mention of subjective judgemental intervention or even of non-quantitative logic here again firmly plants the belief that large-scale consistency checks round the whole model loop were formally carried out to make the scenarios a factual basis for policy. But the adjustment and control of inputs (for example on economic growth rates or energy efficiencies) from the output of, say, required capital investments to achieve a given level of energy use consistent with those economic growth rates simply never happened, even though this is what the claimed consistency on this scale would have involved.

The missing reproducible iteration of the full loop and the apparent lack of clarity and consistency within the accounts of the modelling themselves inevitably weaken the credibility of the scenarios and what was derived from them. It is also important to note how these confused accounts led reviewers to believe that what was actually only arithmetical accounting—a relatively minor technical consistency (part of model validation)—was a large-scale, formalized consistency in the whole energy demand–supply analysis (real-world validity). Instead, the following analysis indicates the great influence of institutional forces upon the analytical process, its content and representation including institutional faiths naturalized into 'reality'. Rather than imagine that analytical designs are planned, implemented, and adapted in a social vacuum, it is worth emphasizing that the cognitive substance of analysis is deeply rooted in its own institutional realities and commitments, which therefore need to be examined in order to evaluate the analysis itself.

COMPLEX MODELS AND SIMPLE RESULTS

Models should be designed for gaining insight and understanding (EiFW, 399).

This is a standard claim for the advantage of models over mental representations of systems. Starting with the input assumptions to the IIASA energy models, a greatly over-simplified analysis of future energy supply can be carried out (using only a hand calculator), as Keepin (1984) has shown. Although this paper-and-pencil analysis entails no equations or dynamic processes, it turns out to reproduce the IIASA energy supply scenarios almost exactly. The unavoidable conclusion is that the major, dynamic results of the scenarios are essentially prescribed in the input assumptions themselves and the apparently extensive analysis performed by the models is equivalent to a back-of-the-envelope calculation. In fact, in many cases, the energy models serve as a simple identity transformation from the inputs to the outputs.

This can be demonstrated by comparing the relationship between the model outputs (scenarios) and the model inputs (assumptions). The idea is to start with the assumptions and to use them in simple calculations to generate rough approximations of the scenarios. Then, by comparing these approximations with the actual scenarios, we should get some idea of the effort of the model's calculations and iterations in producing the scenarios. Thus, in a sense, the input assumptions will be distilled from the model in order to expose the dynamic role of the model itself.

Throughout this discussion, the term scenarios will be understood to denote the published results that were obtained by the IIASA Energy Group from the energy models MEDEE-2, MESSAGE, and IMPACT (MMI) (see Figure 3.1). Meanwhile, for convenience, the simplistic scenario obtained from the input assumptions will be called the scenariette. Note that this analysis is *not* an attempt to design competing scenarios or a new or realistic energy model; the aim is only to understand the effect that the dynamic calculations and iterations performed by MMI have on the assumptions that are fed into MMI.

The analysis presented here is carried out for one particular world region, comprising western Europe, Japan, Australia, New Zealand, and South Africa (called region III in *EiFW*). This region was chosen for several reasons, one of the most important being that it is the only region for which the iterative process of MMI is described in *EiFW* (404–7). In addition, the available data for this region are excellent and voluminous. Finally, region III contains the homelands of virtually all the scientists who developed the demand and supply components of the model loop (MEDEE-2 and MESSAGE). The model's structure and principal assumptions are therefore particularly suited to this region (and most subsequent work with the model has involved applications within region III). Thus, if the value of the model is called into question for region III, it is likely to be even less useful for the other six world regions.

The particular energy forms to be considered are the following:

(1) Primary energy (extraction of resources): oil, coal, natural gas, and uranium.

(2) Secondary energy: electricity generation and liquid fuels.

(Note that natural gas can be placed in either category).

Primary energy

Since oil is a key component of the global energy system, it is a natural starting point. The input data to MMI specify three separate cost categories of this resource, which together define a kind of step function for the cost of oil. Category I is the least expensive, with a unit cost of $62/kWyr. It includes mainly conventional, domestic oil, both existing reserves and those remaining to be discovered. Category II ($103/kWyr) includes some additional, undiscovered reserves, as well as some oil from unconventional sources. Category III is the most expensive ($129/kWyr),[14] consisting of oil from unconventional sources such as oil shales, tar sands, offshore and polar oil, and oil obtained using enhanced recovery techniques.

We now use this input structure to sketch a rough portrait of oil supply for region III. Since conventional oil is the least expensive, we use it first. For simplicity, we will assume in the scenariette that the price of this oil will not change as it is depleted, that is, we assume that the cost of crude oil from domestic reserves remains constant down to the last drop. This is economic sacrilege, but it is acceptable for a rough sketch and it makes things easy: we simply go ahead and use up all the conventional oil first (category I), and only after it has disappeared do we move on to the more expensive unconventional sources. Thus, in the scenariette, the highly simplistic step function (defined by the input cost data) is adopted as the non-linear cost function for oil supply.

In the scenariette we extract as much domestic oil as possible (because it is the cheapest source of oil, by assumption). Thus, the assumed constraint on domestic oil extraction is simply taken to be the domestic oil production curve in the scenariette. The only thing we have to do, is keep a running tab on the cumulative amount of oil extracted—when we pass the 17.48 TWyr mark, we have run out of category I oil (domestic crude), at which time we switch (very abruptly) to category II oil (unconventional); for the high scenariette (this refers to the scenariette obtained from the assumptions of the high scenario, the 'low scenariette' is analogously defined), this happens between 2020 and 2030. We then continue in the same fashion: extract oil at the maximum allowable rate until category II oil is exhausted, then switch to category III and so on.

In addition to domestic oil, there is also imported oil to consider, so we again consult the input assumptions. This time we find a constraint that sets upper limits on the amount of oil that can be imported as a function of time, and this constraint is simply adopted as the curve for imported oil in the scenariette.

This then completes the portrait of primary oil supply, which is displayed in graphical form in Figure 3.2. To generate this figure, the individual data points were plotted and then connected by straight line segments to produce curves. Note that the curves are plotted cumulatively to illustrate the composition of crude oil supply and its evolution over the fifty-year time horizon from 1980 to 2030. Observe the rather abrupt shift from category I to category II oil that occurs around 2020—this is due to the over-simplified assumptions made in constructing the scenariette.

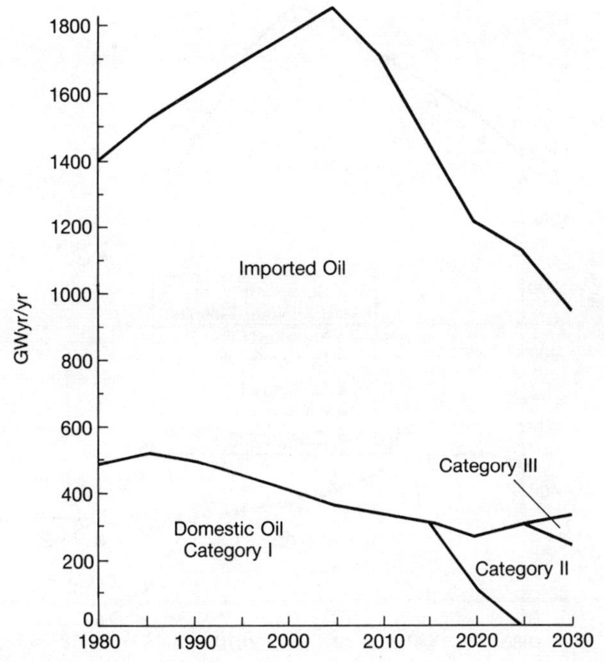

Fig. 3.2 Scenariette for crude oil supply

Region III, high.
The curves displayed here are obtained directly from the exogenous input assumptions to the IIASA energy models.

Now that we have completed this first part of the scenariette, it is interesting to compare it with the results from the published IIASA scenarios itself. To do this, we start with a duplicate of the graph in Figure 3.2, onto which the final scenario results are superimposed by plotting individual data points (which come directly from the MMI computer output listings). The result of this superposition is shown in Figure 3.3. Thus, Figure 3.2 is identical to Figure 3.3, except that some data points have been added; these points are the final scenario results, which are plotted using four different shapes (circles, triangles, squares, and crosses) to distinguish four distinct sets of outputs from MMI. The curves display the scenariette (inputs), and the points display the scenario (output). In Figure 3.3 we see something quite surprising. The data points from the scenario fall almost exactly onto the scenariette curves. There are some minor differences for imports, but these are insignificant.

Scenariettes for other energy resources

The development of similar scenariettes for natural gas, coal, and uranium produces the curves shown in Figures 3.4, 3.5, and 3.6 respectively.[15]

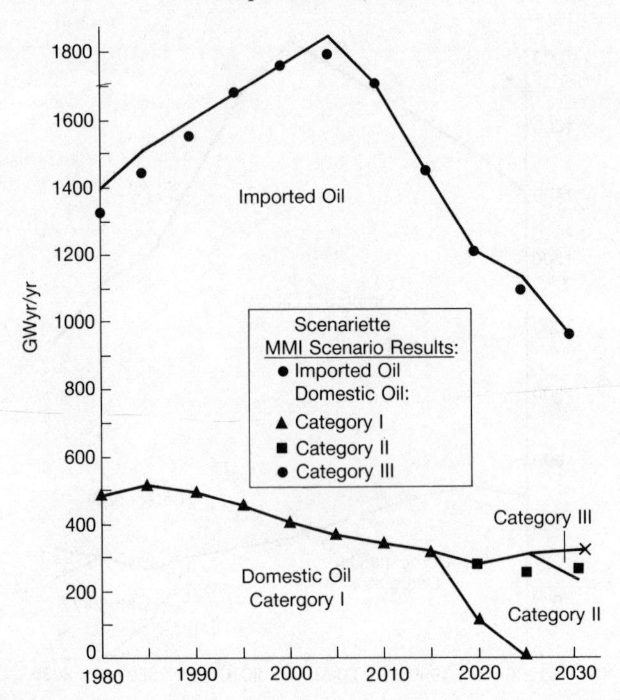

Fig. 3.3 Comparison of scenariette with the IIASA scenario for crude oil supply

Region III, high.
This figure is identical to Figure 3.2 with the addition of the data points, which are the final outputs from the IIASA energy models. Note the agreement between scenariette and scenario.

Again, for comparison, the scenario results are shown as data points, and once again the agreement is essentially perfect. No analysis of any kind was involved in generating Figures 3.4 and 3.5; the curves are plotted directly from the exogenous input listings to MMI, and the points are plotted directly from the output listings from MMI. In some ways these plots look deceptively trivial, which obscures their importance. It is crucial to understand that they are not the result of some curve-fitting exercise. Rather, the data points are the outputs from MMI and the solid curves are the input assumptions to MMI. The fact that they agree perfectly means that, in effect, the scenario results are prescribed exogenously in the input assumptions, and for all its apparent elaboration, complexity, and dynamic calculation, the model itself in fact reproduces these assumptions.

These results have all come out of an analysis of the high scenario. It is quite possible that the entire energy system is operating at maximum capacity in the high scenario, straining every bolt as it were, so that the system comes right up against the constraints. We have therefore also looked at the low scenario, where the strain on the system should be eased con-

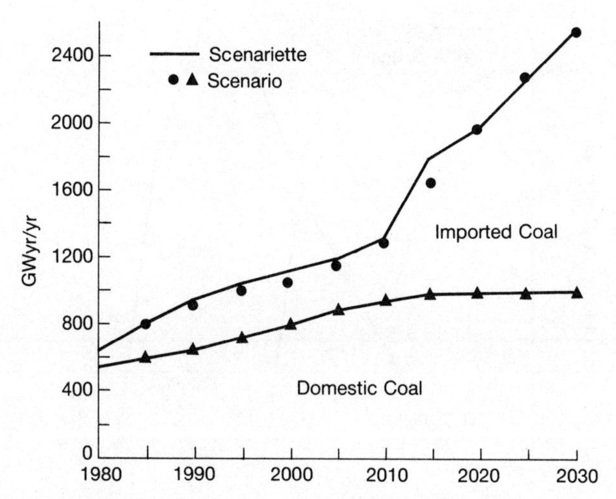

Fig. 3.4 Comparison of scenariette and scenario for natural gas supply

Region III, high.
The curves are inputs to MMI, the points are outputs from MMI.

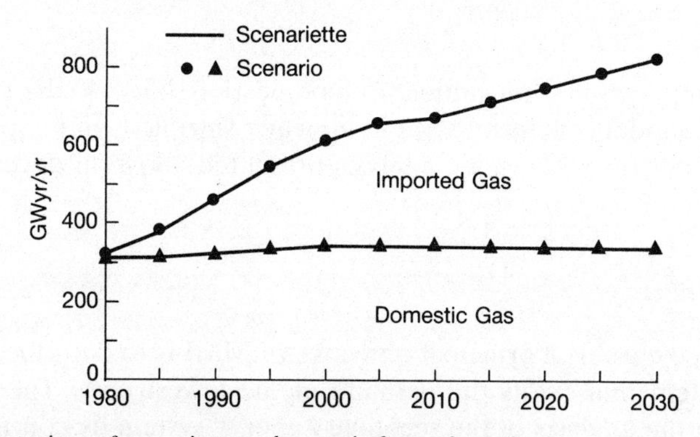

Fig. 3.5 Comparison of scenariette and scenario for coal supply

Region III, high.
The curves are inputs to MMI, the points are outputs from MMI.

siderably (Keepin, 1984). Again, essentially perfect agreement is observed between inputs and outputs in almost all cases.

The principal finding is that both stocks and flows of primary energy sources in the IIASA scenarios are effectively prescribed in the form of exogenous assumptions and constraints. In the schematic diagram of MMI in Figure 3.1, most of these assumptions are contained in the oval labelled P (for primary). Note that this oval lies entirely outside the iterative model loop and that there are no 'major feedbacks' into this oval, indicating that

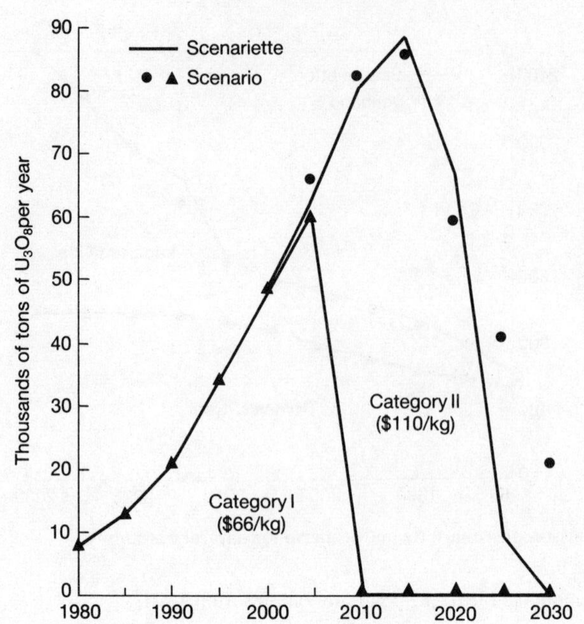

Fig. 3.6 Comparison of scenariette and scenario, for natural uranium extraction Region III, high.

these assumptions are not subject to modification. That is, the key factors driving the model conclusions are in no way controlled. In fact, the model essentially performs the same analysis presented above in developing the scenariette.

Secondary energy

As discussed earlier, a principal objective of MMI is to describe an energy supply system that fulfils the demand at the lowest cost. Therefore, we shall begin the analysis of the secondary energy system by considering the cost assumptions for various secondary energy supply technologies. These are given in Table 3.1, which is reproduced from Table 17–4 in *EiFW* (p. 527). In the following we will present two secondary supply scenariettes; one for electricity and one for liquid fuels.[16]

Electricity generation

Given the objective of cost minimization, we start by looking at the relative cost assumptions for electricity generation. In the last column of Table 3.1, hydroelectric power is found to be the least expensive technology, at $85 per kWyr. Following this, the next cheapest is nuclear power, running from $136 for the light water reactor (LWR) to $143 for the fast breeder reactor

Table 3.1 Cost Assumptions for Major Energy Supply and Conversion Technologies (reproduced from *EiFW*, p. 527 table 17–4)

	Capital Cost (1975$/kW)	Variable Cost (1975$/kWyr)	Final Product Cost (1975$/kWyr)
Electricity Generation			
Coal with scrubber	550	23	154
Conventional nuclear reactor (e.g. LWR)	700	50	136
Advanced reactor (eg FBR)	920	50	143
Coal, fluidized bed	480	36	152
Hydroelectric	620	8.5	85
Oil fired	350	19	256
Gas fired	325	16	216
Gas turbine	170	17	241
Solar central station	1900	28–60	297
Synthetic Fuels			
Crude oil refinery	50	3.7	75
Coal gasification ('high Btu')	480	40	125
Coal liquefaction	480	40	125

(FBR), then comes coal-fired power at $152 to $154,[17] and the remaining electricity sources become increasingly more expensive. Thus, we start with the cheapest source (hydro), take as much as possible, then move on to the next cheapest source (LWR), again taking as much as possible; and continue in this fashion until the demand is met. Thus, to build the scenariette, the technologies are chosen in the order of their cost and each one contributes an amount equal to its supply constraint. This guarantees that when we reach the demand level, we have specified the least expensive supply mix that meets it.

The demand projection is shown in Figure 3.7 by a dashed line. (It is interesting to note in passing that this demand projection entails a 2.7 fold increase in electricity consumption per person living in region III by 2030.) Since the demand is taken as given, a dashed curve is used to distinguish it from the solid curves representing supply, which are the results of the scenariette.

Figure 3.7 displays the scenariette for electricity supply in region III. The area labelled 'coal & other' in this figure is due almost entirely to coal. 'Other' refers to a thin sliver (due to current oil- and gas-fired power plants) which disappears by 2010, and a barely discernible contribution from solar energy after 2020. Figure 3.7 also contains the points plotted from the output listing of MMI. Again, we find that the MMI scenario is identical to the scenariette up through 2010. Notice that after 2015, the data points for LWR and FBR seem to be deflected away from the demand projection as they approach it. During these final fifteen years of the time horizon, coal is being phased out very rapidly, resulting in extensive underutilization of coal-fired capacity. However, MMI imposes an economic

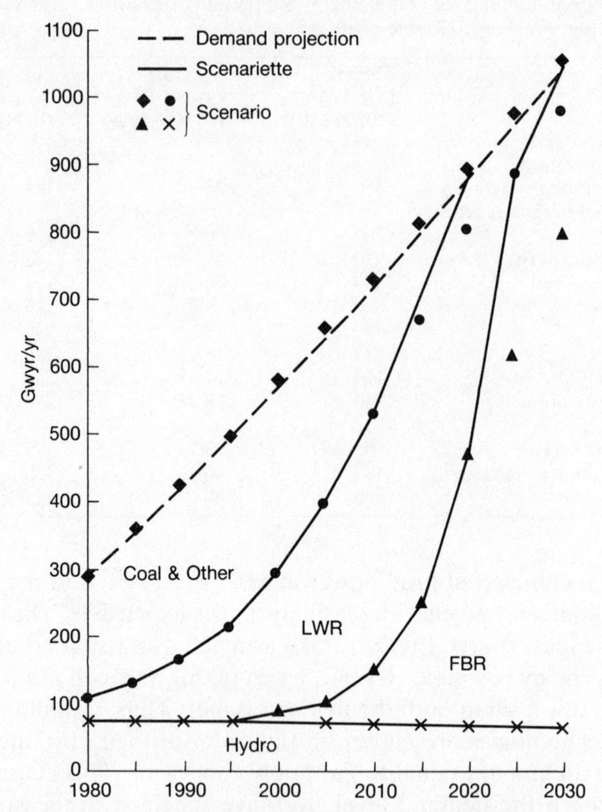

Fig. 3.7 Comparison of scenariette and scenario for electricity generation Region III, high.

penalty for excessive underutilization, so that the rapid decline of coal is attenuated somewhat, producing the observed deflection. This same effect occurs to a lesser extent within the nuclear contribution itself, as LWR gives way to FBR.

Notice that MMI has no knowledge of the physical significance assigned to the particular results that it produces. For example, it might be tempting to conclude from Figure 3.7 that the fast breeder reactor will dominate the future electricity supply. However, this is an assumption supplied *to* the model, and not really a result or conclusion derived *from* the model (as we will see later).

Perhaps the most critical aspect of the global energy system is the supply and international trade of oil. Figure 3.8 shows the high scenario and corresponding scenariette for the global free-market oil supply (excluding the centrally planned economies of regions II and VII). This scenariette was generated by aggregating a few input assumptions (contained in the two

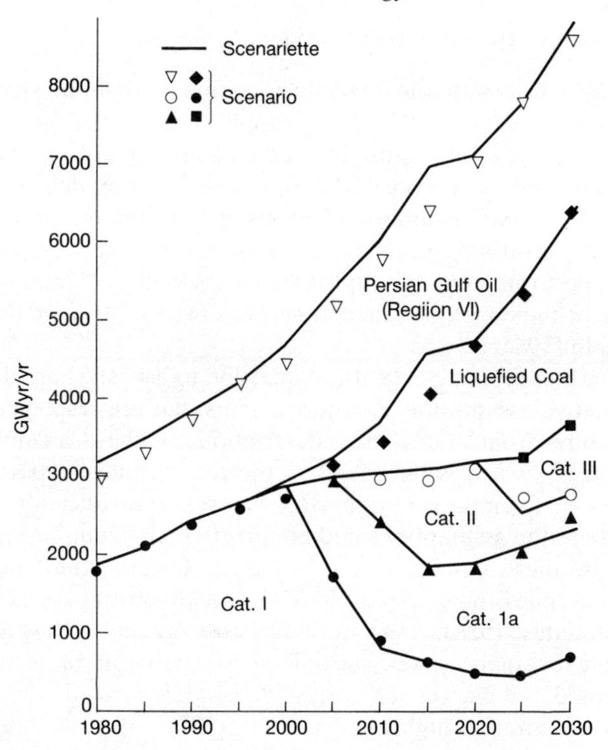

Fig. 3.8 Comparison of scenariette and high scenario for world oil supply
Category I & Ia: Known reserves and some yet to be discovered
Category II: Reserves to be discovered and some unconventional oil
Category III: Unconventional oil (heavy crude, tar sands, oil shales, deep offshore oil, etc.)
Excluding centrally planned economies.

ovals labelled P and S in Figure 3.1), which are exogenous to the model. A variant of Figure 3.8 has been published in *Science* (Häfele, 1980a) and *Scientific American* (Sassin, 1980), as well as in the *Executive Summary* (McDonald, 1981) and twice in *EiFW* (662 and 789).[18] Clearly these results are offered as a key finding from the IIASA energy study. As displayed in Figure 3.8, the differences between scenarios and scenariette are slight, revealing that these important results were essentially exogenous assumptions. This figure is presented in *Science* as evidence for the revealed need to exploit unconventional oil and coal liquefaction.

It should be pointed out that the comparison of scenarios and scenariettes gives similar results for the low scenario as for the high one. This is even true for the electricity sector, despite the fact that the system is not so much under strain as in the high scenario. The results published in Keepin (1984) also suggest that the scenarios and scenariettes exhibit close agreement for other energy regions as well.

Scenarios as objective analysis or considered opinion?

The basic conclusion of the preceding section is that the dynamic and analytic contents of the IIASA energy supply scenarios are directly attributable to assumptions and quantitative judgements that are specified outside the set of mathematical models. In some cases, the models are found to reproduce the input assumptions precisely; in others, they introduce unimportant perturbations to the input structure. At best, the models themselves perform a highly simplistic analysis that is essentially the same as the back-of-the-envelope calculations presented above (and described in full in Keepin (1984)).

In view of these findings, a natural question to ask is: where do the various quantitative assumptions and judgements that are responsible for the scenarios come from? Only brief descriptions of these assumptions are given in *EiFW* (which is described as the full technical report from the study). Almost no empirical evidence or theoretical justification is included to substantiate the assumptions and no quantitative details are included to indicate how these numbers were obtained. Instead, they are candidly referred to as 'guesstimates' (*EiFW*, 531), 'rough average (sometimes consensus) estimates' (528), 'best available assessments' (581), and so on. Thus, whatever analysis was carried out to arrive at these numbers is undocumented and inaccessible: 'these data, while arrived at by averaging many sources, are still highly judgmental' (527). Since the scenarios are largely copies of these assumptions, the conclusion that begins to emerge is that the scenarios are closer to considered opinion than objective analysis. We shall return to this point later, after exploring the robustness of the scenarios.

ROBUSTNESS OF THE IIASA ENERGY MODELS

> . . . All of this leads to a belief (or hope) that the scenarios here are robust, that they can stand up against events whose impacts, in human terms, may be large (*EiFW*, 395).

Robustness is the property that an analysis must have if it is expected to be of some validity in the face of uncertainties in the underlying assumptions and relationships on which the model is structured. To establish robustness, the standard procedure is to perform a detailed analysis of the sensitivity of the quantitative results with respect to variations in the assumed input data. In the case of the IIASA energy scenarios, it has been asserted that 'the sensitivity analysis was done—at length' (Rogner, 1983).

In this section, we explore the sensitivity of the scenarios with respect to certain assumed input data. This senstivity analysis focuses on the energy supply model MESSAGE, because the assumptions and results (from

MESSAGE) represent, in some ways, the core of the energy studies (*EiFW*, 452).

One example of the sensitivity of the IIASA study involves the contribution from the fast breeder reactor in the scenarios. Both the time at which the FBR is introduced and its subsequent rapid expansion are strongly dependent on a two per cent cost advantage that is the result of a small, artificial step in the cost projection for LWR. The temporal location of this step is in turn very sensitive to uncertain estimates of available uranium resources and their costs.

Another difficulty concerns the magnitudes of the contributions from the various supply options, which are based on fixed relative cost assumptions that are presumed to hold for the next fifty years (see Table 3.1).

In the case of the IIASA study, it is found that small changes in the assumptions about relative costs and resource availability can cause the model to produce radically different supply scenarios. This finding is consistent with early internal studies (for a prototype model) that showed tremendous sensitivity with respect to variations in the cost assumptions for technologies and resources. Given the large uncertainties in the future costs of energy resources and supply technologies, this demonstrates the futility of using simple cost minimization linear programming models for describing robust supply scenarios over a long time horizon. Curiously, the earlier work was ignored (Keepin, 1984; Wynne, 1984).

Finally, as explained in *EiFW*, although the assumed costs 'will surely change over time, perhaps dramatically, just one cost estimate for each technology is used here for the entire planning horizon. Sensitivity analyses can test alternative cost estimates' (*EiFW*, 527). However, the documented sensitivity analysis includes only one such test, and this particular test obscures the critical sensitivity to variations in the relative cost structure. In addition, the early work on sensitivity analysis is not cited. Finally, regarding the availability of natural uranium, 'there is a far greater lack of basic information than in the case of fossil fuels' (Häfele, 1983a). Nevertheless, there are no sensitivity tests with respect to the assumed costs and availability of uranium.

The results of a sensitivity test of relative cost changes, which is not undertaken in the study, are summarized in Figure 3.9. Part (a) of the figure displays the electricity supply system of the IIASA low scenario for region I. (Note that this figure shows the actual scenario results, and not a scenariette.) Now suppose the cost of nuclear power is increased by 16 per cent.[19] As revealed by a straightforward calculation (see Keepin, 1984), this produces the greatly altered scenario shown in Figure 3.9b. In this new scenario, coal-fired power accounts for most of the electricity production, while LWR is phased out over the fifty-year time horizon and FBR is never introduced at all.[20] By 2030, the coal contribution to electricity supply reaches 85 per cent (compared to 8 per cent in the original IIASA scenario)

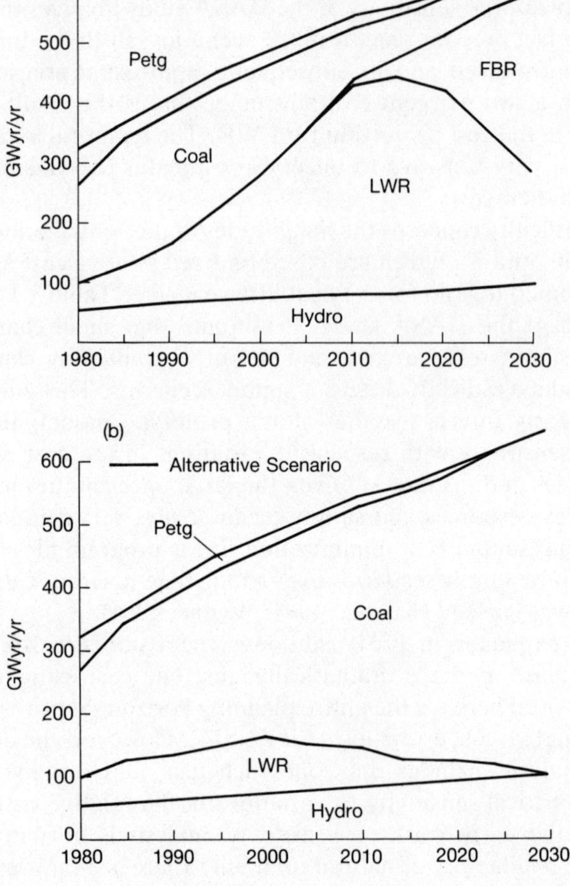

Fig. 3.9 Sensitivity to cost assumptions in IIASA scenario of energy demand for USA and Canada

Region I, low.
(a) Original scenario results for electricity generation; (b) new scenario results, assuming that nuclear costs are increased 16% and that the coal extraction limit is raised 7%.

in Figure 3.9a, and the nuclear contribution almost disappears entirely (compared with 77 per cent in the original scenario). However, it must be emphasized that this is not an alternative scenario to the IIASA one. It is an exercise to test the credibility of the IIASA scenarios via their robustness towards uncertain factors in the system. The main point here is not what actually happens under different cost assumptions, but rather that small changes in these assumptions can produce such tremendously different outcomes from the model.

SUMMARY

Stripped to its bare essentials, the representation of the models in the IIASA energy study has been as follows:

Claim 1. Macro-economic consistency is a big feature of the IIASA scenarios and a major element of their scientific status.

Claim 2. This consistency is achieved by models, and their iteration in the full model loop.

Then we find:

Admission 1. The loop does not exist.

Admission 2. The models anyway do trivial things 'as everyone always knew'. The outputs (that is, the scenarios) come from informal judgements outside the model 'loop'.

So, with these admissions, how *could* the models have done any significant consistency control? The claims are contradicted by their own authors' admissions. This underlines our point that any consistency which may or may not exist in the scenarios is already there or not there before the computer is ever switched on. Assumptions are controlled only by further assumptions; the relatively trivial control for local *arithmetic* consistency does not in itself avoid this circularity, and does not in itself come anywhere near to validating different options. Yet confused representation of the modelling raised it to the higher epistemological status of having objectively controlled, substantive scenario contents and comparative evaluations of policies.

We have seen a slowly emerging, ambivalent, and somewhat roundabout approach towards full acknowledgement of the limitations of the IIASA modelling effort. Within these historical trends, Keepin's discoveries can be seen as a mainstream, albeit somewhat more forthright contribution. They have helped to clarify the full extent and importance of subjective assumptions in the study, whilst at the same time clarifying the actual role of the models. Thus MESSAGE, the 'core of the energy studies reported in this book' (*EiFW*, 402), was found not only to be internally extremely brittle, but totally constrained in its outputs by the externally generated input constraints, some of these undocumented. The fact that MESSAGE was completely tied down by its input assumptions was evidently known to the IIASA ESP team, but the only references to this important limitation are ambiguous, that they were 'often quite tight', or that 'these constraints, taken together, are singular characteristics of the scenarios' (*EiFW*, 402, 527). The full implications of these hints, however, are nowhere spelt out; indeed they are obscured by the repeated prominence given to the formal models themselves. They are referred to as 'real world constraints', but the significance of the fact that they are informal expert judgements and that they completely drive the model outputs and thus the scenarios, is not made clear. Thus, the self-description that, 'The specific approach is to use an optimization procedure to find the combination of energy sources that satisfies the specific temporal sequence of regional final energy demands, subject to (often very dominant) given constraints, while minimizing the discounted total cost' (*EiFW*, 397) does not clarify the fact that there was no optimization within a feasible region defined by the constraints, because the constraints themselves preset the 'solution' at a single point. It was left to Keepin to clarify this and to show just how often and how very (that is, totally) dominant were the constraints, and that some of the more important of these were not documented, even in the 'documentation'.

Thus, the bold assertion in *EiFW* that, 'Computer modelling has at least one other distinct advantage. With a great many interdependent variables to consider in typical systems problems, it is difficult to know which ones are critical, which ones deserve close attention by planners and decision makers. Models can aid in identifying such parameters through for example sensitivity tests and consideration of alternative scenarios' (*EiFW*, 399) does not apply in the case of this study, whether or not it is valid for others. Häfele's earlier advocacy of formalism in analysis and debate as the best way to improve policy[21] is not easily reconciled with later defences of informal judgement.

One of the last two remaining formal models, MESSAGE, is shown to have been virtually moribund. Interestingly enough, the setting of exogenous constraints in a model by its authors so as to exclude subjectively unacceptable outputs was identified and discussed by Nordhaus in an IIASA ESP paper[22] on the Bariloche models following up *Limits to Growth*. The lack of credibility of the outcomes of such exercises was emphasized.

This imprisonment of model outcomes by subjective, external inputs is particularly acute in the case of large complex models because there is so much more data and so many more judgements to check. The IIASA ESP claims that their models are relatively 'simple' and are therefore superior to 'large, monolithic computer models which often suffer from over-complexity and rigidity' (*EiFW*, 400). Yet others, such as Ausubel and Nordhaus (1983), judged the IIASA models to be 'extremely complex' and 'difficult to comprehend, manipulate, change and verify independently'. Each run of MESSAGE, for example, requires the input of approximately 1600 specified constraint variables and 2600 activity variables, even though many of these are zero or constant from one run to the next. Despite having been earlier promised as 'the whole set of input data', the recent supposedly full documentation of MESSAGE in ESP (1982) is selective in reproducing only parts of these input files,[23] so that model reviewers are still unable to gain access to the determinants of the model outputs. Indeed, some of the undocumented parameters involved the most critically sensitive parts of the model. Thus, Keepin (1984: 224) found undocumented *ad hoc* inputs which allow an otherwise inexplicable maximum rate of introduction of the nuclear fast breeder reactor in one region where it otherwise would not have been introduced. The claimed clarity of the so-called simple models, and, as Goldman (1981: 224) put it, 'their ability to travel well' to other modellers, would at least be aided by reproducing the input files themselves rather than a selective description of them. As it turns out from Keepin's scrutiny, the models are analytically simple (indeed analytically non-existent), but still manage to suffer from the 'overcomplexity and rigidity' and lack of clarity which the IIASA team though they had avoided.

The IIASA team's repeated defence against Keepin is that he has utterly misunderstood the craft nature of systems analysis and the fact that the IIASA modelling was only a final framework to check what was a pains-taking process of achieving the model inputs. These, not the modelling, were the real achievements of the study according to the IIASA team (Häfele and Rogner, 1984). Thus Keepin's taking the model inputs as if plucked out of the air is false and unfair. The problem with this defence is that it is contradicted even by the IIASA team's own publications which have repeatedly emphasized the importance of the formal models, even to the extent of misleading other modellers (Wynne, 1984). That analysis, like science in general, is a craft practice involving essential informal judgement as well as more formal procedures of calculation and reasoning is not at issue. The value and importance of institutional mechanisms for peer review and external access is that they help to identify for external policy evaluation the informal judgements and commitments which often involve behavioural and social assumptions, as they did in the IIASA study. Such mechanisms were lacking in the IIASA case, but this is a general problem in policy modelling and analysis.

The completely dominant structure underlying the lack of any degrees of freedom in the modelling is that, given IIASA's primary assumptions about global population, demographic shifts and economic growth, the total primary energy demand always threatens to far outstrip any feasible supply scenarios from what is assumed can only be more capital-intensive, centrally managed supply systems. Within these primary assumptions, there is thus no feasibility space within which to evaluate optimal strategies because most resource and supply options are exploited at the maximum feasible rate. This is true even for the 'low' demand scenario of 24 TWyr per year. Indeed, the range eventually settled upon for the scenarios actually follows a repeated reduction from earlier even higher total demand scenarios.[24] The repeated adjustments downwards, however, remain fully within the framing definition of the whole project set up at the outset, to do with the kinds of technology thought to be realistic and the kinds of corresponding institutions and values thought to be 'natural'. With respect to the latter, it is incidentally worth noting that the 'large-scale consideration' inevitably necessary for a view of global energy supply is unconsciously equated wih centrally managed processes, which is not an inevitable, logical step (Häfele and Sassin, 1977).

One of the major finds of this study is that the IIASA energy supply scenarios are seriously lacking in robustness, particularly with regard to the contribution from nuclear power. This lack of robustness precludes the possibility of drawing reliable conclusions from the scenarios about future energy supply strategies.

A more general insight is that policy modelling, which appears to be purely technical and objective, always contains embedded assumptions about institutional parameters and relationships framing the assumptions about technical constraints and possibilities.[25] These are usually the crucial elements of the analysis. Yet they are often buried from view. It is essential that energy modelling should first accept the inevitability and legitimacy of such dimensions, and then develop the procedures to ensure greater awareness of them and more open evaluation of their implications and comparative merits. In this way technical expertise might emancipate itself from the sterile role of 'revealing' policy answers to a thereby 'de-skilled' policy process, and find a more modest constructive role in helping to clarify the physical–technical implications (including the ignorance) of the range of social assumptions and commitments contending with each other for ascendancy in the real world.

Notes

1. Greenberger (1981: 27)
2. Gass (1981: 1) quoting Judge David Bazelon's editorial in *Science*, 16 May 1980, justifying legal review of agency-scientific decisions to make them fully transparent and accessible. Computer models have been regarded as fulfilling this aim.
3. Tom Long, Comments on paper by G. S. Packer, in F. S. Roberts (ed.) *Energy Modelling: Dealing with Energy Uncertainty*. Proceedings of Second Symposium, Institute of Gas Technology, Chicago, 1981.
4. *Atom*, September 1981: 37–8. *Popular Science*, July 1981: 49, called the study 'perhaps the most impressive of such (energy) studies to date. It was performed by IIASA, in Vienna, using teams of scientists from 20 countries . . . It is by all odds the biggest and perhaps the first truly global and long term examination of the energy situation.' *Bio Science*, 31 (4), April 1981: 292: 'the first truly global and long-range examination of the world's energy future'. *Industrial Heating*, April 1981: 4: 'the first comprehensive and professional survey' (quoting Chauncey Starr). *Dawn*, 19 Aug. 1981: 'unprecedented, detailed analysis . . . analysing options in a quantitative, mathematical form', led by 'a colossus in the form of Prof. W Häfele, . . . a world celebrity'. *Physics in Technology*, Nov. 1981: 'one of the most comprehensive analyses of the world energy dilemma'. *Physics Bulletin*, Nov. 1981: '[Global energy] constraints are examined in detail by computer modelling.'
5. G. E. Brown Jr., 'Can Systems Analysis and Operations Research Help Congress?' *Interfaces* 12 (6), Dec. 1982: 119–25.
6. For example, G. Greenhalgh, *The Necessity for Nuclear Power*, London: Graham and Trotman, 1980. M. J. Chadwick and H. Lindman (eds.) *Environmental Implications of Expanded Coal Utilization*, New York: Pergamon Press, 1982; J. Jaeger, *Climate and Energy Systems: A Review of Their Interactions*, Chichester: Wiley, 1983.

   V. A. Legasov, L. P. Feoktistov, and I. I. Kuzmin, *On Safe Development of Nuclear*

Power, Priroda. A. M. Parry, W. Fulkerson, K. J. Araj, D. J. Rose, M. M. Miller and R. M. Rotty, 'Energy Supply and Demand Implications of CO_2', *Energy* 12, 1982: 991.

7. W. Häfele, 1980a: 156–64. This paper was published also as an IIASA Research Report (RR–81–8), presumably indicating that it was a research paper, and not an expression of policy advocacy.
8. These bodies, over 30 in all, included the UN Environment Programme, the International Atomic Energy Agency, the Siberian Power Institute, Irkutsk (USSR), and the Institute of Energy Economics and Law, Grenoble, France. IIASA's promotion of the study involved the support of seventeen nations' Academies of Science or equivalent member organizations.
9. W. Häfele, 1983a: 3–20. Also the Panel Discussion, 319–28.
10. Häfele (1983a: 325, 1980a: 1980b), Sassin (1980).
11. A. Lovins, 1981: 25–34; D. Meadows, 1981: 17–28. See also Häfele's rejoinder, (1981b).
12. B. Keepin (1984).
13. W. Häfele and P. Basile, 'Modelling of Long Range Energy Strategies with a Global Perspective', in K. B. Haley (ed.), *Operations Research '78*, North-Holland 1979: 493–529.
14. These costs are equivalent to approximately $12.30, $20.40, $25.60 per barrel, respectively, when expressed in 1975 US dollars.
15. The uranium scenariette is obtained in a somewhat different fashion from the other primary energy scenariettes. For details see Appendix E of Keepin (1984).
16. The demand projections for secondary energy in these scenariettes are taken from the endogenous 'Secondary Fuel Mix and Substitutions' procedure, labelled D (for demand) in Figure 3.1. Thus, the present analysis treats these projections as given, and focuses on the supply side of the scenarios.
17. Since these cost about the same and are both coal burning technologies, no distinction is made in the scenariette. It so happens that this distinction was unimportant in the scenario as well.
18. The published figure is somewhat different from Figure 3.8, in part because it probably includes 'constraints for the gradual build-up and depletion of separate oil categories' (*EiFW*, 558). This brief reference (which occurs in the caption of another figure) is the only mention of these constraints—they are not part of the model and we have not found them documented anywhere. In addition, the published figure incorporates specific dynamic estimates of quantities of oil remaining to be discovered, which are also undocumented.
19. Such an increase can easily be envisaged in *any* of the IIASA world regions, for a variety of reasons. For example, the costs of decommissioning nuclear power stations (which could be considerable) are not included in the IIASA cost assumptions. Other factors that could increase the cost are stricter safety regulations, new requirements for waste treatment, tighter emission control standards, legal entanglements, construction delays, etc.
20. In order for this new scenario to be feasible, it is necessary to increase the assumed ceiling on coal extraction in the scenario after 2020 (Keepin, 1984).
21. W. Häfele, 'Hypotheticality and the New Challenges: The Pathfinder Role of Nuclear Energy', *Minerva* 12(1), 1974: 303–23.
22. W. D. Nordhaus, 'World Modelling from the Bottom Up'. Research Memorandum, RM–75–10, esp. pp. 9 and 17. Laxenburg, Austria, 1975.
23. L. Schrattenholzer, 1981: 5.
24. See for example, W. Häfele and W. Sassin, 1977; and Häfele and Sassin, 'Energy Strategies' IIASA RR–76–8, e.g. p. 27.
25. Thus for example already in 1977, the FBR was assumed to be 'feasible not only scientifically, but industrially, commercial feasibility being in sight' (Häfele and Sassin, 1977), even though Nordhaus's (1979: xvi) energy study two years later regarded them as 'unproven for large-scale use'. He, of course, is not the only person to have made the same judgement, nor is Häfele the only one to have made the more optimistic one. The point here is not to say who is correct, but to point out the asymmetry of assuming FBRs as proved, whilst alternatives are treated as unproved. The asymmetry reflects social, institutional commitments, which happen to correspond with FBRs (and centralized large-scale synfuels production, and so on).

Bibliography

Agnew M., L. Schrattenholzer, and A. Voss (1979), *A Model for Energy Supply Systems Alternatives and their General Environmental Impact.* Working Paper WP–79–6, Laxenburg, Austria: IIASA.

Aushubel, J. and W. D. Nordhaus (1983), 'A Review of Estimates of Future Carbon Dioxide Emissions', in US National Academy of Sciences, *Changing Climate*. Washington DC, pp. 153–91.

Basile, P. (1981), Letter to *The Energy Journal*, 2(4), pp. 103–6.

EIA (1982), *Projected Costs of Electricity from Nuclear and Coal-Fired Power Plants*, vol. i, DOE/EIA–0356/1. Washington DC: US Department of Energy Report, Energy Information Administration.

Energy Systems Program (ESP) (1982), *The IIASA Set of Energy Models: Documentation of the Global Runs.* Prepublication Issue, Laxenburg, Austria: IIASA.

Foley, G. and C. Nassim, (1981), *The Energy Question*. Harmondsworth: Penguin.

Gass, S. I. (ed.) (1981), *Validation and Assessment of Energy Models.* US Department of Commerce, National Bureau of Standards.

Goldman, A. J. (1981), 'Reflections on Modelling and Model Assessment', in Gass (ed.) (1981).

Greenberger, M. (1981), 'Humanizing Policy Analysis: Confronting the Paradox of Energy Modelling', in Gass (ed.) (1981).

Grümm, H., D. Gupta, W. Häfele, P. Jansen, M. Becker, W. Schmidt, and J. Seetzen (1966), *Ergänzendes Material zum Bericht 'Kernbrennstoffbedarf und Kosten verschiedener Reaktortypen in Deutschland*'. KFK 366, KFK 466. Karlsruhe, FRG.: Nuclear Research Centre Karlsruhe.

Häfele, W. (1980a), 'A Global and Long-Range Picture of Energy Developments'. *Science*, 209: 156–64.

—— (1980b), 'IIASA's World Regional Energy Modelling'. *Futures*, February, 18–34.

—— (1981a), *Energy in a Finite World: A Global Systems Analysis*, vols. i and ii. Cambridge, Mass.: Ballinger. Volume ii is abbreviated *EiFW* in the text.

—— (1981b), 'Energy in a Finite World—Expansio ad Absurdum? A Rebuttal'. *The Energy Journal*, 2(4): 35–42.

—— (1983a), 'Energy Strategies and Nuclear Power', in G. S. Bauer and A. McDonald (eds.), *Nuclear Technologies in a Sustainable Energy System: Selected Papers from an IIASA Workshop.* Berlin: Springer, pp. 3–20. See also the Panel Discussion, pp. 319–28.

—— (1983b), Lecture presented at IIASA Energy Workshop, Laxenburg, 15 June.

—— and A. S. Manne (1974), *Strategies for a Transition from Fossil to Nuclear Fuels.* Research Report RR–74–7. Laxenburg, Austria: IIASA.

—— and H. H. Rogner (1984), 'A Technical Appraisal of the IIASA Energy Scenarios? A Rebuttal'. *Policy Sciences*, 17: 341–65.

—— and W. Sassin, (1977), 'A Future Energy Scenario'. Paper to the 10th World Energy Conference, Istanbul, Turkey.

IAEA (1982), *Reviews of Nuclear Power Costs around the World*, by L. L. Bennett *et al.* IAEA Report IAEA–CN–42/76, Sept.

Keck, O. (1981), *Policymaking in a Nuclear Program: The Case of the West German Fast Breeder Reactor.* Lexington, Mass.: Lexington Books.

Keepin, B. (1984), 'A Technical Appraisal of the IIASA Energy Scenarios'. *Policy Sciences*, 17: 199–276.

Kok, M. (1983), Private Communication. Sept.

Konno, H. and T. N. Srinivasan (1974), *The Häfele–Manne Model of Reactor Strategies: Some Sensitivity Analysis*. Research Memorandum RM–74–19. Laxenburg, Austria: IIASA.
Kononov, Y. and A. Por (1979), *The Economic Impact Model*. Research Report RR–79–8. Laxenburg, Austria: IIASA.
Landsberg, H. H. (1982), 'Commentary', in W. C. Clark (ed.), *Carbon Dioxide Review*. Oxford: Oxford University Press, p. 364.
Lovins, A. B. (1981), 'Expansio ad Absurdum'. *The Energy Journal*, 2(4): 25–34.
McDonald, A. (1981), *Energy in a Finite World—Executive Summary*. Executive Report ER–81–4. Laxenburg, Austria: IIASA.
McMurry, F. M. (1909), quoted in C. B. Cope *et al.* (eds.) (1979), *The Scientific Management of Hazardous Wastes*. Cambridge: Cambridge University Press.
Marchetti, C. and N. Nakicenovic (1979), *The Dynamics of Energy Systems and the Logistic Substitution Model*. Research Report RR–79–13. Laxenburg, Austria: IIASA.
Meadows, D. L. (1981), 'A Critique of the IIASA Energy Models'. *The Energy Journal*, 2(3): 17–28.
Nordhaus, W. D. (1979), *The Efficient Use of Energy Resources*. New Haven: Yale University Press.
Pearson, C. E. (1974), *Handbook of Applied Mathematics: Selected Results and Methods*. New York: Van Nostrand Reinhold Co.
Perry, A. M. (1982), 'Carbon Dioxide Production Scenarios', in W. C. Clark (ed.), *Carbon Dioxide Review*. Oxford: Oxford University Press, pp. 337–63.
Rogner, H. H. (1983), Colloquium on the IIASA energy models held at IIASA 10 May.
Sassin, W. (1980), 'On Energy and Economic Development'. *Scientific American*, 243(3): 3–13.
—— A. Holzl, H. H. Rogner, and L. Schrattenholzer (1983), *Fueling Europe in the Future: The Long-Term Energy Problem in the EC Countries*. Research Report RR–83–9. Laxenburg, Austria: IIASA.
Schrattenholzer, L. (1981), *The Energy Supply Model MESSAGE*. Research Report RR–81–31. Laxenburg, Austria: IIASA.
—— (1982). Private communication.
Schumacher, E. F. (1964), 'Review of EEC Report on Fuel and Energy by the European Coal and Steel Community'. *The Economic Journal*, March: 192.
Schwarz, B. and J. Hoag (1982), 'Interpreting Model Results—Examples from an Energy Model'. *Policy Sciences*, 15: 167–81.
Suzuki, A. (1975), *An Extension of the Häfele–Manne Model for Assessing Strategies for a Transition from Fossil Fuel to Nuclear and Solar Alternatives*. Research Report RR–75–47. Laxenburg, Austria: IIASA.
—— and L. Schrattenholzer (1974), *Sensitivity Analysis on Hydrogen Utilization Factor of the Häfele–Manne Model*. Research Memorandum RM–74–30. Laxenburg, Austria: IIASA.
Wynne, B. (1984), 'The Institutional Context of Science, Models, and Policy: The IIASA Energy Study'. *Policy Sciences*, 17: 277–320.
Zalai, E. (1982), *Foreign Trade in Macroeconomic Models: Equilibrium, Working Paper, Optimum, and Tariffs*. WP–82–132. Laxenburg, Austria: IIASA.

PART III

Negotiated Futures

4

Energy Forecasting in West Germany: Confrontation and Convergence

Hans Diefenbacher and Jeffrey Johnson

INTRODUCTION

THE energy situation of the Federal Republic of Germany (FRG) changed radically during the 1970s. Energy was a rather unimportant political topic before 1973.[1] It had been completely self-evident to the public that enough (imported) energy in the form of oil from the Middle East would be available to satisfy the enormously growing needs of the West German economy.[2] The initial doubling of the crude oil price and the temporarily 'car free' Sundays in 1973 destroyed this complacent attitude.[3] Nothing has remained self-evident on the energy scene since then.

Energy policy is not characterized by a broad consensus regarding the objective of 'away from oil' and a major political struggle over the new energy system and the path leading to it (Meyer-Abich and Dickler, 1982: 221).

Three areas of energy planning and forecasting can be identified before 1973 and the switch to the social consensus of 'away from oil':

(1) government and private planning with respect to the introduction of nuclear energy;
(2) planning and forecasting activities of the (publicly controlled) electric utilities in pursuit of their economic management objectives;
(3) planning and forecasting activities of the (privately owned) oil industry and the (publicly controlled) coal industry.

The beginning of the oil price inflation in 1973 brought also with it an inflation of energy forecasts. The competiton between different scientific, business, and political groups led not only to the development of forecasting methodologies and to the multiplication of forecasts, but also to the perfection of the forecast as an instrument of politics.

Before going into some of the details of this, we shall first present some information on the energy situation and on major forecasting actors. We then present the major forecasting results, ordering them by their year of completion or publication. There are two reasons for this approach. First, the forecasting mistakes have been in no area as drastic as in the energy area. A doubling of the electricity consumption every decade since the end of World War II led experts in 1970 to expect an eight to tenfold increase in the consumption of electricity by 2000 and even a consumption thirty times higher by 2020 (see Rudzinski, 1983). Secondly, this approach reveals an interesting pattern in the relationship between forecasts and concurrent changes in energy consumption at the time of their publication.

We link forecasting results to the authors of the forecasts, discussing not only forecasting results but also methods and intentions. We complement this description with an analysis of the events around the creation of the Enquête Commission, a parliamentary commission charged with an investigation of the future of nuclear energy policy. The work of this commission demonstrates in exemplary form the link between politics, forecasting methods, and the implications of forecasting results. We conclude with methodological and theoretical remarks that should be considered as working hypotheses until further investigation of the politics of forecasting can provide additional knowledge of this little studied topic.

However, the reader should keep in mind the political context within which the forecasting events described here had been taking place. It was not a peaceful process that led to the creation of the Enquête Commission. There were frequent confrontations between demonstrators and the police at nuclear power plant building sites, and there was a permanent challenge of political and scientific decisions with respect to nuclear power in the administrative courts by the political and scientific representatives of 'citizen initiatives' (*Bürgerinitiativen*).[4]

The challengers of the nuclear energy policy were also often seen as part of the wider challenge to state authority then taking place. They and the providers of scientifically and technically sound arguments had to brace public and emotional campaigns. This conflict, partly fought with forecasts and models, was all but a genteel debate at an academic energy forecasting conference.

THE WEST GERMAN ENERGY SITUATION

Primary energy consumption (PEC) peaked in 1979 at 408 mtce.[5] It then fell for three consecutive years by a total of 11 per cent before showing again a slight increase of less than 1 per cent in 1983. The post-war period up to 1973 is characterized by a massive expansion of the energy supply and an equally impressive substitution of coal by oil and gas in all sectors of the economy. Electricity more than tripled its share of the energy market, reaching 15 per cent in 1983 (see also Meyer-Abich and Dickler, 1982).

However, the middle of the 1950s witnessed the emergence of the fear of a future energy gap, the '*Energielücke*'. This fear was fuelled in part by the coal shortages between 1949 and 1953 and the assumption that oil imports could not be adequately increased. This assumption turned out to be wrong by the end of the 1950s. But the research centre of the energy industries at the Technical University in Karlsruhe forecasted in 1955 the onset of an energy crisis in 1975, peaking in the year 2000 with unforeseen consequences, a forecast which rested on the postulated fixed relationship between energy consumption and GNP (Radkau, 1983: 113 f.). Radkau points out that this fear of a future energy crisis was an important, although not a decisive, argument behind the nuclear energy euphoria in the mid-1950s.

Economic growth was considered during the entire post-war period as an important source of political stability. Deliberate policies to maintain growth were started in 1967 with the passage of the Economic Stabilization Act (the *Stabilitätsgesetz*) (Meyer-Abich and Dickler, 1982: 223). The German model of economic development required the gradual substitution of increasingly expensive coal with ever cheaper oil. But this substitution, on the one hand, substantially increased dependence on imported energy resources rising from 8 per cent in 1960 to a peak of 60 per cent in 1977. The risks inherent in this development were perceived well before the onset of the oil crisis and protection was sought in the promotion of a nuclear energy research programme (Meyer-Abich and Dickler, 1982: 223). The reduction of domestic coal production, on the other hand, led to unemployment problems in the coal-mining areas. The political costs were kept in check thanks to the general economic growth and the increasing use of incentives to help make the electricity industry burn domestic coal.[6]

The energy consumption growth rate began to slow down as of 1973, that is, antecedent to the oil price shock. Oil had by that time already begun to lose market share. (Its share peaked in 1971 at 60 per cent.) The period after 1973 is identified with energy conservation activities. But the specific energy use of nearly all industries has in fact been declining since 1950 despite low and relatively weakening energy conservation efforts. These have begun to be exploited in response to the increased energy costs. But significant further progress is obviously possible when one considers that a full 63 per cent of primary energy consumption is lost for the provision of end-use services in the course of transformation and distribution (Ruske and Teufel, 1980: 10).

FORECASTS AND FORECASTING INSTITUTIONS

A large number of institutions have at one time or another become engaged in the business of forecasting energy supply and/or demand. Five main groups can be distinguished in general by using the criteria of the institutions' links with the interest groups participating in the energy discussion (see Figure 4. 1):

(1) electricity utilities;
(2) other energy companies;
(3) economic research institutes with identifiable energy divisions, working for utilities, other energy companies and government;
(4) research institutes, with mainly an ecological or system dynamics orientation;
(5) government and government-related institutions.

These institutions taken together have dominated the scientific and public debates about the future energy system of the FRG. We discuss below each of these groups, focusing on their reasons for becoming engaged in the energy debate and on the forecasts they have been publishing.

Energy forecasting activity was heavy after 1973 (see Figure 4.2). There was no year between 1973 and 1981 without the publication of a new energy forecast (see Table 4.1); 1980 was the peak year with the final report of the parliamentary Enquête Commission. This event certainly closed the debate, if only temporarily and out of exhaustion of the participants and the modellers.

The picture of all these forecasts suggests a process of convergence of establishment and alternative forecasts towards an increasingly low energy future (see Figure 4.2).

Electricity utility and other energy companies

For two reasons, electricity utilities have played a role in the forecasting history of West Germany. First, increasing power plant size and construction lead times have lengthened the utilities' planning horizon. Secondly, the saturation of traditional electricity markets has led the utilities to try to penetrate other energy markets. This has required all energy forecasters to take explicit account of substitution possibilities.

Power plant size has grown parallel to the size of the utilities, and they in turn have been accompanied by increasing concentration in the industry. This complex of development has increased the time needed for the planning and building of base-load power plant to now more than ten years, especially in the case of nuclear plants. Plant life of about twenty-five years therefore imposes a planning horizon of thirty-five years if not substantially more. These plans make economic sense only if the plants' full utilization is assured. Longer-term as well as accurate forecasts have therefore become of importance for the decision making and the economic well-being of the utilities. The suppliers of the utilities, such as the reactor builders (now only KWU) and the coal industry, have similarly been forced to second-guess the utilities over ever longer time periods.

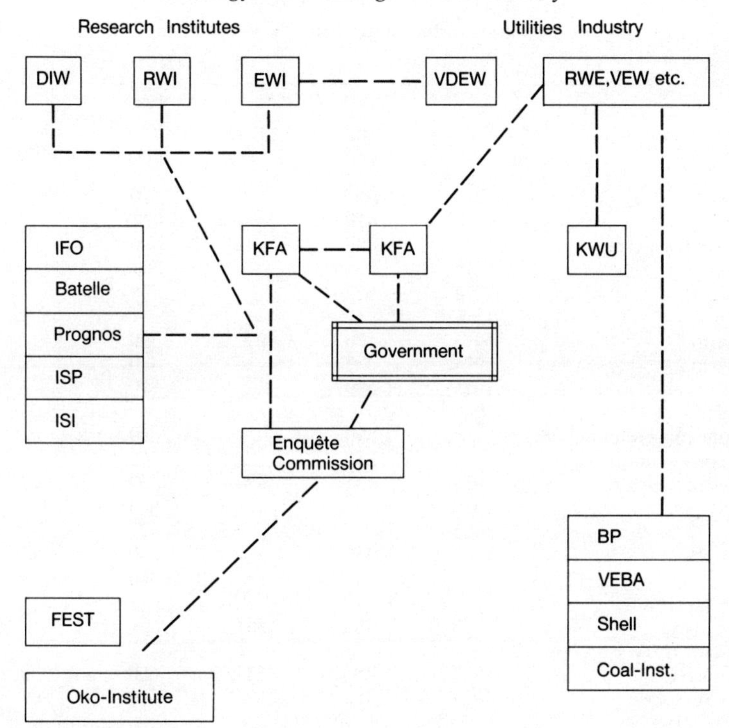

Fig. 4.1 Forecasting institutes and their links in 1980

Utilities
Rheinisch-westfälische Elektrizitätswerke (RWE)
Vereinigte Elektrizitätswerke Westfalens (VEW)
Vereinigung der Deutschen Elektrizitätswerke (VDEW)

Industry
British Petroleum (BP) VEBA AG
Shell AG Kraftwerksunion (KWU)

Research Institutes associated with the Utility Sector
Energiewirtschaftliches Institut an der Universität zu Köln (EWI)
Forschungsstelle für Energiewirtschaft an der Universität München

Economic Research Institutes with Energy Economic Divisions
Deutsches Institut für Wirtschaftsforschung, Berlin (DIW)
Rheinisch-westfälisches Institut für Wirtschaftsforschung, Essen (RWI)
IFO-Institut für Wirtschaftsforschung, München (IFO)
Prognos-AG, Basel

Independent Institutes for Systems Analysis
Battelle-Institut, Frankfurt/M
Institut für angewandte Systemforschung und Prognose (ISP), Hannover
Institut für System–und Innovationsforschung (ISI), Karlsruhe

Government Financed Institutes
Kernforschungsanlage Jülich (KFA), Abteilung STE
Kernforschungszentrum Karlsruhe (KFK)

Church Institutes
Forschungsstätte der Evangelischen Studiengemeinschaft (FEST), Heidelberg

Ecology Orientated Institutes
Institut für angewandte Ökologie (Öko-Institut), Freiburg/Darmstadt
IFEU-Institut für Energie und Umweltforschung (IFEU), Heidelberg
Gruppe Ökologie (GÖK), Hannover

Table 4.1 Prognoses and Scenarios of the Future Energy Needs of the Federal Republic of Germany

Institute	Publ. *Year*	1985 Prim.En. *(mtce)*	Electr. *(TWh)*	2000 Prim.En. *(mtce)*	Electr. *(TWh)*
Prognos	1973	(800)		1350	
BMWi	1973	610		(920)	
Deutscher Bundestag	1974	600		900	
BMWi	1974	555		(760)	
KFA Jülich	1974	550		760	
Bossel					
Scenario 1	1974	520		700	
Scenario 2	1974	460		440	
BMFT	1975	590		890	
Fichtner	1975				
without conservation		590		860	
with conservation		550		760	
KFA Jülich	1975	485	538	760	960
Ziesing	1976	494		683	
Bossel	1976				
1		450		540	
2		440		400	
DIW, EWI, RWI:					
reference case	1977	496	561	—	—
2. Fortschr. (DIW)					
reference case	1977	482.5	534	600	900
alternative case	1977	482.5*	534*	560	800*
Deutsche BP AG					
Variant 1	1977	480	525	625	975
Variant 2	1977	—	—	475	—
ISP					
reference scenario	1977	437	439	584	687
conservation scenario	1977	418	—	495	552
FEST					
Scenario 1	1977	555	—	760	—
Scenario 2	1977	445	—	530	—
Shell AG	1978	—	—	—	600
AUGE	1978	440	485	604	690
Deutsche Shell AG					
Scenario evolution	1979	464	—	513	648
Scenario disharmonies	1979	430	—	435	518
VEBA AG					
high variant	1980	450	415*	550	635*
low variant	1980	435	431*	500	578*
Enquete Commission					
Path 1 (high)	1980	—	—	600	749
Path 2	1980	—	—	445	382
Path 3	1980	—	—	375	317
Path 4 (very low)	1980	—	—	345	293
KWU					
Path 1	1980	—	—	—	850
Path 2	1980	—	—	—	750
Path 3	1980	—	—	—	650

Table 4.1—Continued

Institute	Publ. *Year*	1985 Prim.En. *(mtce)*	Electr. *(TWh)*	2000 Prim.En. *(mtce)*	Electr. *(TWh)*
Öko-Studie					
Scenario coal and gas	1980	—	—	310	264
Sc. extrapolation	1980	—	—	293	264
Sc. solar and coal	1980	—	—	298	264
3. Fortschreibung	1981	416–433		469–497 (1995) 499–534 (extrap. for 2000)	
VDEW	1981	—	—	—	734
Deutsche Shell AG	1982	—	—	336–435	—
Prognos	1984				
Scenario 1	1984	378.4		338.7	
Scenario 2	1984	380.3		369.4	
Scenario 3	1984	382.8		381.4	

* Domestic consumption excl. of own consumption, energy for pumped storage, losses and exports.
Sources:
FEST (1977), Krause *et al.* (1980), w/o author (1983). Exact references are given in the Bibliography at the end of the chapter.

The gains in economies-of-scale of ever larger central power generation units—which are apparently rapidly diminishing—are purchased at the price of increasing uncertainty. This uncertainty has been countered in the past with the provision of expensive overcapacity. Its costs were passed on to the electricity consumers through higher prices (Ford and Youngblood, 1982). This development provided another reason for the improvement in the quality of forecasts. Furthermore, this building strategy could increasingly be justified in the face of rising popular resistance with reference to high quality forecasts only. The paradox of this situation is, however, that the development requiring this quality jump in forecasts is also creating a context in which forecasting is becoming more difficult.

Declining growth rates in electricity consumption and, in their wake, growing generating overcapacity, and the (technically biased)[7] conception of power-plant economics forced the utilities to look for new markets, not only in order to match capacity with demand, but to find the quality and flexibility of demand which would allow using base-load capacity and which would answer to the future uncertainty. But this turned an electricity forecasting exercise into a differentiated energy forecasting problem, thereby providing an opening for policy evaluation and scenario analyses.

Oil companies have also been very active in the forecasting business. The oil price developments had of course changed the parameters on which

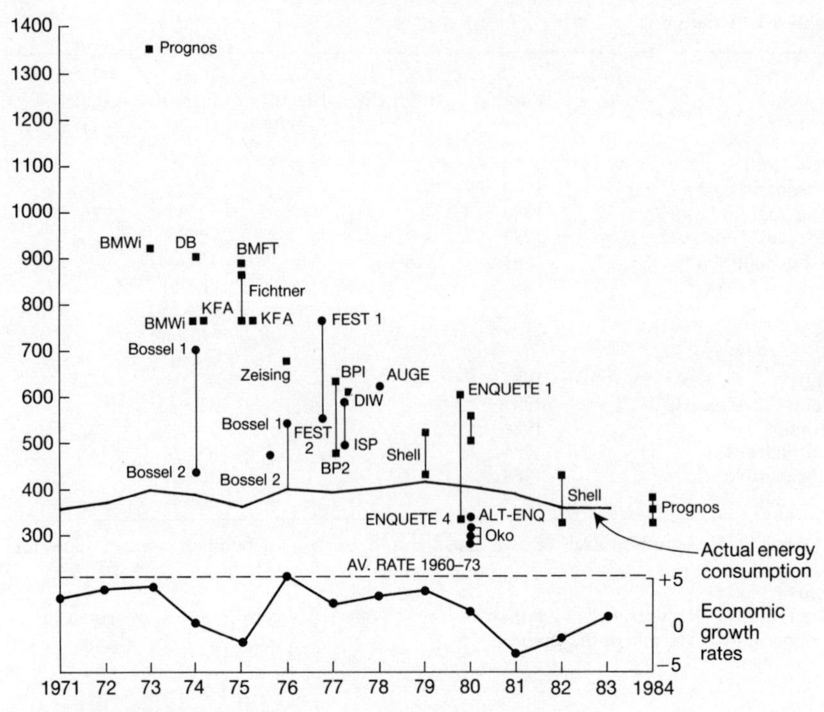

Fig. 4.2 W. Germany
Forecasts of primary energy consumption in year 2000 (in MTCE)

Prognos: Prognos-AG Basel
BMWi: Bundesministerium für Wirtschaft
DB: Deutscher Bundestag
KFA: Kernforschungsanlage Jülich
Bossel: Institut für Systemtechnik und Innovationsforschung
BMFT: Bundesministerium für Forschung und Technologie
Fichtner: Fichtner Beratende Ingenieure GmbH
Ziesing: Ziesing, H. J.: Der künftige Energieverbrauch in der Bundesrepublik Deutschland
DIW: Deutsches Institut für Wirtschaftsforschung
BP: Deutsche BP AG
ISP: Institut für angewandte Systemforschung und Prognose
FEST: Forschungsstätte der Evangelischen Studiengemeinschaft
AUGE: Arbeitsgruppe Umwelt, Gesellschaft, Energie der Universität Essen
Shell: Deutsche Shell AG
VEBA: VEBA AG
ENQ: Enquête Commission
ENQ-ALT: Ecological Scenario of the Enquête Commission
Öko: Institut für angewandte Ökologie

they had based their plans. Forecasting had always been part of their normal planning procedure. But these forecasts began now to be made public in an attempt to influence public opinion and, through it, government decisions.

The most interesting forecasts were made by BP (in 1977), Shell (1978, 1979, and 1982), VEBA (1980), VDEW (1981), and KWU (1980) (see Figures 4.2 and 4.3). The interest lies not so much in their methodology as

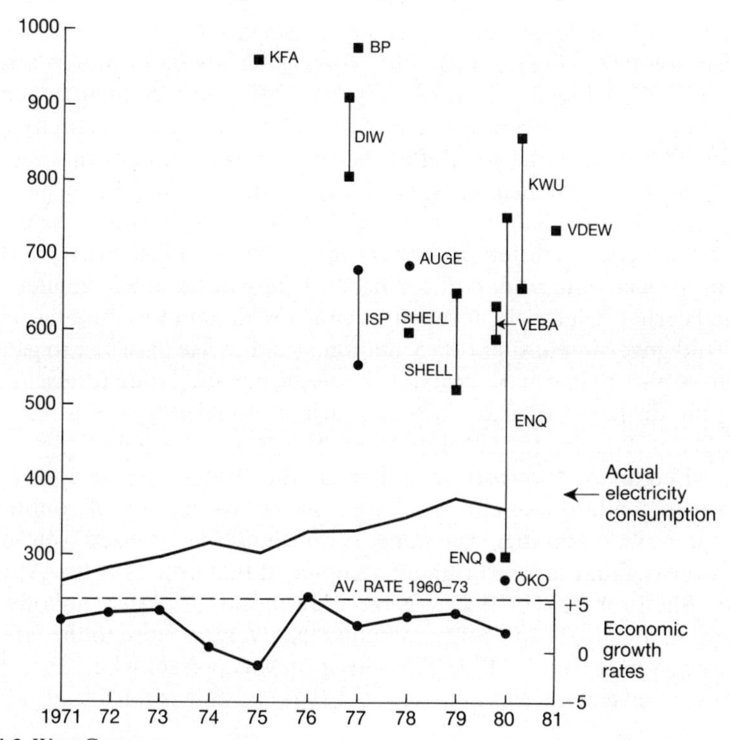

Fig. 4.3 West Germany
Forecasts of electricity consumption in year 2000 (in TWH)

KFA: Kernforschungsanlage Jülich
DIW: Deutsches Institut für Wirtschaftsforschung
BP: Deutsche BP AG
ISP: Institut für angewandte Systemforschung und Prognose
AUGE: Arbeitsgruppe Umwelt, Gesellschaft, Energie der Universität Essen
SHELL: Deutsche Shell AG
VEBA: VEBA AG
ENQ: Enquête Commission
ÖKO: Institut für angewandte Ökologie
KWU: Kraftswerksunion
VDEW: Vereinigung der Deutschen Elektrizitätswerke

in their assumptions and results. The forecasts were based on standard, sectorially disaggregated models with fixed coefficients, and assumed high rates of economic growth along with other assumptions on the general economic and demographic development. This method has obvious structural implications. The energy price level, for instance, does not affect the level of economic activity. Energy policy influences energy demand only through its effects on technological developments. Supply is whatever earlier decisions with respect to future capacity have made it to be. The model simply forces demand to adjust to available supply and resolves the problem of the modal split.

A comparison of the forecast results reveals the much greater uncertainty about future electricity developments (Figure 4.3) than the forecasts for PEC (Figure 4.2). PEC forecasts converge towards today's consumption level while electricity forecasts by energy companies remain substantially above current consumption. (The scarcity of electricity forecasts for 2000 made in the years after 1980 possibly falsifies this analysis. But this absence is in itself an interesting observation, reflecting the socio-political struggles over nuclear power.) Electricity forecasts of course reflect the competition between different energy carriers (above all electricity and oil) over future shares in a market for heat which will be much smaller than assumed earlier. Electricity forecasts are the instrument to assure oneself of public and government support to maintain (in the case of oil) or to gain (in the case of electricity) market shares. Ecologically orientated forecasts, reflecting the dislike of their backers for nuclear electricity, even foresee the possibility of reducing the absolute level of electricity consumption.

The oil company forecasts are relatively moderate. The BP forecast of 1977 was a standard one and was based on the assumption of continuous economic growth and increasing energy consumption. It fitted well to the other forecasts that the government accepted at that time as valid. Already the first Shell forecast in 1978 was quite a sensation because nobody had expected figures from an energy company that were so close to the alternative forecast prepared by ISP. The sensation was perfect when Shell published even lower forecasts in its second (1979) as well as third study (1982) (Figures 4.2 and 4.3).

Shell is a company known world-wide for its sophisticated planning department with a feeling for socio-political developments in its markets. But oil companies in general had to face during the 1970s the fact that their expansion was likely to be over. Their strategic goal instead became assuring survival and market share. Changed relative prices and power shifts suggested the need for, among other things, a reduction in refining capacity. The resultant lay-offs in a situation of good profits could be justified with relatively moderate energy forecasts.

Other energy companies followed only reluctantly this trend of lower forecasts. VEBA, with large interests in oil and electricity, published a forecast close to the one of ISP, but three years later and after PEC had again started to fall. The VDEW study (1980, Figure 4.3) concentrated on analysing the development of the heating market. Its purpose was to push the use of electricity for space heating, thus using the official slogan of 'away-from-oil' to buttress its own interests as an electricity utility. Realization of its forecast would have required a doubling of electricity production between 1980 and 2000. The KWU forecast (1980, Figure 4.3) predicts an electricity demand level similar to the VDEW study. Understanding this forecast in the narrow, but traditional meaning of the term as the prediction of an independent, unaffectable variable, would have suggested the building of one or two nuclear power stations per year. One should know that KWU is the only remaining commercial reactor builder in West Germany.

Of course, even low forecasts from energy companies are not all that a quick glance at forecast energy consumption levels suggests. The Shell forecast of 1982 contains a substantial downward revision of its two perspective—scenarios in all but name—with the high perspective now assuming PEC in 2000 equal to the low perspective of the 1979 study. This high perspective, assuming a 25 per cent increase in consumption compared to 1982, is called *Strukturwandel* (structural change) and assumes the societal capacity to manage a benign transition to new energy structures. The low perspective, foreseeing a reduction of 1982 PEC by about 10 per cent is called *soziale Disharmonien* (social disharmonies). It assumes social conflicts over the employment problem, social disintegration due to changing attitudes to work and due to the development of an informal sector, and the rigid defence of the economic interests of the formal, increasingly bureaucratic sector. It is social conflict that keeps economic growth and energy consumption low, not deliberate energy policy as in the other low forecasts (of, for example, the Enquête Commission or the Öko-Institut).

Alternative forecasting institutes

Two alternative energy forecasting exercises merit special attention. The ISP study of 1977, while not the first alternative low-energy-growth forecast, is the first to use an alternative methodology, system dynamics. The *Energiewende* study of the Öko-Institut in Freiburg (the Institute for Applied Ecology) in 1980 is the ecological movement's vision of the future. Developing policy-based scenarios, it suggests that future energy consumption can be reduced to the levels prevalent at the end of the 1960s, and can do so with renewable instead of nuclear energy. This study also pioneered looking towards the time horizon of 2030 and it played an important role in the work of the Enquête Commission discussed later (Krause *et al.*, 1980).

The ISP system dynamics model

ISP, the Institute for Systems Analysis and Forecasting, had been founded by Eduard Pestel, the author of the second Club of Rome report (Pestel and Mesarovicz, 1974). Pestel was director of ISP at the time the energy study was initiated. The institute's methology is system dynamics, and its so-called 'Deutschland-Modell' was the first energy model based on this methodology (Pestel *et al.*, 1978a and b).

This energy model consists of a number of submodels for energy, economic production, the labour market, population, and education. Submodels are connected in such a way that a model providing data inputs into another model does not itself rely on data inputs from the latter model. The energy submodel calculates first final demand for the three sectors of industry, services, and households using the developments in the economic and the population submodels. Final energy demand is converted into PEC through a conversion matrix which itself can be changed with the help of a number of 'scenario' variables. Thus different energy systems can be expressed and their consequences explored. Another original feature of the energy submodel is the disaggregation of the electricity sector with the help of a load curve. This allows explicit determination of base, middle, and peak load capacity needs and allows therefore the evaluation of different power station configurations.

This apparatus produced forecasts which at the time were at the lower end of the forecast spectrum, especially for the near future (see Figure 4.2). But the conservation 'scenario' still foresaw an electricity consumption in 2000 about 60 per cent above actual 1977 consumption. Nevertheless, the gap between establishment forecasts and the ISP electricity forecast was more substantial than for the energy forecasts (see Figure 4.3) and this explains in part the signal effect of this study.

The 'Energiewende' of the Öko-Institut

The Öko-Institut in Freiburg (the Institute for Applied Ecology) had over the years become centrally involved in providing technical expertise to the various citizen initiatives that were challenging nuclear power plant projects and related installations. The experience of this challenge in political, and above all judicial, institutions led to the decision to oppose an in-house energy analysis to the industry and government forecasts which were used to justify the nuclear power programme. Given the institute's alternative perspective on the present and the future economic, social, and energy situation in the FRG, this analysis could not rely on a traditional forecasting methodology.

The *Energiewende* published in 1980 (Krause *et al.*, 1980) innovated in two areas: it systematically used the scenario methodology and it used the concept and the calculation of energy services as starting point of the analysis.

The *scenario methodology* demonstrates that the future energy demand is no longer the result of current trends and a passive acceptance of their extrapolation into the future. Rather it results from the full exploitation of currently existing and expected technologies for an efficient production and use of energy. Implicitly this analysis suggests the taking of policy decisions to bring about this efficient production and use pattern.

To emphasize the feasibility of this low-energy future, the analysis relied on growth assumptions which were 'unrealistic' from the perspective of the study's authors: they assumed increasing welfare (measured by GNP) and growing energy services (which means an increase in the use of energy-consuming machines and appliances, not an increase in total energy consumption).

An *energy service* is defined as the effect, the result of using an energy flow: for instance, the heating of a room to a certain temperature or the transportation of a person over a certain distance within a certain time.[8] The *Energiewende* postulates that increased energy services in 2030 could be provided with a PEC equal to 60 per cent of current consumption. Other main results of the study were as follows:

1. It is feasible to supply the (reduced) energy needs in 2030 without oil or nuclear energy. Nuclear energy is not necessary in any of the three scenarios (compared to a 15 per cent contribution in 1980). Both scenarios 'solar and coal' and 'coal and gas' show possibilities for an oil-free economy by 2030. 'Solar and coal' demonstrates a totally domestic supply of energy.
2. Energy demand may be reduced to a value that is below today's demand even though the energy services provided will increase. This effect will be achieved by more efficient energy use.
3. Energy consumption tends to increase until the energy saving methods are successively introduced. By the year 2000, the energy demand is on a downward trend; the absolute minimal value is attained only much later. This result is even more significant, considering that only today's technologies were assumed (for instance, no photovoltaic technologies).

Energiewende made the point that future energy demand is not a technical or economic necessity but instead is the consequence of individual and political action to reach *the* energy-efficient society. The study purports to show that there is no need for technological solutions requiring increasing imports of oil or the use of nuclear power.

The energy programme forecasts of the federal government

We pointed out at the beginning that the federal government was not very much engaged in energy policy-making before 1973 and that even after that time it played a rather passive role in this area, letting market forces push the system in the desired direction. The government started to elaborate its first energy programme in 1973 and has updated it and its forecasts already three times since then. This programme is however more an information and co-ordination tool than a firm statement of government policy.

The government has, however, always been involved in supporting two energy resources. The domestic coal industry has been subsidized and protected mostly for employment and regional development reasons. The government also became an early and enthusiastic backer of nuclear technology, mostly for industrial policy reasons.[9] The government has, therefore, definitely an interest in the valorization of coal and nuclear power. The government is therefore not a neutral provider of energy forecasts. The actors linked to the nuclear power industry have always produced the highest energy forecasts. Government forecasts have consequently acquired a double character: they are forecast or prediction and policy

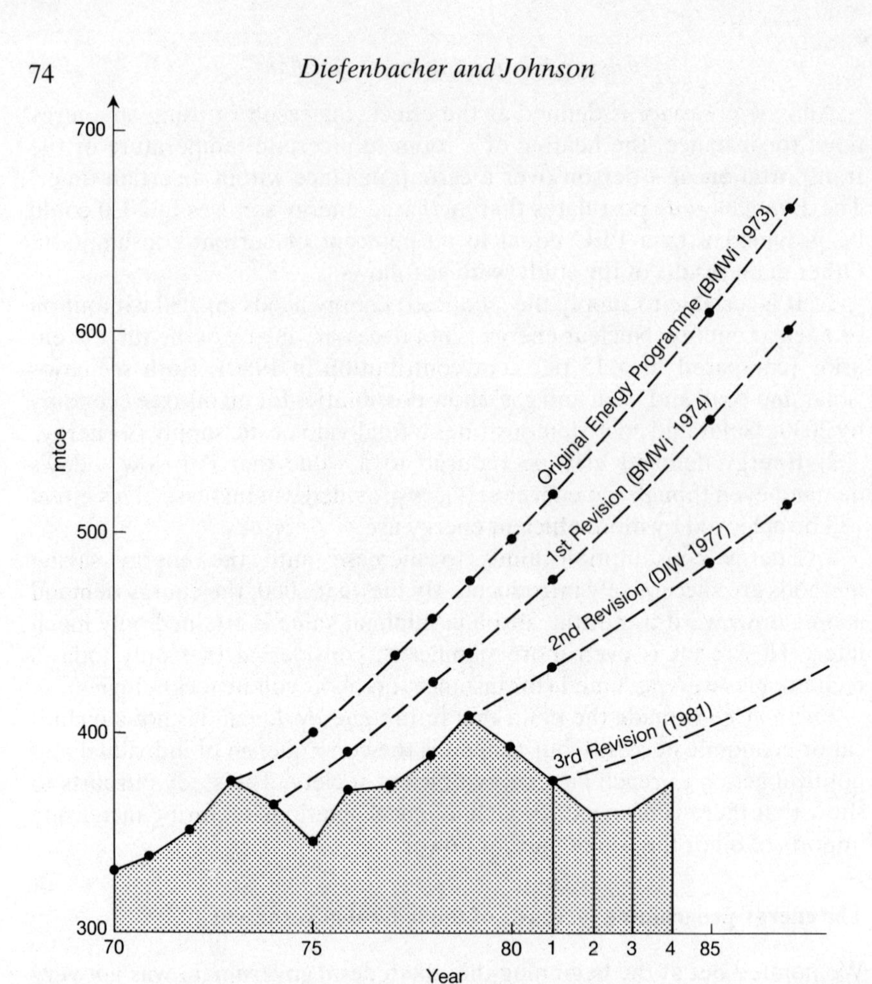

Fig. 4.4 Comparison between actual primary energy consumption and the forecasts of the Federal Energy Programmes

statement at the same time. The energy programmes are instruments of orientation for and of co-ordination between government agencies, energy companies, and industry. They also serve to legitimize government decisions made in the past as well as those the government would like to take in the future. This explains the expansive nature of the four forecasts published by the government so far (identified as BMWi (1973, 1974), DIW (1977), and 3. Fortschr. (1981)) (see Figure 4.4).

The government has always relied on academic research institutes with an energy expertise (of the traditional kind) to prepare its energy forecasts. DIW and RWI are the two economic research institutes with separate energy divisions, and the EWI at the University of Cologne is *the* energy institute in Germany. These three institutes prepared a common forecast for the original energy programme. The government published its version alongside this forecast in 1973. The procedure was the same for the first and second revisions. The third revision failed to include the institute's forecast although the government admited its plausibility. However, the institutes had added a third, lower case to the original two cases when it became clear during the year that actual energy demand growth was below the one assumed initially. But this adjustment was not welcomed by the government.

Both DIW and RWI used econometric models with detailed energy sectors to elaborate their forecasts for the original governmental energy programme. Their economic growth assumptions concurred with government views although the rates used of between 2 and 2.7 per cent proved later to have been wishful thinking (see Figure 4.2). None the less, the forecasts indicated a stagnation of final energy demand in all sectors. But PEC was thought to continue to increase at approximately half the rate of GNP growth. 'The clearly stronger increase of PEC compared to final energy consumption is above all explained with the above average growth in electricity production and the concomitant increase in transformation and transportation losses.'[10]

The scenario analysis of the Enquête Commission[11]

The creation of the Commission of Inquiry into the Future Nuclear Energy Policy of the West German Parliament (Enquête Commission) was one of the most important events in the history of West German energy policy.[12] The outcome of this inquiry and the recommendations made by the commission were to a large extent dependent on the commission's evaluation of the different energy demand forecasts available at the time. Surprisingly enough, the commission came up with a set of ideas on energy policy which differed in many aspects from previous assumptions and decisions. For the first time in West German energy policy-making, an official parliamentary commission admitted to having doubts about the future use of nuclear energy. The commission's final report of 25 June 1980 states (Altner, 1982: 54):

> Provided that the measures listed below concerning technological development, energy conservation, and greater use of renewable energy resources are vigorously undertaken and pursued, it should be examined in about the year 1990 whether
> — a definitive extension of the use of nuclear energy with breeder reactor systems and all they entail will prove to be necessary,
> — or whether the use of nuclear energy is to remain restricted to nonbreeder reactors,
> — or whether it is possible to even forego the use of nuclear energy in the future.

The commission's formation, organization, and work procedures

To understand the genesis of these startling recommendations it is important to look at the political process leading to the creation of the commission and at the work procedures it adopted. The Enquête Commission was specifically instructed by the parliament to examine the possibilities and consequences of using *and* of abandoning nuclear energy. Yet all the parties represented in parliament at that time were officially in favour of nuclear energy.[13] But since around 1976, sizeable minorities in each of them began to become sceptical about the future use of nuclear energy. Official party information therefore made increasing use of the formula of nuclear energy 'as little as possible and as much as necessary'. But the majorities were always voting in such a way as to make it clear that they thought nuclear energy was possible *and* necessary in considerable quantities.

The anti-nuclear movement had by that time already succeeded in substantially delaying the further expansion of nuclear power generation. Above all, a large group of politicians had become reluctant to push the building of new nuclear power plants. They preferred to gain time and to avoid, at least in part, taking responsibility for the future course of action. This situation helps explain the events around the fast breeder reactor (SNR 300) in 1978. The construction licence for the third building stage was by that time long overdue. The constitutional court had decided that fast breeder reactors could be built on the sole legal basis of the Nuclear Power Law. It was therefore upon the government of Nordrhein–Westfalen to grant the construction licence. This government was at the time formed by a coalition of SPD and FDP, similar to the federal government. The government of the *Land* was however divided on this issue and preferred to pass on its power to the federal parliament.

In the meantime, the FDP party congress had decided against the completion of the SNR 300 in Kalkar and asked for the constitution of a commission to evaluate risks and benefits of a further expansion of nuclear power. However, the FDP ministers in the federal government threatened to resign unless the party's members in parliament disavowed this decision. The FDP parliamentary group therefore abstained when parliament voted on this issue on month after the FDP congress. In this way a three part compromise came about:

(1) continuation of the construction of the SNR 300;
(2) deferral of the decision to commission the SNR;
(3) constitution of the parliamentary Enquête Commission to evaluate the fast breeder option.

The political discussion around the formation of the commission led to the appointment of critics of nuclear energy, a first in the history of parliamentary commissions of inquiry. These discussions also produced a remarkable broadening of the commission's mandate. Its task was now not only to judge the merits of the fast breeder reactor, but also to estimate the technical risks and the social consequences of the nuclear energy option as such. In addition, it also was to investigate the possibilities for and the consequences of terminating the use of nuclear energy.

Both developments, the appointment of critics and the extension of the mandate, were clearly a concession to the sizeable minorities who, within the large parties, were against nuclear power. This constellation of forces also led to further innovations with respect to the operation of the commission. It received a secretariat and a scientific staff whose members were chosen personally by the individual commission members. The commission were also the first one to acquire the right to call on experts critical of nuclear power.

The commission itself consisted in the end of fifteen members:

seven members of parliament: Gerlach, Gerstein, Stavenhagen from the CDU/CSU; Reuschenbach, Schaefer (*), and Ueberhorst (*), as chairman, from the SPD; and Laerman from the FDP;
eight outside experts: Altner (Öko-Institut and FHS Koblenz) (*), Birkhofer (GRS and TU München), von Ehrenstein (University of Bremen) (*), Häfele (IIASA, now KFA Jülich), Knizia (VEW), Meyer-Abich (GH Essen) (*), Pfeiffer (DGB) and Schäfer (TU München).

Five of these fifteen members were more or less opposed to nuclear power (here marked with (*)). The others had precedingly revealed themselves as more or less strong advocates of the nuclear path.

The commission organized its work in seven working groups with between two and five commission members, including always the chairman, actively participating (Conrad, 1982):

(1) Future energy demand and the necessity of nuclear energy (energy paths).
(2) Societal implications of energy-supply patterns and risk-benefit-assessments of various energy sources (safety risks).
(3) Criteria for the acceptance of nuclear energy and other energy sources (criteria).
(4) Recommendable decisions on the fast breeder reactor technology, especially on the SNR 300 (fast breeder).
(5) Acceptable suggestions for nuclear waste management (back end of the fuel cycle).
(6) Acceptable suggestions on alternative fuel cycles, including the results of INFCE.
(7) Possible contributions of the West German nuclear energy policy towards non-proliferation (non-proliferation).

This organization of the commission's work provided—at least in principle—opportunites for intensive discussions and confrontations of viewpoints. There were twenty-two plenary meetings and twenty-four reunions of the working groups. The commission also contracted for research, organized expert hearings, and undertook information trips. A large number of working papers were written, discussed, and considered for the final, 600 page report.

The Enquête results

The results of the first working group merit of course our special attention. It too has to be stressed here from the beginning that the investigative methodology and the presentation of the findings were rather revolutionary for the work of a parliamentary commission. This difference is especially striking when compared to the government's own energy programme and its subsequent revisions, shown in Figure 4.4. An official body, backed with the prestige of the parliament, demonstrated for the first time through the use of energy path analysis that action alternatives do exist and what their likely consequences are for energy consumption. This demonstration swept away all arguments which used predictions of a generally rapidly rising energy demand to justify a further expansion of energy production capacity as the natural response to exogenously given developments, reflecting somehow immutable, 'natural' laws.

Figure 4.5 summarizes the basic features of the four energy paths elaborated by the commission. The 'official' scenario, Path 1, 'assumes high availability of energy and aims at a massive increase in the use of nuclear energy—165 GWe, 84 GWe alone from fast breeder reactors. However, to reach the intended goal, . . . the quantity of coal used is supposed to double until the year 2030 and no substantial savings of oil are foreseen' (Altner, 1982: 54). The declared aim is to achieve a continuously high economic growth rate.

Path 2 already assumes lower economic growth rates. But it still relies on a substantial expansion of nuclear energy, at least when compared to the present situation. Over a hundred nuclear reactors would be required in this case and 54 GWe would still have to come from fast breeder reactors. Reinforced energy conservation measures assure the lower share of nuclear energy compared to the first path, but these would not rely on tough administrative regulations and controls. The use of nuclear energy and energy conservation measures are responsible for the clear reduction in the consumption of gas and oil.

Paths 3 and 4 both assume a phasing out of nuclear reactors between 1990 and 2000. Both paths assume the same economic growth rate as path 2. The non-nuclear future is reconciled with economic growth through the implementation of severe energy conservation measures and an important use of renewable energy resources. The economy would undergo substantial structural change. The fourth path is especially systematic in the application of these strategies. It envisages therefore, in addition to the phasing out of nuclear energy, a large reduction in the consumption of oil and gas. This part is almost identical with the scenario presented by the Öko-Institut in its *Energiewende* study (Krause *et al.*, 1980 and above).

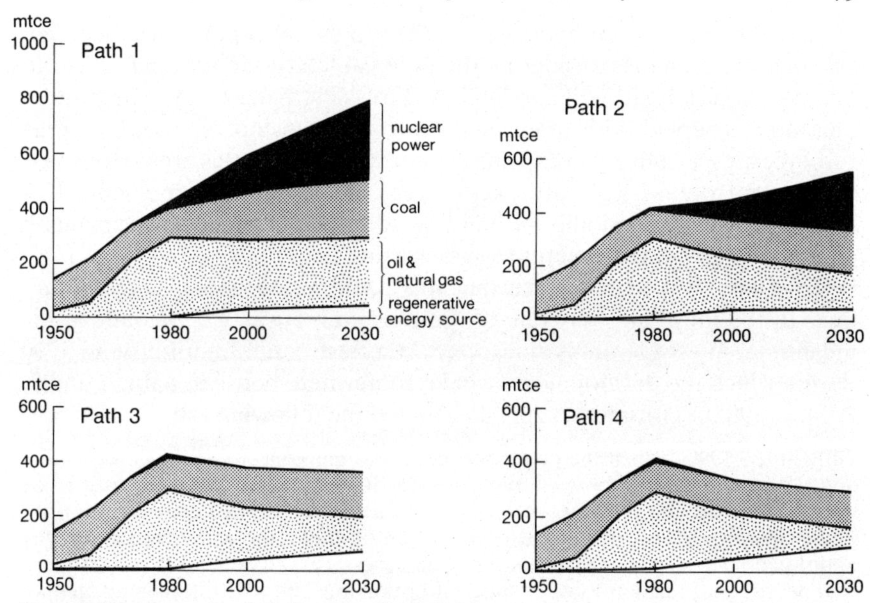

Fig. 4.5 The four paths of the Enquête Commission
All four paths result in the same energy service for consumers.

The impact on the public of this way of looking at the energy problem was and still is enormous. The calculations of the critics of the government's energy policy were for the first time made acceptable. They in fact were presented as variants of a future societal development that were to be taken seriously. All the official predictions and scenarios of the preceding years—which were all variations of Path 1—lost thereby their sheen of inevitability. They were shown to be only an extreme variant of the whole range of possible societal development paths—a variant which in addition would not even manage decisively to reduce the dangerous dependence of the country on imported oil, and this despite massive expansion in the production of coal and in nuclear generating capacity.

The contrast between Paths 1 and 2, on the one hand, and Paths 3 and 4, on the other hand, faithfully reflects both the polar opposites in the political sphere with respect to the energy problem as well as the composition of the Enquête Commission itself. On the one hand, continued high growth and welfare are to be assured through an ever-growing energy use. On the other hand, the search is on for another quality of growth and welfare characterized by a 'decreased demand for end-use energy through economical consumption (decentralized use of coal and of renewable energy resources, conservation technologies)' (Altner, 1982: 54). The conflict over the future of nuclear energy, and over the solution of the energy problem more generally, contributed decisively to making the quality of life, of growth, and of welfare a central part of today's public debate.[14]

That the commission had to conclude a political compromise does not surprise when one is considering the political landscape in which it had to operate. Basically the commission opted for increased energy conservation measures, coupled with intensified efforts to develop renewable energy resources, especially by increasing support for R&D in this area. However, the nuclear option was to be kept open, at least for the moment. This encompasses the possibility of building additional nuclear power plants if energy demand developments should warrant it.

It is important to recognize that this compromise is only a *temporary* one. In the long run, only one of the two alternatives will prevail. In the meantime, West Germany finds itself in a 'test period' until around 1990 during which the developments will lie somewhere between paths 2 and 3. Altner (1982: 5) formulates this situation in the following way:

> Only in the 1990s, when the policy recommendations concerning the conservation and utilization of renewable energy resources have been implemented, is it to be decided whether the previous use of nuclear energy was a mere episode or a prior phase leading to its definite expansion. Until then, the construction of any additional nuclear power plant that may be necessary is to be guided by consumption needs, and, furthermore, the nuclear option must be scientifically and technologically guaranteed.

This means that until the final choice in the 1990s, a 'fair competition' between the two main lines of societal development has to be guaranteed. Whether this will be so might merit a later investigation.

CONCLUSION

Here we would like to draw the readers' attention to the main lines of the forecasting development traced in this chapter.

1. Energy forecasting in West Germany has experienced a great change of forecasting methods during the last few years. The more or less superficial procedures of extrapolating past growth rates were first supplanted by econometric and system dynamic model analysis. These then gave way to scenario analyses which make political decisions with respect to energy supply the normative basis of analysis.

2. It is increasingly realized as a result of forecasting exercises that all developments in the energy sector depend on political decisions by governments.

3. The forecasts made at the beginning of the 1970s were continuously reduced over time, albeit with reluctance and great hesitation. These large corrections and the increasing discrepancy and conflict between individual forecasts contributed to discrediting forecasting in general. This has especially been the case because it was possible to link individual forecasts to political positions.

4. The large downward revisions of the PEC-forecasts have not been accompanied by a corresponding change in forecasted electricity consumption. The latter has at best been slightly revised downward. This reflects quite clearly the political problem and the use of forecasts to solve it. Some of the forecasts are designed to justify the construction of new (nuclear) power plant capacity.[15]

5. Forecasts of future electricity demand are obviously important for the investment decisions of the utilities. Yet their misallocation of resources due to wrong forecasts are apparently more serious than in other energy branches. The oil companies, for instance, began quite clearly in the energy crisis to revise their planning and to put it on a more realistic footing.

Notes

1. There was a coal policy before 1973, but it was as much an employment and regional development policy as an energy policy. Coal as well as energy policy in general followed—and still follows today—the fundamental approach 'to allow market forces to adjust to new situations'. The government follows the 'piecemeal approach, that is, energy policy by individual intervention, without explicitly taking into account the long-run aspects of energy supply and demand, the total system, or the interrelations among the different energy sectors' (Schmitt, 1982: 140, 144).
2. This is somewhat simplified. The public discussion starting in the late 1950s about the future of the German coal industry and the best ways of protecting it against foreign oil and coal included expressions of concern about the security of oil imports (see e.g. Wessels, 1964). This aspect of the problem was, however, possibly overemphasized because it legitimized state subsidies for domestic coal production despite obvious contradictions to the dominant economic philosophy (Peters, 1971: 53). State intervention in favour of coal failed because the government was never able or willing to regulate oil companies. The switch from coal to oil was basically accepted by the government in 1966–7.
3. The federal government forbade driving on six consecutive Sundays in the winter of 1973–4. This was an exceedingly grievous intervention in view of the Germans' identification of the right to drive with personal freedom.
4. On the special role and strong position of the administrative courts see Atz (1983), Thompson (1983: 305), and, more generally Reichel (1981).
5. A temporary fall of 3 and 5 per cent in 1974 and 1975, respectively, was quickly compensated by renewed consumption growth up to 1979.
6. Domestic coal production had been protected since the end of the 1950s with tariffs, a tax on heating fuel and investment subsidies. The subsidizing of the use of coal for the production of electricity began in 1965 with the first Electricity Production Act (*Verstromungsgesetz*) and has been repeatedly improved since then. The utilities agreed in 1980 to use a minimum amount of coal for the next 15 years (Schiffer, 1982).
7. The BMWi, the federal Ministry of Economics, issued the so-called 300 MW-Erlässe in 1964. These guidelines suggested that only power plants with at least 300 MW capacity should be built and that they should be considered a priori (!) to be economic. The largest power plant under construction at the time was a 300 MW unit. In 1974, similar guidelines were issued in favour of 1,300 MW nuclear and 700 MW coal-fired power plants.
8. A given energy service can almost always be provided by different technologies which in turn require different qualities and quantities of energy inputs: a room can be kept at 19°C with a large heating system and little insulation or a small system and a lot of insulation. The total energy input into the system can then be calculated as

$$\sum_i E_i = \Sigma\, ES_i \,.\, e_i$$

where *ES* is the energy service, *e* the specific energy consumption per unit of *ES*, and *E* the energy consumption. Different technology combinations have different specific energy consumptions (*e*). By introducing energy efficient technologies (i.e. technologies with low *e*), it is possible to supply increasing *ES* while both secondary and primary energy use is reduced.

9. See Radkau (1983) for an analysis of the nuclear power history in the FRG, and Keck (1981) more particularly on the breeder development.
10. Deutscher Bundestag, 1983, TZ.19. Translation by authors.
11. This section is based on the reports by Conrad (1982), Altner (1982), and Radkau (1983).
12. Such parliamentary inquiries are an established tool predating the foundation of the Federal Republic of Germany. These inquiries look into significant social issues that are likely to have great structural significance.
13. The Social Democrats (SPD) with 43 per cent of the seats in parliament were forming a coalition government with the Liberal Democrats (FDP, 10 per cent), while the Christian Democratic Union (CDU) and its Bavarian sister organization, the Christian Social Union (CSU) formed the opposition with 44 per cent of the seats.
14. An increasingly intense debate is taking place within the trade unions on the issue of 'qualitative growth'.
15. Nuclear has been consciously set in parentheses. Opinion about the type of new power plant capacity has recently begun to shift in response to changing electricity price calculations.

Bibliography

Altner, G. (1982),'Prospects for Transition to a Nuclear Free Future'. *Anticipation*, No. 29: 54–60.

Anon. (1980), 'Energieprognosen für die Bundesrepublik Deutschland für das Jahr 2000'. *Atomwirtschaft/Atomtechnik*, Nov.: 567.

—— (1982), 'Energieflussbild der Bundesrepublik Deutschland'. *Energiewirtschaftliche Tagesfragen*, 32 (3).

—— (1983), 'Die Elektrizitätswirtschaft in der Bundesrepublik Deutschland im Jahre 1982'. *Elektrizitätswirtschaft*, 82 (21).

Abeitsgemeinschaft Energiebilanzen (1984), *Energiebilanzen der Bundesrepublik Deutschland*. Frankfurt: Verlags- und Wirtschaftsgemeinschaft der Elektrizitätswerke.

Atz, H. (1983), 'Bundesrepublik Deutschland: Wenig Lärm um Wilhemshaven', in H. C. Kunreuther, J. Linnerooth, *et al.*, *Risk analysis and Decision Processes: The Siting of Liquefied Energy Gas Facilities in Four Countries*. Berlin: Springer Verlag, pp. 40–77.

AUGE—Arbeitsgruppe Umwelt, Gesellschaft, Energie der Universität Essen (1978), *Technologien zur Einsparung von Energie. Studie im Auftrag des BMFT*, Part III, Summary. Essen: Gesamthochschule.

Bossel, H., R. V. Denton, P. V. D. Hijden, W. Hudetz (1974), *Dialogprogramm zur Entwicklung und Überprüfung von Langfristkonzepten für das Energieversorgungssystem und Anwendung auf die Bundesrepublik Deutschland.* Hannover: Institut für Systemtechnik und Innovationsforschung (ISI).

BP AG (1977), *Energie 2000—Tendenzen und Perspektiven*. Hamburg.

BMFT—Bundesministerium für Forschung und Technologie (1975), *Auf dem Wege zu neuen Energiesystemen*, Part I. Bonn.

BMWi—Bundesministerium für Wirtschaft (1973), *Das Energieprogramm der Bundesregierung*. Bonn.

—— (1977), *2. Fortschreibung des Energieprogramms der Bundesregierung*. Bonn.

—— (1981), *Daten zur Entwicklung der Energiewirtschaft in der Bundesrepublik Deutschland*. Bonn.

—— (1982), *Daten zur Entwicklung der Energiewirtschaft der Bundesrepublik Deutschland im Jahre 1981*. Bonn.

Bundesregierung, Presse- und Informationsamt (1977), *Bulletin Nr. 30: Grundlinien und Eckwerte für die Fortschreibung des Energieprogramms*. Bonn.

Conrad, J. (1982), 'Future Nuclear Energy Policy—the West German Enquête Commission'. *Energy Policy*, 10 (3): 244–9.

Deutscher Bundestag (1973), *Energieprogramm der Bundesregierung*. Bonn: Bundestags-Drucksache VII/1057.

—— (1974), *Protokoll der Energiepolitischen Anhörung des Deutschen Bundestags vom 06.11.1974*. Bonn.

—— (1974b), 1. *Fortschreibung des Energieprogramms*. Bonn: Bundestags-Drucksache VII/2713.

—— (1977), 2. *Fortschreibung des Energieprogramms*. Bonn: Bundestags-Drucksache VII/1357.

—— (1980), 'Zukünftige Kernenergiepolitik, Bericht der Enquêtekommission des Deutschen Bundestags'. *Zur Sache*, No. 1/2, 1980; Bonn.

—— (1983), Bonn: Bundestags-Drucksache IX/983.

DIW/EWI/RWI (1978), *Die künftige Entwicklung der Energienachfrage in der Bundesrepublik Deutschland und deren Deckung—Perspektiven bis zum Jahre 2000*. Essen.

ESSO AG (1977), *Energie nach Mass*; *Hinweise zur sinnvollen Energie-Verwendung in der Haustechnik*. Hamburg (2nd edn.).

FEST—Forschungsstätte der Evangelischen Studiengemeinschaft (1977), *Alternative Möglichkeiten für die Energiepolitik—ein Gutachten*. Heidelberg: FEST, Texte und Materialien, Series A.

Fichtner–Beratende Ingenieure GmbH (1975): *Abschätzung der Energieeinsparungsmöglichkeit innerhalb der Bundesrepublik Deutschland*. Stuttgart.

Ford, A. and A. Youngblood (1982), 'Simulating the Planning Advantages of Shorter Lead Time Generating Technologies'. *Energy Systems and Policy*, 6, (4): 341–74.

Hampicke, U. (1983), 'Einsparung von Energie in der Industrie', in: K.-M. Meyer-Abich, *et al.* (eds.), *Energiesparen: die neue Energiequelle*. Frankfurt (2nd edn.): Hanser.

Keck, O. (1981), *Policy-Making in a Nuclear Program: The Case of the West German Fast Breeder Reactor*. Lexington, Mass.: Lexington Books.

KFA Jülich (1974), *Programmstudie Sekundärenergiesysteme* (Summary). Jülich: KFA. Jül/1148/SE.

—— (1975), *Einsatzmöglichkeiten neuer Energiesysteme*. Jülich: KFA.

—— (1977), 'Die Entwicklungsmöglichlkeiten in der Energiewirtschaft in der Bundesrepublik Deutschland—Untersuchung mit Hilfe eines dynamischen Simulationsmodells'. *Angewandte Systemanalyse, No. 1*.

Krause, F., H. Bossel, K. F. Müller-Reissmann (1980), *Energiewende—Wachstum und Wohlstand ohne Erdöl und Uran*. Frankfurt: S. Fischer.

KWU AG (1980), *Möglichkeiten und Grenzen der Ölsubstitution*. Erlangen: KWU.

Meyer-Abich, K.-M. and R. A. Dickler (1982), 'Energy Issues and Policies in the Federal Republic of Germany'. *Annual Review of Energy*, 7.

OECD, Nuclear Energy Agency (1975), *Uranium Resources, Production and Demand*. Paris: OECD.

Pestel, E. and M. Mesarovicz (1974), *Menschheit am Wendepunkt—2. Bericht an den Club of Rome*. Stuttgart: Deutsche Verlags-Anstalt.

—— R. Bauerschmidt, M. Gottwald *et al.* (1978a), *Das Deutschland-Modell. Herausforderungen auf dem Weg ins 21. Jahrhundert*. Stuttgart: Deutsche Verlags-Anstalt.

—— R. Bauerschmidt, K. P. Möller, W. Oest (1978b): 'Das Deutschland-Modell, Teil 2—Energie bis 2000'. *Bild der Wissenschaft*, 2.

Peters, H. R. (1971), *Grundzüge sektoraler Wirtschaftspolitik*. Freiburg: Rombach.

Prognos AG (1984), *Die Entwicklung des Energieverbrauches in der Bundesrepublik Deutschland und seine Deckung bis zum Jahre 2000*. Basel: Prognos.

Radkau, Joachim (1983), *Aufstieg und Krise der deutschen Atomwirtschaft, 1945–1975*. Reinbek: Rowohlt.

Reichel, P. (1981), *Politische Kultur der Bundesrepublik*. Opladen: Leske und Budrich.

Rudzinski, Kurt (1983), 'Sonne und Wind ergänzen die Kernkraft—Fehlprognosen zum Energiebedarf'. *Frankfurter Allgemeine Zeitung*, 7 December.

Ruske, B. and D. Teufel (1980), *Das sanfte Energie-Handbuch*. Reinbek: Rowohlt.

Schiffer, H. W. (1982), 'Die Elektrizitätswirtschaft in der Bundesrepublik Deutschland' *Zeitschrift für Energiewirtschaft*, 4: 228–42.

Schmitt, Dieter (1982), 'West German Energy Policy', in Wilfrid Kohl (ed.), *After the Second Oil Crisis*. Lexington: D. C. Heath, pp. 137–57

Shell AG (1978), *Der Beitrag des Mineralöls zur künftigen Energieversorgung—Prognosen erfordern heute schon Entscheidungen*. Hamburg: Shell.

—— (1979), *Trendwende im Energiemarkt—Szenarien für die Bundesrepublik bis zum Jahr 2000*. Hamburg: Shell.

Thompson, M. (1983), 'Postscript: A Cultural Basis for Comparison', in H. C. Kunreuther, J. Linnerooth *et al.*, *Risk Analysis and Decision Processes: The Siting of Liquefied Energy Gas Facilites in Four Countries*. Berlin: Springer Verlag, pp. 288–325.

VDEW (1980), *Wärme 2000—Die volkswirtschaftliche Bedeutung der elektrischen Energie auf dem Wärmemarkt*. Frankfurt: VDEW.

Veba AG (1980), *Energieprognose für die Bundesrepublik Deutschland 1980–2000*. Düsseldorf: Veba.

Wessels, T. (1964), 'Die Sicherheit der nationalen Versorgung als Ziel der nationalen Wirtschaftspolitik'. *Zeitschrift für die gesamte Staatswissenschaft*, 120: 602–17.

Ziesing, H. J. (1976), 'Der künftige Energieverbrauch in der Bundesrepublik Deutschland', *Der Bürger im Staat*, 26: 17–23.

5

The Dutch Energy Scenario Game: Corporatist Search for Consensus

Reinier de Man

INTRODUCTION

THIS chapter provides a dense description and analysis of the complex interaction between politics and energy forecasting in the Netherlands. The second section describes the history of policy and forecasting in the Dutch post-war period. Energy policy and energy forecasting did not play an important role until 1974, and were largely confined to single fuel supply policies. Energy forecasting was not much more than the making of some incidental extrapolations when needed. On the other hand, economic forecasting by means of econometric models is a respectable tradition in this country. The CPB, the Central Planning Bureau, plays a predominant role in this field. Energy forecasting, therefore, became a special task of this institution too, once energy problems became a central concern. In the early seventies, like in other industrialized countries, growing anti-nuclear and environmental opposition arose in the Netherlands. This opposition produced its own alternative forecasts, using its own assumptions and its own methodologies. Subsequently, the environmentalists, seeking influence in official policy networks, succeeded in getting financial support from the government for the development of a low energy scenario. In this process, however, they lost a part of their autonomy. Government support implied an outside control over important parameters of the scenario calculation. The detailed historical account of the development of this low energy scenario and the political factors which have played a role is followed by a political analysis focusing on the period between 1976 and 1984. It can be summed up that the 'Dutch energy scenario game' has contributed to the formation of consensus between the opposition and the establishment. But this game does not seem to have been politically effective, and the resulting 'negotiated reality' is, in fact, almost unrealistic.

THE WEIGHT OF THE PAST: THE ECONOMETRIC PLANNING TRADITION

Econometric forecasting in the Netherlands has a long history and is strongly institutionalized. The Government's institute, the 'Centraal Planbureau' (Central Planning Bureau), plays a predominant role in legitimating all kinds of policy decisions. Directly after World War II, it had become by law responsible for preparing the Central Economic Plan once a year (Van den Beld, 1979). Tinbergen, the founder of the Dutch econometric school, made a decisive contribution during the formative years of the CPB. Econometricians and their econometric models became the basis of the CPB, which has developed many different econometric models over time. The VINTAF model came to play an important role in the energy debate. VINTAF, made public in 1975, has been designed for mid-term economic forecasting, and is able to describe the substitution of labour by capital.

Energy forecasting was not a concern of the CPB until recently. In 1981, CPB constructed an energy model, which translates sectorial developments into energy demand. The sectorial developments are obtained by judgement on the basis of the macro-economic forecasts calculated by VINTAF. That is, the disaggregated input to the energy model is not machine-based.

The CPB periodically produces forecasts for the Central Economic Plan and the Macro-Economic Survey (since 1961 on), or whenever such forecasts are needed for government policy-making. Usually, no other institutes are involved: the CPB and its econometric models hold a strong monopoly position in legitimating government policy. This role of producing authoritative forecasts implies an emphasis on certainty rather than uncertainty. It is a good tradition in the Netherlands not to present error margins, but to provide single, unambiguous forecasts to the policy-maker. As Van den Beld, the former director of CPB, has said: 'Obviously the CPB does not say that there could be alternatives to one or more relations. This would be a return to the "discussion economy", and that is exactly to be avoided with the CPB's working method' (Van den Beld, 1979: 66)[1]

THE EVOLUTION OF ENERGY POLICY AND FORECASTING

The history of energy forecasting in the Netherlands may be divided into three distinct periods (See Table 5.1):

1. *1957–1974*. An energy policy did not yet exist. There were rather a number of disjointed fuel policies. Energy forecasting was confined to pretty crude extrapolations reflecting the high growth expectations of those days. The energy situation was changing quickly: coal was replaced rapidly by imported oil and domestic natural gas, and the total energy demand grew at an unprecedented rate.

Table 5.1 The development of Energy Forecasting in the Netherlands

	I 1957–1974	II 1974–1982	III 1982–1984
High/low forecasts	high growth rates	ultra low and very high	moderate growth expectations
term	mainly short term	focus on 2000	focus on 2000
methodology	rough extrapolations, macro	technological and econometric, disaggregated and macro	coupled to existing econometric and technical models
energy policy	disjointed supply-side policies, high nuclear	demand and supply conservation vs. nuclear and coal	lower on the agenda, new nuclear plans
economic environment	high GNP growth, full employment	stagnating growth, rising unemployment	moderate industrial recovery
political environment	technological optimism, broad consensus	technological optimism questioned, failing consensus	reindustrialization is the basis for a new consensus

2. *1974–1982*. In 1974, not long after the 'energy crisis', the first official document on energy policy was published. Strong anti-nuclear opposition had already formed, and a lively debate on alternative energy futures took place between 1974 and 1982. The time horizon, in 1974 still being 1985, shifted to the year 2000. Technological models on a highly disaggregated basis were used to criticize the official macro-econometric forecasts. The government gave financial support to the low energy forecasters within the framework of the Public Energy Debate. The expected growth in energy demand declined steadily during this period.

3. *1982–1984*. The end of the Public Energy Debate also meant the end of the forecasting debate. There is much more consensus on the general shape of the energy future: no one expects high growth rates in the near future.

Phase 1: nuclear dreams, 1957–74

During the 1950s, many in the Netherlands expected the coming of the nuclear energy age. These expansive nuclear scenarios prevailed until the early 1970s. An ambitious nuclear energy programme was presented to parliament in 1957. The CPB presented a forecast of energy demand for the year 1975 based on fairly simple extrapolation techniques.[2] In 1957, two major determinants of the later energy demand were still unknown: the discovery of huge domestic gas reserves, which would make the Netherlands a net energy exporter for some time, and the availability of cheap

Middle-Eastern oil. For this reason, the forecast proved much too low. On the other hand, the planned introduction rate for nuclear power was grossly overestimated. The 1957 paper assumed that in 1975 about fifty per cent of all electricity would be produced by nuclear plants. This implied an installed capacity of 3000 MWe. This capacity will probably not be realized before the turn of the century. At present (1984) it is only 531 MWe.[3]

The Slochteren (Groningen) gas was discovered in 1959. This had a dramatic impact on the Dutch energy situation and energy policy. Initially, the gas reserves were estimated rather conservatively. Therefore, planned sales were systematically below actual sales. In 1967, the gas sales forecast for 1972 was 30 billion m^3 (of which 12 billion m^3 were to be exported). In reality, sales reached 58 billion m^3, and 25 billion m^3 were exported. This bias in forecasts was not only due to a lack of geological knowledge, but was also related to underlying political motives (Odell, 1973a and 1973b).

The penetration of natural gas into the Dutch economy evidently caused a competition between natural gas and nuclear energy. Nevertheless, the nuclear plans remained essentially unchanged for a while. It took some time to recognize the full impact of the new energy abundance. This is beautifully demonstrated by the government paper on nuclear energy, published in 1971.[4]

The general energy policy framework of this government paper was extremely sketchy. The development of total primary energy demand or the competition between the various fuels were non-issues. The increasing demand for electricity and the need for nuclear energy were treated in complete isolation. The forecasting technique was based on rough extrapolations. The annual growth in electricity demand at that time was about 10 per cent per annum. This growth was extrapolated, allowing for a somewhat slower growth after 1980. As a result, total electricity capacity was supposed to increase from about 9 GW in 1970 to 70 GW in 2000. The nuclear component would be 2 GW in 1980, 14 GW in 1990, and 35 GW in 2000. This government paper did not reflect the changed situation due to gas abundancy, although the situation was correctly known at the time. But when this paper was discussed in parliament in October 1974, there remained hardly any relation between the world discussed in the paper and the world outside, on account of the international oil crisis, the growing anti-nuclear movement, and the technical and economic problems of nuclear programmes in other countries (UK, USA). Soon after this paper, another government paper, in fact the first Paper on Energy Policy, was discussed in parliament. This paper signalled the end of the nuclear dreams.

The 1974 government paper on Energy Policy,[5] was the first paper on overall energy policy ever presented to parliament. It was made by the new government, a coalition of the Socialist (PvdA) and Christian Democratic parties, and it reflected the growing doubts on nuclear energy. These doubts were most prominent in the PvdA and the other left wing parties. But they were also shared by many Christian Democrats. The view on nuclear planning was therefore much more moderate than in 1971. In principle, 3 GW of nuclear capacity was to be constructed, but the final decision was postponed until the results of three studies became available. In addition, it would depend on a public inquiry into the choice of reactor sites. Two energy scenarios were presented in this paper, based upon an unspecified methodology. The first scenario represented an 'unchanged policy', and the second scenario with a 'changed policy' accounted for some energy conservation and the proposed moderate nuclear programme. The most striking feature of these scenariors (for 1980 and 1985) was the increasing dependence on growing oil imports.

In the parliamentary debate on this 1974 government paper, the left wing political parties asked the minister whether the CPB could develop a model by which the relations between energy demand, economic growth, and employment could be simulated. The minister said it was worth studying, but that it would take some time.[6]

Phase 2: the protracted debate, 1974–82

The period after the 1974 government paper on energy policy is characterized by the government's indecision over nuclear energy, a powerful anti-nuclear and environmental movement, and a steady decline in economic growth. The official government forecasts were heavily criticized by the anti-nuclear movement, which produced its own counter-forecasts. The result was a lively debate about forecasts and forecasting methodology. A comparison of the forecasts for the year 2000 leads immediately to two interesting observations (see Fig. 5.1). First, we see an overall trend of rapidly decreasing forecasts combined with a sharp reduction of band width between upper and lower forecasts. This growing consensus is not only the result of the shortening forecasting horizon, but also of negotiation processes between the parties involved. A second point of interest is the position of the environmentalists. In 1977, Potma's low energy scenario was far outside the 'credibility range' of established energy forecasting. In 1982, the low energy scenario, the CE scenario developed by Potma within the context of the developed Public Energy Debate, is close to the establishment forecasts.

This process of energy modelling ending in forecasting convergence began in 1974, when the new coalition government of Socialist and Christian Democrats commissioned an outline of a national energy research programme from LSEO, the Steering Group on Energy Research. LSEO needed some long term energy scenarios on which they could base their thinking. The 1974 government paper on Energy Policy gave only forecasts until 1985. LSEO, therefore, developed its own forecasts in 1976. These

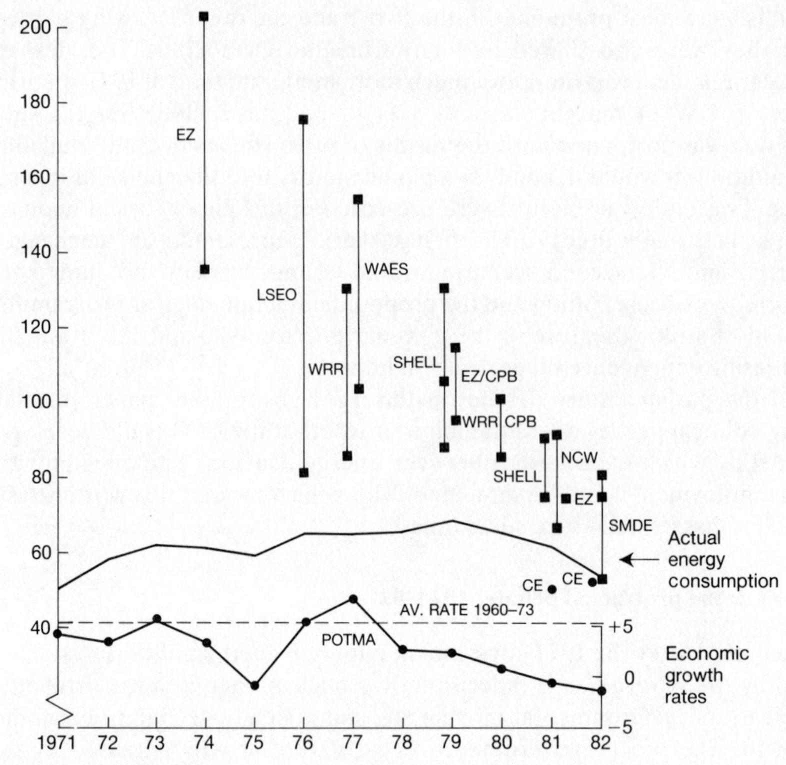

Fig. 5.1 The Netherlands
Energy demand forecasts for year 2000 (in MTCE)

EZ: Ministry of Economic Affairs
LSEO: The Steering Group for Energy Research
POTMA: Leader of the alternative forecasting group
WRR: The Scientific Council for Government Policy
WAES: Workshop on Alternative Energy Strategies
SHELL: Shell Nederland
EZ/CPB: Ministry of Economic Affairs/Central Planning Bureau
NCW: Nederlands Christelijke Werkgeversverbond
CE: Centre for Energy Conservation
SMDE: Steering Group for Public Energy Debate

forecasts were very simple extrapolations on the basis of explicit assumptions. This was quite the contrary to the lack of specification in the 1974 government paper.[7] No doubt, this forecasting procedure reflected the dominance of technical experts (instead of econometricians) in the LSEO. Scenario 1 was a straightforward extrapolation of energy demand. In scenario 2, the growth rate of energy demand was supposed to decline gradually. Scenario 2 was the actual basis for the recommendations made by the LSEO.

In anti-nuclear and environmental circles, the awareness was growing that there were large unexploited possibilities of conserving energy and using renewable energy sources. The ideas of Amory Lovins in the USA about soft energy paths and the energy analyses of Peter Chapman (and later of Gerald Leach) in the UK inspired the Dutch critics to develop their own low energy scenarios (Lovins, 1974 and 1976; Chapman, 1974; Leach, 1979). Theo Potma became the prominent Dutch low energy advocate when he published, in 1977, a scenario which he felt LSEO had forgotten (Potma, 1977a and 1977b). This 'forgotten scenario', projecting a declining energy demand, became a symbol of the low energy proponents.

This development went parallel to an increasing awareness in the public at large of long-term problems. The Scientific Council for Government Policy (WRR) had been set up in 1972 as an advisory body for general government policy. Its special task was to advise on long-term problems and on the stimulation of future research (Huisman/Van der Sluijs, 1981). This meant a broadening of attention as compared to the then still prevailing technical/economic tone of most forecasting work. The first important publication (1977) was a long-term study of 'the next 25 years'.[8] It included two energy forecasts based on different economic growth assumptions. Three years later, the WRR again presented two energy scenarios within the framework of a 'policy-orientated' survey of the future. This was a specification of the OECD Interfutures study for the Dutch situation. It was meant as a stimulus for discussion on alternatives.[9]

In 1979, the second official government paper on energy was published.[10] It presented a number of forecasts calculated by the CPB. The methodology was not made clear. The forecasts reflected the uncertainty about the contributions of nuclear energy and coal, both of which depended on public decisions. The decision-making process was halted, while awaiting the results of the public energy debate, which the government had organized under pressure from parliament.[11]

In this same period, Shell Nederland developed long-term energy scenarios on behalf of their corporate planning (De Bruyne, 1979). These scenarios covered a wide spectrum of possible developments. The key variable in these scenarios was the social and politial development. The scenarios published in 1979 were named 'Business as Usual', 'Frustration and Conflict', and 'Realism and Restraint'. Other scenarios, such as the 'Kafka scenario', were never officially published. In 1981, Shell Nederland published again two energy scenarios.[12] The forecast energy demand was lower than in the former scenarios, and no growth in nuclear energy was assumed to take place.

In the framework of the Public Energy Discussion, four energy scenarios were developed. Onc of them was the further elaboration of the former 'forgotten scenario', and the other three scenarios represented the more established views. The highest 'establishment' scenario (the IH scenario) was even lower than the low scenario in the 1979 government paper on energy policy.

Three aspects of this whole complex of the public energy debate merit a more careful analysis. We first discuss Potma's 'forgotten' scenario'. We then trace the history of and struggles around the CE scenario, which was evolving out of the 'forgotten' one. We then look at all three scenarios of the Public Energy Debate, before discussing the responses to them of various societal forces.

The 'forgotten' scenario

LSEO had presented its two exploratory energy scenarios in 1976. Potma, part of the growing environmental movement, published an additional scenario in 1977 as a critical reaction to the LSEO work. He called it the 'forgotten' scenario, that is, the scenario which had been forgotten by the LSEO. Forecast energy demand in the year 2000 would be only 40 mtoe in contrast to the 80 mtoe in the low LSEO scenario. A number of energy conservation techniques and changes in lifestyle would thus result in decreasing energy demand between 1977 and 2000. This scenario, like the two LSEO scenarios, was not much more than a rough sketch. But Potma, supported by an extensive environmentalist network, wanted to give the 'forgotten scenario' a solid, scientific basis. In this way he would have stronger arguments to support criticism of the official high growth views. The environmentalists founded their own research organization, the *Centrum voor Energiebesparing* (CE, the Energy Conservation Centre). This organization submitted a research proposal to the National Steering Group for Environmental Research (the LASOM) in July 1978.[13] The LASOM is an advisory body to the Ministry of Health and Environment. The proposal was supported by the National Platform for Environmental Organizations (LMO), by an important labour union, and also by four prominent Dutch economists. The proposal asked for research funds to work out a low energy scenario along the lines suggested earlier by the 'forgotten scenario'. The LASOM's attitude was not negative. In fact, such research could have helped the Ministry of Health and Environment in its resistance to the energy policy of the Ministry of Economic Affairs, which was likely to have grave environmental consequences. But the very fact that the latter ministry has central responsibility for energy policy proved fatal to this low energy study. The Ministry of Economic Affairs refused to let the Ministry of Health and Environment finance any energy studies at all. LASOM, at the end of 1979, refused to fund this low energy research (Hueting, 1983), in order to avoid a serious inter-departmental conflict.

The CE scenario

In the early 1970s, the growing public controversy over nuclear energy was dominating the political scene. Earlier nuclear projections had already been reduced considerably, but much doubt remained on the acceptability and necessity of the 3000 MWe nuclear capacity announced in the 1974 government paper. This doubt was shared by left-wing and centre political parties, as well as by the extra-parliamentary opposition. The final decision on nuclear plant construction was therefore postponed several times. The public inquiry into the choice of plant sites caused many difficulties. The anti-nuclear movement succeeded in subverting the participation procedure by infiltrating most of the public hearings (see: Abma, 1981). The discussions at the inquiry emphasized more the general energy policy questions than the specific issues of plant-site choice. The government, at the time unable to reach a clear decision on nuclear energy, decided to stop the inquiry. An alternative had to be found. The alternative was the 'Public Energy Debate'.

The Public Energy Debate

The idea emerged in the 'intermediary' societal network which had formed between the moderate opposition and the moderate nuclear proponents. This network, in which the Churches played a key role, was a meeting place for the élites of the environmental movement of the labour unions and of civil servants and members of parliament. The idea was finally accepted by the government, most probably influenced by the outcome of the British Windscale Inquiry.[14] There remained, however, much ambiguity about the actual form of such a debate. The government asked the General Energy Council (AER) for an opinion. This advice was published in 1978,[15] at the same time as the *Initiatiefgroep Energiediscussie*, a group of people from this 'intermediary' network, put forward a more elaborate proposal.[16] The government proposal for the organization of the Public Debate, published in 1979,[17] assumed that the second government paper on energy policy would be the basis for the debate.

The government asked for comments on their proposal. Many organizations with a critical attitude towards the official energy policy reacted in September 1979.[18] The important national environment organization, *Natuur en Milieu*, for example, wrote:

> We are not happy with the plan to start the Debate on the basis of the three volumes of the second Energy Paper . . . These are too narrow a basis for discussion. It would be more correct to put the government standpoint into an information package together with other standpoints and scenarios representing the most characteristic examples from the wide range of ideas in the energy field. All these standpoints and scenarios should be worked out to the same level. This should be made possible by government resources. In this framework we would like to remark that we regret that no decision has yet been reached on financing the so-called 'forgotten scenario' of Mr Potma, for which support is asked from the National Steering Group for Environmental Research.

After these comments had been received, the Parliamentary Commission on Nuclear Energy invited several organizations to explain their standpoint. The *Initiatiefgroep* began co-ordinating, in the first months of 1980, the political pressure on formulating more energy scenarios. They finally succeeded in convincing a number of left-wing and Christian Democratic members of parliament of the importance of alternative scenarios.[19] Many motions were formulated and discussed at the public session of the Parliamentary Commission on Nuclear Energy during February 1980 and March 1980.[20]

Motion No. 25, proposed by a Christian Domocratic member of parliament asked the government to support the elaboration of more energy scenarios. It mentioned that it was not yet clear whether the additional scenarios should be worked out by the CPB or not. The Minister for Economic Affairs was strongly opposed to such an exclusion of the CPB:

> I do not understand why the CPB should be by-passed, since this institute is the pre-eminent Dutch institute, at the service of government, to forecast long-term economic developments. Outside the energy field, we do not involve other institutes either. I will not allow any distrust in the CPB's work.

The member of parliament defending this motion answered that he had not wished to exclude the CPB, but that he had wanted to give other models a fair chance, without interference from the CPB. He had doubts as to whether all aspects of the energy problem could be done justice by the CPB models. Another deputy emphasized the need for 'absolutely comparable and equivalent presentations of different options for the energy future'. The Minister of Economic Affairs reacted: 'I should think that this would imply that the calculations should be carried out by the CPB. Otherwise, one would never realize this equivalence.' The motion was supported by a majority in parliament despite the minister's objections.

The parliamentary majority for this motion was a success for the critics and a problem for the minister. In April 1980, the minister asked the General Energy Council (AER) for advice on the possibilities and constraints of energy models and scenarios, on the assumptions regarding energy, economics, and environment, on the models to be used and on the organizational set-up of the scenario studies. The AER advice was published in June 1980.[21] The AER considered the scenario concept and stressed its limited relevance. The value of scenarios, according to the AER, was not dependent only on the quality of the models, but also 'on the acceptability of the assumptions at many levels of society'. The AER advised developing a scenario based on low or zero economic growth and to base it on the 1978 proposal for a low energy study made by the Energy Conservation Centre (CE) to the Environmental Research Steering Group (LASOM). The AER based its choice for the CE on the observation that there existed massive support from the organizations which had founded the CE, the environmental platform LMO and the *Werkgroep Energiediscussie*. The latter was a network uniting a variety of organizations with a critical attitude towards existing energy policy. The AER, however, queried some basic assumptions of the CE proposal. It remarked that technical and economic models were already available in the establishment institutions of CPB and ESC, the Energy Research Centre at the government-financed Technical Energy Research Institute ECN. It also pointed out that 'these models contain an important part of the factors to be studied'. The AER stressed 'the need for comparability, which could be achieved in the short term by working out both scenarios by means of the CPB energy model. . . . the Council takes due note of the CE's wish to deviate from this model . . . if its elaboration will turn out to be impossible on the basis of the model'.

A commission would be constituted to control the comparability of the two scenarios. In addition, the AER suggested that 'the commission shall give recommendations on the usefulness and presentation of the scenarios for the Public Debate'.

The AER also published a list defining the input and output parameters which were to be taken into account by both scenarios (that is, the CE scenario and the offical scenario). It indicated those parameters that were 'free', that is, open to the scenario builder's choice, and those that were 'fixed', and the same for both scenarios. This list was produced by a special commission made up of members from the AER, the Ministry of Economic Affairs, CE, and the Energy Research Council (REO).

Preliminary scenario results

Two different economic models were used for the elaboration of the CE scenario. The first model relates energy demand to economic development. For this purpose, CE made use of the CPB energy model. The second model calculates the economic effects of the proposed energy policy. CE did not use the CPB VINTAF model, but the SECMON model of the Foundation for Economic Research (SEO) of Amsterdam University instead. This was contrary to government preference. The main reason for not choosing VINTAF was that environmental circles greatly distrusted the CPB, because of its role in legitimating the established economic policy. CE searched, therefore, for a less 'contaminated' institute. The SECMON model has a structure different from the VINTAF model. SECMON is a sectorial model, in which macro-economic developments are estimated on the basis of sectorial developments. VINTAF does not calculate these sectorial developments. The CPB estimates them by a judgemental procedure on the basis of the macro-economic development (Hueting, 1983).

It was not easy to translate the CE's assumptions into parameters for the SECMON model. A spokesman for SEO estimated that 80 per cent of the SEO's time on the scenario project was spent on translating CE's wishes into model inputs. It was impossible, however, to incorporate all assumptions into the econometric model. In fact, SEO recognized that it was simply impossible to use a historically based econometric model to calculate developments of a hypothetical, alternative economic order. SEO, of course, could not openly confess its failure.

Preliminary results from the CE scenario analysis attracted considerable public attention in September 1981. An important national newspaper published a few results from an internal report,[22] which made clear that the official scenario of the Ministry of Economic Affairs, the EZ scenario, implied a huge increase in unemployment. In the confused debate which followed and in which the Steering Group for the Public Energy Debate, the Energy Conservation Centre, the Commission on Energy Scenarios, the Foundation for Economic Research, the General Energy Council, and the environmental organizations were all involved, voices were heard demanding that further publication of results should be prevented. This never happened, as there were no formal grounds to do so. The commotion was mainly caused by the fact that the attention focused on the shortcomings of the established policy.

In reaction, the critics of the CE scenario tried to prove its low quality. They argued that there was not yet any CE scenario and that the SECMON model was not operational at all. However, the accusation that incomplete and non-operational models were used was heard from both sides. There were rumours that the VINTAF model had never been used in its totality and that large parts of the CPB energy model did not even exist. The Energy Conservation Centre (CE) was not very pleased about this premature publication, but its supporters from the environmental organizations played the game, aiming at maximum political effect. From that moment on, unemployment figures were the central issue in the struggle against established energy policy. It was the explicit strategy of the director of the CE to use the unemployment issue to attract political attention.

The interim reports

The interim reports of CE and SEO were published in December 1981.[23] But a major change in assumptions had occurred shortly before the publication date. It was decided to use lower future labour productivity growth rates in the service sector models. This assumption reflected the ideas of Hueting, a welfare economist who was influential at the Energy Conservation Centre. He argued that, from the perspective of welfare economics, the negative environmental effects of continued production growth should be counted as negative production. The optimal production development implied, then, a lower labour productivity than under conventional economic assumptions.

It was the suddenness and the way in which the decision was taken that created resistance on the part of some of the actors involved. There were some objections on the part of SEO, and the AER scenario commission objected very vigorously. Members of this commission felt that an analysis of the service sector was beyond the limits of the scenario analysis as defined by the AER. But at that time they could only accept what had happened.

The interim report of CE gave five preliminary conclusions:

(1) A reduction of energy consumption by 20–25 per cent can be reached in the period 1980–2000, which is a reduction by 35 per cent compared to an unaltered policy.
(2) There will be a positive effect on the balance-of-payments in spite of large investments in energy conservation.
(3) Pollution can be reduced by 30–50 per cent.
(4) There will be a shift in the expenditure pattern towards more housing, public transport, and consumer-directed services.
(5) Unemployment will be reduced by over 25 per cent from the 1980 level by additional social and economic policies, while the official reference scenario foresees an increase of almost 150 per cent to 600,000 unemployed.

The CE tried to use the employment issue to get support for its energy scenario!

The CE used the CPB energy model to derive estimates of aggregate energy demand from economic developments. However, the CPB model structure was not clear for those who had to prepare the input. At the start of the CE scenario study there was considerable confusion deriving from the fact that only minor parts of the CPB energy model were known to CE. At the end of 1980, only a list of exogenous parameters and a small number of model equations had been published by the CPB. This meant that CE had to feed parameter values into a model which was not yet understood. In addition, CE and the Ministry of Economic Affairs had different opinions about possibilities and costs of energy conservation. The values used by CE were the result of a compromise between both parties after consultation of the scenario commission.[24]

The Steering Group's scenarios

Organizational details of the Public Energy Debate had been revealed for the first time in 1979. But the first stage of the procedure did not begin until July 1981. This lag resulted from the political difficulties in composing the Steering Group and from the problems of formulating the alternative energy scenario. On 3 July 1981 the Steering Group was formally set up by government. Tasks until then belonging to the Ministry of Economic Affairs and the General Energy Council (AER) were assigned to this new institution. This happened at a time when many preparations for the Public Debate had already been made. The Steering Group would have preferred to start with a clean slate, but was confronted with the two energy scenarios under development: the CE scenario and the scenario of the Ministry of

Economic Affairs. The group decided, however, that two additional energy scenarios should be worked out, the IH (Industrial Recovery) and the AD (Labour Distribution) scenarios.

The motives of the Steering Group The Steering Group started life with the desire to develop new scenarios. The premature publication of the CE scenario provided an additional argument for this course of action. The CE scenario suggested that the environmental path would also guarantee low unemployment, while the high energy path of the government, in contrast, promised nothing but continued misery. The Ministry of Economic Affairs, of course, objected to this situation, but the Steering Group also found it unacceptable. This comparison was considered too cheap a success for the environmentalists, especially because the government scenario did not contain energy policy measures after 1980, or even policy measures designed to control rising unemployment.

The Steering Group concluded that the two scenarios worked out so far were no basis for fair comparison. The CE scenario was a normative scenario, incorporating a multitude of policy measures, whereas the government scenario was not intended to reflect a normative policy option. Hence, the Steering Group decided to develop other scenarios to replace the government scenario. The latter was renamed the 'Reference Scenario' under pressure from the Economic Affairs Ministry, which had no inclination to defend it. The ministry obviously did not want to be stuck with having to defend a weak position in the public energy debate. This decision coincided with a change of ministers after the 1981 government crisis. Many believed at the time that the withdrawal of the government scenario had been caused by this change (V. d. Hoevem, 1982; Hueting, 1983). But any minister would have refused to take responsibility for such a weak scenario, which, after all, was not official policy.

The absence of the interested parties In order to formulate an alternative to the government scenario, the Steering Group sought normative orientations that would gather sufficient political support behind them. Two such orientations were found. One was the idea of reindustrialization put forward by the government's Reindustrialization Commission (the Wagner Commission). The other was the idea of work redistribution, a concept put forward by an economist and supported by the labour unions. Both orientations were subject to much Public Debate and the Steering Group decided to base two alternative energy scenarios on them: an Industrial Recovery (IH) and a Labour Redistribution (AD) scenario. It would have been natural that the groupings supporting these orientations were going to work out the scenarios. The Steering Group indeed contacted the most important employers' organization, VNO, and the Reindustrialization Commission as well. But both were unwilling to take any responsibility for setting up and defending an energy scenario. As a consequence, the Steering Group itself became responsible for the IH and AD scenarios.

The reactions from the environmentalists The environmentalists had hoped to present an attractive case to the politicians by stressing the contrasting unemployment figures. The withdrawal of the government scenario was, therefore, highly unwelcome. They even called it a falsification of the debate and blamed the Ministry of Economic Affairs for walking away from the debate. The Energy Conservation Centre (CE) wrote:

> The public discussion originated because the government wanted nuclear plants and the public did not. In any case therefore the government is a party. When the Public Debate was being set up, this was made clear by the claim of the Economic Affairs Ministry to discuss exclusively its own standpoint . . . But what happens? The Steering Group agrees to the withdrawal of Economic Affairs as a discussion partner in 1981, because the new Minister Terlouw wants this . . . The responsible authority and party will not be present in the Public Debate. A sordid affair. The government walks away. . . . What should the public do with four scenarios? Only confusion will increase. Therefore it is better to reduce the choice to the original two policy directions.[25]

The CE had good political reasons for keeping the situation simple and clear. But their option was potentially in conflict with principles of state law. If the ministry had acted as a 'party' in the debate, it would have defended a policy in public before parliament had been consulted. This would have been quite unusual.

The final scenarios

The three scenarios IH, AD, and CE, which in the end attained prominence in the discussion, are all (also) calculated for a non-nuclear future.[26] Projected energy demand is low compared to the numbers in the 1970s forecasts (see Figure 5.1).

The Shell BU scenario of 1979, which has a normative background similar to the IH scenario, still forecast 132 mtoe for the year 2000. The low-demand Shell RR scenario with 88 mtoe was even higher than the high IH scenario with its 81 mtoe. The environmentalists, however, went slightly upwards: the 'forgotten scenario' of 1977 projected 40 mtoe for the year 2000; the CE scenario gives 53 mtoe.

The most striking feature of the scenarios, however, is that nuclear energy is not a key parameter anymore, although this issue was the reason for the Public Energy Debate. Lower energy demand growth eliminated the need for either coal or nuclear energy. Even the IH scenario can be realized without nuclear energy, although this requires a corresponding shift to coal of about 6 per cent of total consumption. Yet nuclear energy policy had been waiting for the results of the Public Energy Debate since 1978.

The unemployment issue, which had been central in December 1981, has also disappeared. The Steering Group simply discarded this issue from the agenda by imposing an employment policy valid for all scenarios, assuming a reduction in unemployment to 200,000.

The institutional response to the scenarios

The scenarios were published in January 1983. Important reactions were forthcoming in March 1983. The Scenario Commission published its comments, and a conference on the CE scenario revealed the response from different institutional actors.

The scenario commission's comments As has been mentioned, a commission had been set up to control the quality of the scenario studies. This commission had been formed in November 1980 and had become a part of the Steering Group for the Public Energy Debate as of March 1982. Its main task was supervision of the scientific quality, of the plausibility of assumptions, and of comparability and relevance of the scenarios for the Public Energy Debate. The final report was published in March 1983.[27] The commission had concluded that the scenarios were not very comparable. In particular, the CE scenario was not comparable with the AD and IH scenarios:

> The resulting non-comparability of the scenarios is so large that the commission doubts the possibilty of using these four scenarios together in the Public Energy Debate. [But there are] . . . some other comparability problems . . . The most important one is that the comparison of scenario outcomes does not give a fair impression of the different policy options. The scenario outcomes are not only caused by policy differences, but also by differing assumptions on the autonomous behavioural changes and on technical energy conservation maxima.

The Scenario Commission was very critical of the relevance of the scenarios. It regretted that nuclear energy had not been considered explicitly in the scenarios, especially as the nuclear controversy had given rise to the Public Debate in the first place. The commission made a number of critical remarks on the scientific quality of the scenarios. It signalled the problem of the two different econometric models. Only the CE scenario had made use of the sectorial SECMON model. The other scenarios had been based on VINTAF. VINTAF had been validated to a greater extent than SECMON, but the commission recognized the advantages of SECMON's sectorial structure. About the CPB energy model, used in all scenarios, the commission remarked that the model had not yet been published. It had, therefore, not yet been exposed to the critique of the scientific community.

The commission found the assumptions of the CE scenario not very plausible: 'These assumptions make the CE scenario outcomes vulnerable and less plausible. . . . It would be misleading to conclude that the CE scenario proves the realizability of the CE's goals.' On the basis of its critical analysis, the commission concluded that the scenarios could only serve a very limited function. The commission was above all convinced that the scenarios could not fulfil their political function of helping to bring about a consensus on societal goals and on policy action.

The CE scenario congress The political function of the scenarios was, of course, very important for the environmentalists. For this reason, CE organized a congress on 1 March 1983, where many organizations were represented.[28] The chairman of CE said in his opening address:

> The 'forgotten scenario' has evolved to the CE scenario, which is a crystallization point for the environmental organizations. We see this as an exercise of special importance for discussions in our country, related to more than only energy questions. The inert policy networks, to which the established economists also belong, dare not participate in such a discussion. The economists of the Rotterdam University have even withdrawn from the Public Debate!

The prominent Dutch economist Pen had some reservations about the CE scenario. He said that the environmentalists' reluctance towards reindustrialization was unnecessary. Industrial policy and the protection of the natural environment were not contradictory. Professor Weitenberg from the CPB was also not particularly impressed by the low energy scenario. But the most negative critique came from the representative of the employers' organization NCW. In his opinion, industry needed reindustrialization, deregulation, and reduction of costs, not the type of future the CE scenario implied. The representatives of the Churches, labour unions, and environmental organizations supported the CE scenario.

Phase 3: the new reality, 1982–84

The Dutch energy scenario game was played between 1980 and 1983. It resulted in a growing consensus on the expected energy situation in the year 2000. The differences between the views of the 'establishment' and those of the 'critics' have never been so small, at least when measured in terms of aggregate energy demand. However, the differences are much greater in terms of normative orientations. The now dominant reindustrialization option[29] is contrary to the type of future the environmentalists envisage. The near consensus on outcomes, therefore, is a mere coincidence. The economic growth rate, attacked by the environmentalists in the early seventies, has been declining without any deliberate policy. As the 'forgotten scenario' evolved towards the final CE scenario, the economic environment continued to change drastically. Ironically, the argument for slower economic growth has been evolving into an argument for economic recovery. Nine years passed between the emergence of the first ideas for a low energy strategy and the final publication of the CE scenario. As a result, the scenario had lost much of its actuality when it was finally presented to the policy-makers. The CE scenario has not had any important influence in the political parties. Shifting responsibility for the debate, including the construction of the energy scenarios, from the government to the Steering Group placed the debate outside parliamentary politics. The final Steering Group report,[30] with strong arguments against further nuclear power planning, will most probably have no substantial impact on future energy policy. In this context, the CE scenario was only supported by the smaller left-wing political parties. In 1984, preliminary sketches of a pro-nuclear energy policy were publicly circulating and political negotiations to reach a consensus on the issue are on their way. On 11 January 1985 the government proposed to parliament the construction of two additional nuclear power plants of between 900 MW and 1300 MW, and this despite signs of continued public hostility to this course of action.

The public energy policy discussion has ended in a political vacuum. It seems unlikely that energy scenarios will again come to play an important role in the near future. In the 'no-nonsense' climate of the 1980s, long-term scenario exercises are increasingly regarded as impractical and unreliable tools for decision-making.[31]

THE POLITICS OF ENERGY FORECASTING

The history of energy planning seems to follow a very haphazard course. However, a certain logic can be found, despite the contingent nature of many influencing factors. The dominant political culture in the Netherlands is an important explaining factor. The use of forecasting models can be related to the emphasis in Dutch politics on consensus formation and the nature of the negotiation and consultation processes.

The politics of accommodation

In fact, it is impossible to speak of the political culture in a country. There are always a variety of cultures, which can be found in different institutional environments and which are competing against each other. This explains why the dominant political culture in a country may suddenly change from one to the other extreme. This is to be kept in mind when discussing the dominant Dutch political culture. This culture, historically connected to the need for integration between the different 'pillars' of society,[32] is characterized by an emphasis on consensus formation and the avoidance of open political conflict. The process of consensus formation implies a corporative style of policy formation, involving extensive policy networks with representatives from all relevant interest groups and often tiresome and time-consuming consultation procedures. This culture is a dominant factor in Dutch politics, but its significance is dependent on the specific institutional settings. It is most strongly interconnected with the Christian Democratic parties, but it is also a significant element of the Social Democratic political culture. Almost the opposite style of decision-making culture is found in industrial circles. Here open conflict is avoided less, decision procedures are preferably kept simple by confining the number of participants and by setting unambiguous deadlines. The traditional corporatist culture has come into conflict with this new 'no-nonsense' orientation. The latter is gaining influence in many parts of Dutch society. The Public Energy Debate, however, is a clear expression of the traditional type of politics. A great variety of interest groups participate and there is a stress on reaching consensus. Essentially, the energy scenario game is played by two parties, the environmentalists and the establishment, meeting an ever-changing arena of temporary institutional constructions for the Public Energy Debate.

The development of the arena

Figure 5.2 indicates the most important changes in the arena during the scenario game. Before 1980, the environmental organizations were fairly isolated from the official institutions. In the framework of the Public Energy Debate, however, the ties between the two parties were growing. Figure 5.2 gives two structures for that period. The first represents the situation before the institution of the Steering Group, and the second represents the later situation, where the role of the Ministry of Economic Affairs (EZ) has been taken over by the Steering Group. The fourth structure in Figure 5.2 gives the 1984 situation after the Public Energy Debate. The temporary organizations no longer exist, and the formal ties between the environmentalists and the establishment are reduced accordingly.

The nature of the negotiation and consultation processes

In the negotiation processes between the environmentalists and the establishment, the former—actually consisting of a variety of different groups and organizations—acted as a unity. Only one of the many possible energy scenarios was opposed to the established energy policy. The great emphasis which has been placed on the comparability of the different scenarios illustrates the dominant consensus culture (De Man, 1983). To be useful as a tool for discussion and policy, the scenarios had to be comparable and, therefore, based upon the same forecasting models. The high status of econometric models in the Netherlands, seemingly a technocratic feature of the political system, is an expression of the need for consensus, characteristic of the prevailing style of corporatist decision-making. The models provide an unambiguous reality for all parties involved. The environmentalists had to accept this reality in order to be taken seriously in the debate.

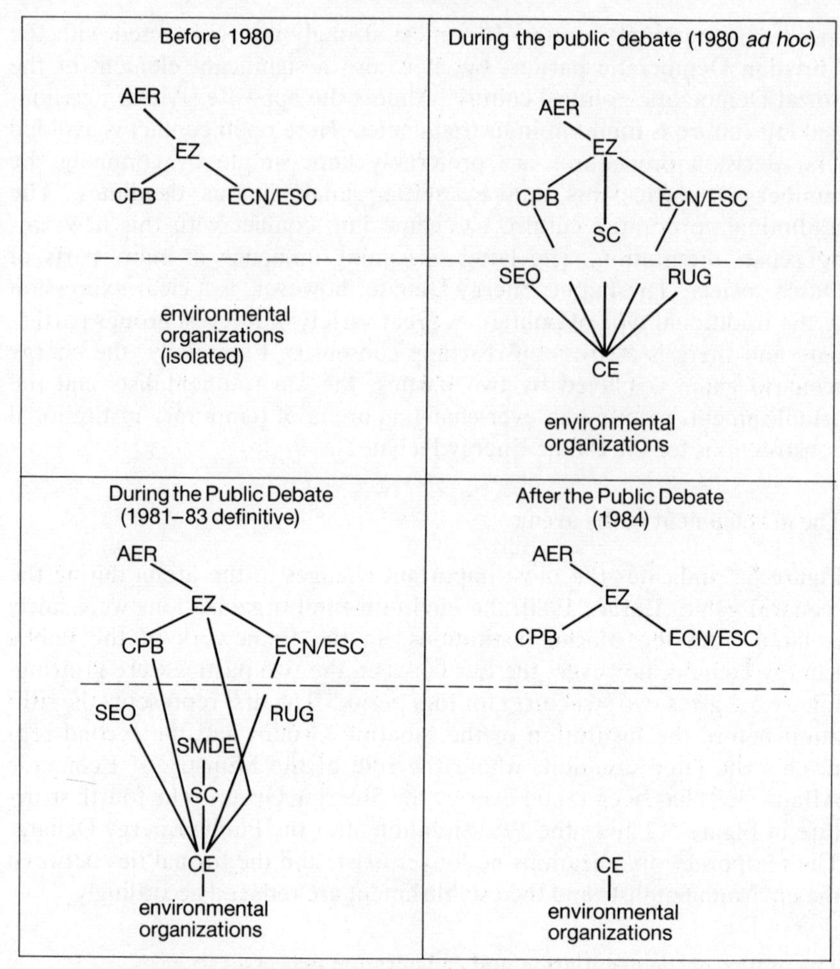

Fig. 5.2 The development of the Dutch energy forecasting arena

EZ: Ministry of Economic Affairs
AER: General Energy Council
CPB: Central Planning Bureau
ECN: Energy Research Centre
ESC: Energy Study Centre
SMDE: Steering Group Public Energy Debate
SC: Scenario Commission (in the 1980 situation a commission of EZ, later a commission of SMDE)
CE: Centre for Energy Conservation
SEO: Foundation for Economic Research
RUG: Groningen University

They managed to make their calculations without the CPB VINTAF model, but they used another econometric model instead.

There is always a basic dilemma in such negotiation situations between the preservation of one's own identity and the pay-off of co-operation (De Man, 1984). The conflict between the environmentalists and the establishment is basically over the legitimation bases of political decisions. Traditional economics, in the view of the environmentalists, helps to legitimate political decisions, which are fundamentally wrong because they lead to unacceptable effects. Thus, traditional economic thought is no longer a valid legitimation base, according to this vision (Offe, 1983; Bons, 1982). In the practical, political situation, however, the environmentalists must make concessions when searching to gain influence. In the Netherlands, they have played the game of making econometric calculations. This may have been a positive contribution to the legitimacy of the low energy scenario, but at the same time they have lost something of their own identity.

CONCLUSION: THE OUTCOME OF THE SCENARIO GAME

The energy scenario game has been played in a policy arena which was largely isolated from the rest of society. The negotiation processes have been time-consuming. Both the isolation and the large time-span have contributed to low political efficacy. The arena had no effective ties with the actual centres of policy-making, and when the final scenarios were presented to the policy-makers, the political climate had changed so radically that the scenarios were hardly recognized as important. The energy problem was much lower on the political agenda. Other issues, like the international arms race and the high unemployment situation, were now competing for attention. The Public Energy Debate had the one important effect of postponing the decision-making for some time.

The energy scenario game has no doubt contributed to an increased consensus. The difference between the establishment and environmentalist scenarios have never been so small in terms of forecast, aggregate energy demand. But what degree of reality do these scenarios have? We have seen how important parameters were simply dictated by a commission, and how the unemployment level was bureaucratically fixed at 200,000. To satisfy the need for consensus and comparability, bureaucratic procedures have replaced sound analysis. The resulting pseudo-certainties run counter to the intrinsic uncertainty of the parameters of any energy policy. The price of oil on the world market, for example, is such a parameter. This was not a 'free parameter', and was, therefore, not seriously considered.[33] Thus, the increased consensus has failed to produce an increased sense of reality.

Notes

I have received useful comments from Eric-Jan Tuininga, Leo Jansen and others on an earlier draft of this report.

1. All quotations in this chapter have been translated from Dutch.
2. *Nota inzake de kernenergie*. Parliamentary Report 4727 No. 2 and Appendix: 'Nota van

het Centraal Planbureau'. The Hague, 1957. Actual PEC in the mid-seventies was about 70 per cent above the forecast level, after another acceleration of consumption growth after 1965.

3. Stuurgroep Maatschappelijke Discussie Energiebeleid, *Het Tussenrapport*. The Hague 1983, p. 80: Borssele 477 MWe, Dodewaard 54 MWe.
4. *Nota inzake het kernenergiebeleid*. Parliamentary Report 1971–2, 11761, No. 2.
5. *Energienota*. Parliamentary Report 1974–5, 13122, No. 2.
6. Parliamentary Report 1974–5, 13122, No. 6. 'Voorlopig verslag', Chapter 4: Potential Lines of Development (potentiële ontwikkelingslijnen).
7. Parliamentary Report 13250, No. 4: *Interimrapport van de Landelijke Stuurgroep Energie Onderzoek, Energie 1976*. The Hague, 1976.
8. Wetenschappelijke Raad voor het Regeringsbeleid, *De komende vijfentwintig jaar, een toekomstverkenning voor Nederland*. Rapporten aan de Regering No. 15, The Hague, 1977.
9. Wetenschappelijke Raad voor het Regeringsbeleid, *Beleidsgerichte toekomstverkenning, deel 1: een poging tot uitlokking*. Rapporten aan de Regering No. 19. The Hague, 1980.
10. *Nota Energiebeleid, Deel 1/Algemeen*. Parliamentary Report 1979–80, 15802 No. 2.
11. *Opzetnota, maatschappelijke discussie over de toepassing van kernenergie voor elektriciteitsopwekking*. Parliamentary Report 15100 No. 2. The Hague, 1979.
12. Shell Nederland, *Long-Term Scenarios for the Netherlands*. Executive summary, Central planning, and functional services, Mar. 1981.
13. See Brief van de Initiatiefgroep Energiediscussie aan de Vaste Kamercommissie Kernenergie, 15 February 1980, (Appendix 2).
14. The Windscale Inquiry was held between 14 June and 4 Nov. 1977. The outcome was a 'nuclear apologia': Patterson, 1978.
15. *Advies van de voorlopige algemene energieraad inzake een brede maatschappelijke discussie over kernenergie*. The Hague, 7 June 1978.
16. *Meedenken, meedoen, democratisch beslissen over (kern) energie*. Initiatiefgroep Energiediscussie. Amsterdam, June 1978.
17. *Opzetnota*, op.cit. (n. 11 above).
18. *Letters*: Brief Initiatiefgroep aan de Vaste Kamercommissie voor Kernenergie, 20 Sept. 1979; Brief Natuur en Milieu aan deze Commissie, 26 Sept. 1979; Brief FNV-Federatiebestuur aan deze Commissie, 28 Sept. 1979.
19. Meeting of the Parliamentary Commission for Nuclear Energy, 'Vaste Kamercommissie voor de Kernenergie' on 26 Nov. 1979.
20. Parliamentary Report 15100 (1979–1980) Nos. 19, 20, and 22–40.
21. *Energiescenario's*. The General Energy Council's Advice to the Minister of Economic Affairs, 18 June 1980.
22. *De Volkskrant*, Sept. 1981.
23. H. Y. Becht and T. G. Potma, *Het CE-scenario, een realistisch alternatief*. Delft: Centrum voor Energiebesparing, Dec. 1981. W. Driehuis et al: *Zuinigheid met vlijt, een voorlopige economisch-technische uitwerking van een scenario voor Nederland ontworpen door het Centrum voor Energiebesparing*. Amsterdam. Stichting voor Economisch Onderzoek, Dec. 1981.
24. Becht and Potma, op.cit.
25. W. van Dieren (ed.), *Groene economie*. Delft: Centrum voor Energiebesparing, 1983.
26. Stuurgroep Maatschappelijke Discussie Energibeleid, *Het Tussenrapport*. The Hague, Jan. 1983.
27. 'Adviescommissie scenario's', *Eindrapport*. The Hague, Mar. 1983.
28. *Is het CE-scenario haalbaar?: commentaar op commentaar naar aanleiding van de brede maatschappelijke discussie over het CE-scenario op 1 maart 1983*. Delft: Centrum voor Energiebesparing, May 1983.
29. See the reports of the Government Reindustrialization Commission: H. van Dellen (ed.), 1984.
30. Stuurgroep Maatschappelijke Discussie Energiebeleid: *Eindrapport*, The Hague 1984.
31. The Reindustrialization Commission, for example, is opposed to 'the continuing confusion which has arisen in the public discussion about the quantification of the effects of

alternative policy programmes. This will not contribute positively to a climate which favours a policy of economic recovery.' (From the June 1981 Report.)
32. Lijphart speaks of the 'consociational society' (Lijphart, 1975).
33. In this climate of 'negotiated realities', the Rotterdam economist Peter Odell has never been taken seriously by either of the two parties: his perpetual attacks on the idea of 'scarcity', underlying both the established policy and the environmentalist strategy, have not played any role in the national energy debate, although his predictions and analyses have proved right more than once.

Bibliography

Abma, E. (1981), 'Kernenergie als maatschappelijke splijtstof: een analyse van een protestbeweging', in P. Ester and F. L. Leeuw (eds.), *Energie als maatschappelijk probleem*. Assen; van Gorcum.

Beld, C. A. van den (1979), 'Het Centraal Planbureau: Zijn invloed, zijn macht en zijn onmacht', in W. M. van den Goorberg, T. C. M. J. van de Klundert, A. H. J. Kolnaar (eds.), *Over macht en wet in het econmisch gebeuren*, Leiden: Stenfert Kroese.

Bons, C. P. (1982), *Left and Right in a Melting Pot?* Leiden: Leiden University, Sociological Institute.

Bruyne, D. de (1979), 'Nederlandse energie tot 2000: vooruit met halfgas', in *SMO-informatief* 79/1.

Chapman, P., *et al.* (1974), 'The Energy Costs of Fuels', *Energy Policy*, 2 (3): 231–43.

Dellen, H. van (ed.) (1984), *Een nieuw elan, de marktsector in de jaren tachtig*. Deventer: Kluwer.

Hoeven, E. van den (1982), 'Scenario's in de energiediscussie', *Economisch-Statistische Berichten*, 20 Jan.

Hueting, R. (1983), 'Results of an Economic Scenario that gives Top Priority to Saving the Environment and Energy instead of Encouraging Production Growth', *Symposium on Economic Growth and the Role of Science*, Stockholm, 9–11 Aug.

Huisman, H., and H. van der Sluijs, (1981), 'Toekomstverkenningen van de WRR: Planning of verkenning?', in J. van Doorn and F. van Vught, *Nederland op zoek naar zijn toekomst*. Utrecht: Het Spectrum.

Leach, G., *et al.* (1979), *A Low Energy Strategy for the United Kingdom*. London: Science Reviews.

Lijphart, A. (1975), *The Politics of Accommodation, Pluralism and Democracy*. Berkeley: University of California Press, 2nd edn.

Lovins, A. B. (1974), *Technical Bases for Ethical Concern*. London: Friends of the Earth.

—— (1976), 'Energy Strategy: The Road Not Taken', *Foreign Affairs*, Oct.

Man, R. de (1983), 'Energy Models and the Policy Process: The Dutch Scenario Game', *Simulation and Games*, Dec.

—— (1984), 'De onderhandeling over energiescenario's voor Nederland', *Beleid en Maatschappij*, Nov.

Odell, P. R. (1973a), 'Indigenous Oil and Gas Developments and Western Europe's Energy Policy Options', *Energy Policy*, June.

—— (1973b), 'The So-Called Dutch Gas Shortage: Unanswered Questions and an Alternative Hypothesis'. Unpublished Research Paper, Rotterdam, Mar.

Offe, C. (1983), *Legitimation Problems in the Nuclear Energy Conflict*. Paper presented at SOMSO-meeting on the nuclear energy debate, Utrecht, 24 June.

Patterson, W. C. (1978), 'The Windscale Report: A Nuclear Apologia', *Bulletin of the Atomic Scientists*, June.

Potma, T. G. (1977a), *Energiebeleid met minder risico*. Amsterdam: Milieudefensie (Dutch Friends of the Earth).

—— (1977b), 'Het vergeten scenario', *Bêta*, 1 Jan.

Appendix

The development of Forecast Total Primary Energy Demand for the Year 2000 (mtoe)

2000 forecasts						
EZ	a	1974		135.1		206.0
LSEO	b	1976	sc. 1,2	80.0		175.0
Potma	c	1977	sc. '3'	40.0		
WRR	d	1977	B, A	85.0		133.0
WAES	e	1977	D, C	107.0		155.0
Shell	f	1979	RR, FC, BU	88.0	106	132.0
EZ/CPB	g	1979	L, H	97.0		118.0
WRR	h	1980	GG, NG	84.0		101.0
Shell	i	1981	LG, HG	73.0		96.0
EZ	j	1981	RS		74	
NCW	k	1981		64.55		96.3
CE	l	1981			55	
SMDE	m	1982	CE, AD, IH	53.0	75	81.0
CE	n	1982			53	

References

a. Energie nota
b. Parl. rep. 1975–1976 13520 no. 4, *Interimrapport van de Landelijke Stuurgroep Energieonderzoek: 'Energie 1976'.*
c. T. G. Potma, 'Het vergeten scenario', *Bêta*, 1-2-1977.
d. Wetenschappelijke Raad voor het Referingsbeleid, *De komende vijfentwintig jaar,* The Hague, 1977.
e. Workshop on Alternative Energy Strategies, *Energy Supply Demand Integrations to the Year 2000,* 3rd technical report 1977.
f. D. de Bruyne, 'Nederlandse energie tot 2000, vooruit met halfgas' in *SMO-Informatief* 79/1 (1979).
g. *Nota Energiebeleid*, Part 1/Algemeen, Parl.rep. 1979–1980 15802 no. 2.
h. Wetenschappelijke Rad voor het Referingsbeleid, *Beleidsgerichte Toekomstverkenning*, Part 1: Een poging tot uitlokking, The Hague, 1980.
i. Shell Nederland, *Long-term scenarios for the Netherlands*. Executive summary, Central planning and functional services, March 1981.
j.and m. Stuurgroep Maatschappelijke Discussie Energiebeleid, *Het Tussenrapport*, The Hague, 1983.
k. Nederlands Christelijk Werkgeversverbond.
l. and n. Centrum voor Energiebesparing, Delft: Het CE sceneario een: realistisch alternatief.

Abrreviations (English Translation)

EZ	Ministry of Economic Affairs
LSEO	National Steering Group on Energy Research
WRR	Scientific Council for Government Policy
WAES	Workshop on Alternative Energy Strategies
CPB	Central Planning Bureau
NCW	Protestant Employers' Association
CE	Energy Conservation Centre
SMDE	Steering Group of Public Energy Debate
RR, FC, BU	Realism and Restraint Frustration and Conflict Business as Usual
L, H	Low, High
GG, NG	Moderate Growth, No Growth
LG, HG	Low Growth, High Growth
RS	Reference Scenario
CE, AD, IH	Energy Conservation Centre Labour Re-distribution Industrial Recovery

6

United Kingdom Energy Policy and Forecasting: Technocratic Conflict Resolution

Reinier de Man

INTRODUCTION

THIS chapter describes and analyses the development of energy forecasting in the United Kingdom between 1965 and 1984 in the context of the ever-changing energy policy and the broader economic and political situation. The development considered here may be divided roughly into three distinct periods.

(1) *1965–76*: A period in which high energy forecasts dominate the energy policy scene. Energy policy is almost entirely supply orientated. Forecasting technology is little developed, especially in the policy institutions.

(2) *1976–81*: Low-energy forecasts attract considerable attention. Forecasting methodology is heavily discussed both inside and outside government. Attention is shifting gradually from the supply side to a more equilibrated view of both supply and demand. Ultra-low-energy scenarios are developed with financial support from the government.

(3) *1981–84*: A new consolidation is occurring. The critique of the government's past forecasting practice is to a considerable extent absorbed. Consequently, high energy scenarios almost disappear. There is much more interest in energy conservation, and forecasting methodology is altered accordingly. But the claims of the ultra-low energy forecasters are at the same time seen to be unrealistic.

These three periods describe a process of reinstitutionalization. Periods (1) and (3) are more or less stable states, whereas period (2) is the transition period to which the bulk of analysis below will be directed.

Table 6.1 UK Energy Forecasting: Three Periods

	1 'Good old days'	2 'Confusion and change'	3 'A new consensus'
High/low forecasts	very high forecasts	discussion: ultra-low vs. very high	moderate forecasts
policy orientation	supply	demand vs. supply, high conservation vs. high nuclear	supply and demand, moderate conservation, more moderate nuclear
methodology	rough economic (macro) forecasts	disaggregated technological vs. macro-economic	macro and disaggregated, more technological forecasting, no single methodology
short/long term	predominantly short term	longer term	importance of long-term recognized, but also intrinsic uncertainty stressed

DEVELOPMENTS IN ENERGY FORECASTING

The overall development of longer-term energy forecasting is summarized in Figure 6.1, where forecasts for total primary energy demand in the year 2000 are shown as a function of the year in which the forecast was published. A division has been made between 'establishment' forecasts and the more unconventional forecasts presented by critics of the government's energy policy. The low energy forecasting debate really took off in 1979, when Leach presented his energy scenarios, which demonstrated how considerable economic growth could be combined even with a declining demand for primary energy. (G. Leach, *et al.*, 1979). As a consequence, the official forecasts were subjected to repeated downward revisions, although there existed at the time considerable opposition towards the scenarios and the methods used by Leach.

This made *Energy Policy's* reviewer remark (P. Coyne 1979), '. . . within 2 years the IIED study will itself be the conventional wisdom, and within 5 years we will be dismissing it as dismally conservative'. Figure 6.1 indeed shows that the wide range of forecasts presented by the Central Electricity Generating Board (CEGB) in 1982 includes the whole range of Leach's forecasts of 1979. But to assume that Leach's energy scenarios are already conventional wisdom, is, as will be shown in more detail below, to indulge a misleading half-truth. The official forecasts did indeed systematically decline over the last decade, but the reasons why are distinct from the arguments presented by the low energy forecasters: only a minor part of the decline of official forecasts refers to the effects of an active conservation policy. The greater part is caused by the protracted economic

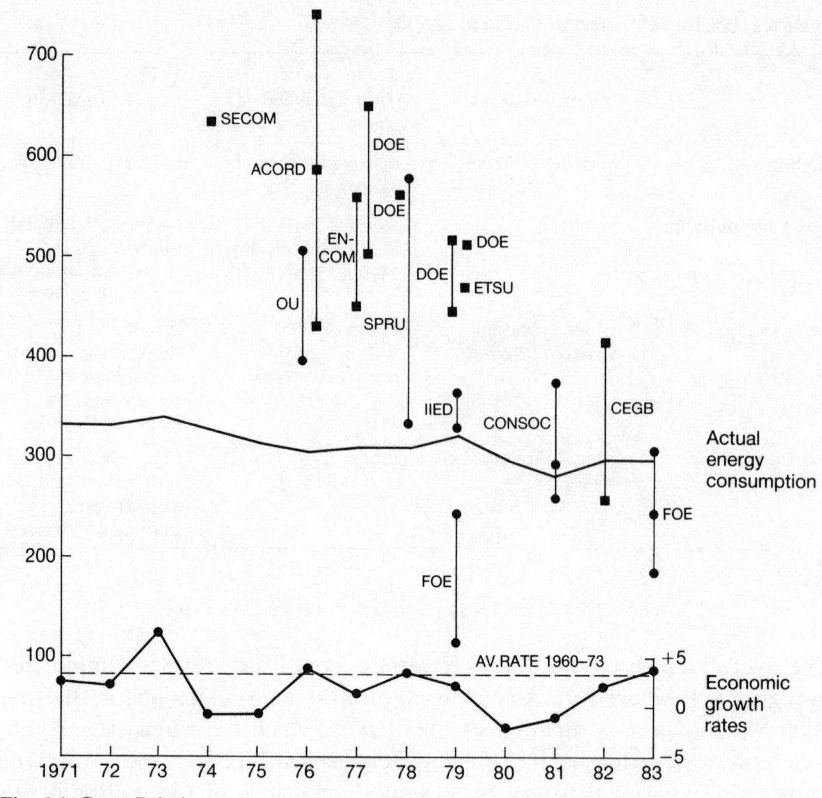

Fig. 6.1 Great Britain
Primary energy demand forecasts for year 2000 (MTCE)

SECOM: Select Committee on Science and Technology (Estimates for year 2000 are based on trend extrapolations)
OU: Open University, Energy Research Group
ACORD: Advisory Council on Research and Development for Fuel and Power
ENCOM: Energy Commission
DOE: Department of Energy
SPRU: Science Policy Research Unit
ETSU: Energy Technology Support Unit
IIED: International Institute of Environment and Development
FOE: Friends of the Earth
CONSOC: Conservation Society
CEGB: Central Electricity Generating Board

depression of the late seventies and the early eighties. If eventually there is a strong economic recovery in the second half of the present decade, energy forecasts will no doubt increase again.[1]

The British energy policy context

The UK energy situation of the late seventies and the early eighties is characterized by ample availability of energy resources, over-capacity of electricity plants, and the absence of substantial growth in energy consumption due to economic stagnation. This ample availability of energy derives from Britain's endowment of coal, oil, and gas resources. Coal reserves, whilst not as cheap as, for instance, Australia's, are large, with their resource lifetime measured in hundreds rather than tens of years. Oil and gas reserves are such that the UK should remain broadly self-sufficient in these resources until the next century. This endowment has ambivalent consequences for the UK. On the one hand, it acts as a 'cushion', making the UK less vulnerable to external risks, but, on the other hand, it has a negative effect upon exchange rates and the competitiveness of exports.

An equally important feature of the UK energy policy context is Britain's strong involvement in nuclear energy. After World War II, an important national nuclear industry was built up and many nuclear reactors were built. Britain has greater experience in the nuclear energy field than any other country, and the safety record of the nuclear industry is excellent. The British nuclear industry, however, was not able to compete on the international market on the basis of its own reactor designs. Therefore it came into serious trouble in the late sixties. Notwithstanding the nuclear industry's present weakness, the historically evolved government institutions on nuclear energy remain powerful actors in the energy policy (D. Burn, 1978; R. Williams, 1980).

The ample availability of indigenous energy resources in combination with low growth expectations for the British economy makes the British energy problem rather peculiar. There is no need for meeting large additional demands. The problem of dividing the total primary energy demand between the different supply sectors gives energy policy in Britain an important political dimension, involving competition between the different (nationalized) fuel industries, notably between coal and nuclear energy for electricity generation (NCB versus UKAEA and CEGB) and between gas and electricity (from coal or nuclear). The rapid penetration of North Sea gas into the British energy economy, made possible by low prices, threatens the vested interests of coal and nuclear energy. An example of failure in energy forecasting, even for the relatively short term, is found in the 1965 National Plan.[2] Although overall energy demand was forecast pretty accurately (forecast for 1970: 337 mtce; actual: 336.7 mtce), the contribution from natural gas was heavily underestimated (forecast for 1970:1.5 mtce; actual: 17.9 mtce).

This situation of abundance and interfuel competition sets the stage for UK energy policy in the seventies and the eighties. In the longer term, however, the exhaustion of North Sea oil and gas reserves will cause a dramatic change. There will be a transition towards more imports, other fuels (such as synthetic and natural gas, breeder reactors, renewables) or conservation will be more important. This is crucial for the very long-term projections (mainly for the year 2025), which will be discussed further below.

Phase 1: The good old days, 1965–1976

Energy policy in this period, between the launching of the second nuclear programme in 1964[3] and the Report on Nuclear Power by the Royal Commission on Environmental Pollution in 1967,[4] was largely supply orientated. Energy policy then was fuel policy, and until 1974 there was no Department of Energy. Externalities of energy production were not yet recognized as important; a strong environmental movement had not yet developed. Long-term energy forecasts did not yet play an important role in the government's fuel policy. This policy was related to the transition from a single-fuel (coal) energy economy to a multi-fuel (coal, nuclear, oil, and gas) economy. Expectations about the contribution of nuclear energy were high during the first part of this period. The coal industry was expected to contract drastically (P. L. Cook and A. J. Surrey, 1977; M. V. Posner, 1973).[5] These expectations turned out to be wrong for three reasons. First, serious technical and economic problems, due to considerable planning mistakes, meant the end of nuclear optimism. Second, the development of North Sea Oil and gas created a strongly competitive situation, especially between nuclear energy and gas. Third, rising energy prices made British coal less uneconomic than expected. The coal industry's contraction, therefore, could be slowed down. It could not be stopped, however.

The 1965 National Plan contained energy forecasts for the different fuel sectors for the year 1970. The considerable underestimation of the contribution by natural gas has already been mentioned. The nuclear and hydropower contributions, on the other hand, were substantially overestimated: 16.5 mtce were forecast for 1970, 12.2 mtce were actually consumed. In the same year, a White Paper on Fuel Policy was published.[6] Two years later, a second White Paper on Fuel Policy was presented,[7] including forecasts for 1970 and 1975.

The penetration of natural gas was now accounted for, but an utterly unrealistic view of the contribution of future nuclear power still prevailed: it forecast 35 mtce for the year 1975 (actual 12.9 mtce!). Even at that time, such a forecast was not realistic. The signs of the developing AGR planning disaster were visible for those who were able to see (D. Burn, 1978). But it would take some time before official belief in the nuclear power programme was replaced by a more realistic assessment. The 1967 White Paper implied the commissioning of about 1200 MWe capacity each year between 1970 and 1975. But the AGR nuclear programme ran into such serious problems that the first station ordered, Dungeness-B, would not supply electricity until 1976. Although the 1967 White Paper based itself on the practice of long-run marginal cost pricing, the reality has been different. Gas prices were kept low (J. Chesshire and C. Buckley, 1976),[8] bringing about a rapid substitution of coal-based town gas. Longer-term considerations did not play a decisive role.

The period after the 1967 White Paper on Fuel Policy until the 1973–4 'energy crisis' showed a haphazard state of energy policy. For a great part, this period coincides with the 1970–4 Conservative Government. The abundance of imported oil did not make energy policy a top priority. The coal industry's problems remained. The CEGB found itself at the centre of differences between the fuel industries, arising from growing interfuel competition (Cook and Surrey, 1977). The CEGB, for example, preferred diversification to exclusive dependence on coal, that is, on the NCB. On the other hand, the NCB needed more certainty than the short-term forecasts of the 1976 White Paper could offer. Their strategic planning, especially manpower planning, needed a longer time horizon. NCB's campaign to assure the future of coal resulted in conflicts between the NCB and the Ministry of Fuel. As a result of this NCB–CEGB antagonism, the two industries took substantially different views of the future of coal and oil. The government's policy towards development of the North Sea was more reactive than active, and was made by incremental policy-making rather than by active planning. Between 1967 and 1974 there was no serious attempt to explore the longer-term energy future of the UK.

Certainly, the new Labour Government in July 1974 was more in favour of planning and setting long-term goals, but the state of the art of energy forecasting in the UK was still poorly developed. In 1974, the National Coal Board published its Plan for Coal.[9] The planning targets for coal output in 1985 and 2000 were set at 150 and 170 mtce. The government agreed with NCB's plan.[10] Thus, the decline of the UK coal industry was expected to turn into gradual recovery. For the new Labour Government, energy conservation became, at least in verbal statements, an area of major concern. On 9 December it launched its Energy Saving Programme and more action, mainly exhortation and information programmes, followed, (W. C. Patterson, 1978a). But it took considerable time before any substantial conservation policy was formulated. In 1975, the House of Commons Select Committee on Science and Technology presented its report on energy conservation.[11] Longer-term energy forecasts prepared by the Department of Energy were presented here in a highly condensed form: only one page of scenario results for the years 1975, 1980, and 1990, based on an unspecified computer model. More revealing than these forecasts was the first recommendation of the committee: 'Henceforth the government should consider the extent to which increases in energy demand should be met by investment in additional supply capacity or avoided by investment in energy conservation measures.' This recommendation can be considered the trigger for a political and economic dispute which would be at the centre of the low-energy debate for years to come.

Phase 2: confusion and change, 1976–1981

During this period, energy policy and energy forecasts were heavily discussed both inside and outside government. The energy scenarios presented by Gerald Leach in early 1979 formed the climax in the low-energy debate. Accordingly, the following text has been divided into three parts: the events before the Leach scenarios, the Leach scenarios, and the evolution of the (low) energy debate after 1979.

Towards Windscale

As has already been demonstrated, electricity forecasts, the basis for capacity planning, were wildly inaccurate in the sixties and early seventies. From 1961 onwards, demand had been invariably overestimated, usually by some considerable margin. For example, in 1966 the CEGB forecast that maximum demand for electricity in England and Wales would be 55 GWe in 1972. The 1982 evidence of the CEGB to the Sizewell Inquiry suggests that the board did not believe that this level would be reached until about 2030. The CEGB forecasts were thus a ready target for a critique of their methods. The Open University Energy Research Group (OUERG) published such a critique in 1976.[12] The director of this group, probably best known for his work on energy analysis, had earlier presented scenarios in his book *Fuel's Paradise* (P. Chapman, 1975)[13] (P. Chapman *et al.*, 1974: 231–43). The OUERG work presented three significant innovations in forecasting methodology:

(1) The use of scenarios: in fact, unlike then current practice, the scenarios were developed to reflect policy choice rather than intrinsic uncertainty.
(2) The forecasting time horizon was stretched to 2020, much farther than in most previous exercises.
(3) The scenarios were based on technological rather than conventional econometric methods.

The next substantive contribution came from the Advisory Council on Research and Development for Fuel and Power (ACORD).[14] The objective was to assess what research and development on energy sources was necessary in order to ensure that demand could be safely met at reasonable cost. The ACORD team was primarily made up of UKAEA staff under the direction of the then Chief Scientist at the Department of Energy, Sir Walter Marshall. Except for the 1967 White Paper, the then most recent government forecast for UK energy demand (!), ACORD, had no government forecasts on which to base its thinking. For this reason, ACORD itself developed a set of seven scenarios, each representing a possible view of the future energy economy. These, however, were 'not intended to represent preferred views'. The methodology for generating these demand forecasts was not specified, and there was considerable dissatisfaction amongst the independent members about the level of the forecasts.[15]

About the same time, the Secretary of State for Energy, Tony Benn, called a National Energy Conference at Church House in Westminster. Tony Benn's wish for a public debate on energy policy was clearly related to his own worries about nuclear energy and his difficulties with changing the course of his own department in this respect.[16] At this conference, the growing opposition against the established energy policy was expressed by many different groups.[17]

The September 1976 Report by the Royal Commission on Environmental Pollution (RCEP) is a splendid illustration of the lack of realism in the forecasts of nuclear energy at the time.[18] It used two scenarios, one described as 'official', which was later disowned by the Department of Energy as being 'a heavily nuclear projection made by the United Kingdom Atomic Energy Authority', and one developed by the Royal Commission, which used less nuclear power, a so-called 'alternative scenario'.[19] The UKAEA forecast 104 GWe of installed nuclear capacity in the year 2000 against an RCEP forecast of only 24 GWe. The Department of Energy believed that capacity would come to lie anywhere between 40 GWe and 135 GWe (Pearce, 1982: 37).

The alternative scenario of the RCEP presented a challenge to the high nuclear forecasts. In the view of the commission, such a nuclear future was undesirable, mainly because of the problems of a 'plutonium economy'. Ironically, the commission did not state that such a future was simply impossible, but just undesirable (Williams, 1980: 278–85; Pearce, 1982). Starting in 1980, the construction of the planned 90 GWe of nuclear capacity would have meant the construction of one nuclear plant every five and a half weeks! The commission had presented an alternative to what could never have become practical policy. Although it was generally believed that the RCEP report had dealt the nuclear industry a severe blow, it can be shown that the Royal Commission's 'alternative' was well within the limits set by the policy of the Department of Energy at the time the report was published.

The Department of Energy finally produced its own forecasts in 1977 in the *Energy Policy Review*.[20] This document represented little advance on the ACORD and Royal Commission forecasts. It was entirely supply-orientated, and contained only two pages of very general discussion of conservation. The methodology of demand forecasting was very crude, and the assumptions about future demand developments remained at a very aggregate level.

By this time, however, Tony Benn, the Secretary of State for Energy, announced the setting up of an Energy Commission. The latter had already been presaged at the Church House Conference, but Tony Benn clearly did not succeed in opening up the energy debate, especially the nuclear debate, in the way he preferred. The fuel industries and the trade unions were dominating the commission, and thus were in a position to frustrate Tony Benn's intention (D. W. Pearce *et al.*, 1979: 196–7). The commission failed to have a substantial impact on energy policy and was abolished by the new Conservative Government in 1979. One of its effects, nevertheless, was to elicit a further set of forecasts from the Department of Energy. These were discussed at the Energy Commission, and finally published in February 1978 as a Green Paper,[21] followed shortly by a paper on forecasting methodology.[22]

The turning point of Windscale

The Windscale Inquiry provided a major impetus for the development of energy forecasts in the UK. In late 1976, the government announced it was to have this inquiry into the application of British Nuclear Fuels Ltd. for the reprocessing of spent oxide fuel rods at Windscale. The inquiry was held at Whitehaven from 14 June until 4 November 1977 (W. C. Patterson, 1978b). Many objectors found it impossible to discuss the desirability of nuclear fuel reprocessing at Windscale without considering the overall context of British energy policy. The Labour Government gave permission to the inspector to cover this broader range of issues related to general energy policy. But the major part of the arguments on this subject proved to be poor and sketchy. Only Gerald Leach, who was already working on his energy study with financial help from the Ford Foundation, and Peter Chapman of the Open University had something substantial to offer. Their alternative forecasts and their criticism of official forecasts were given considerable attention. A lower demand for electricity would result in a lower need for reprocessing capacity. The inspector, however, based his conclusions only on the official Department of Energy forecasts, not because they were necessarily better, but because they were produced by people with official responsibility (Pearce *et al.*, 1979: 153–4).

The other critics, such as the Friends of the Earth, the Centre for Alternative Technology, and the Conservation Society, did not present much by way of reasoned scenarios. Although the outcome of the Windscale Inquiry was very disappointing for the anti-nuclear critics, and only helped to polarize the issue further (Pearce *et al.*, 1979) its indirect effect was a small but unprecedented success for the advocates of an unconventional energy future. Tony Benn was willing to give financial support to these people in order to improve the quality of their low energy scenarios. He succeeded in his difficult task to persuade his own department, which was then, and still is today, suspicious of 'funding the opposition'.[23] Funds were made available to work out the scenarios of the groups mentioned above, with the exception of Gerald Leach (IIED), who had his own funding.

In the organizational set-up, a central role was given to the Energy Technology Support Unit (ETSU) of the Department of Energy, and Professor David Pearce from the Department of Political Economy at the University of Aberdeen was asked to act as a go-between 'unorthodox' energy advocates and the establishment, and to advise ETSU on social, economic, and political aspects of the scenarios. ETSU itself did not carry out any low energy scenario work, but invited the different groups to do so.

The groups themselves, however, did not construct the scenarios either. This task was given to more 'respectable' institutions in order to assure a minimal degree of acceptance. The contract for the Conservation Society and the National Centre for Alternative Technology was, therefore, subcontracted out to Richard Lecomber at the University of Bristol. Initially, the Conservation Society people caused problems by stating that all quantification was basically wrong, but, finally, they agreed in recognizing their ideas in the Lecomber scenarios.

Friends of the Earth also failed the acceptance test, even though they had already been working earlier together with Amory Lovins on low energy strategies. Their contract was finally given to Earth Resources Research (ERR), the research arm of the Friends of the Earth.

Although it seemed natural to include a scenario by OUERG, this did not happen because Peter Chapman, whilst still influential at the Windscale Inquiry, had already diverted his attention elsewhere at the time.[24]

The 1978 Green Paper on energy policy was the first official statement on energy policy since the 1967 Fuel Policy document. It was a first attempt to present a systematic revision of energy demand forecasting. It represented a considerable advance on previous forecasts, being far more explicit in its assumptions and methodology.

However, the balance of discussion was still very much supply orientated. This was also made clear by the document on forecasting methodology, which was published shortly after the Green Paper. This paper on methodology showed that the Department of Energy used a set of loosely coupled submodels rather than applying one energy model in order to produce its forecasts. The most extensive modelling was on the different fuel-supply sectors, whereas energy demand was treated far less thoroughly. The modelling of energy conservation, in particular, presented peculiar problems to the model builders, 'because there is very little by way of a historic data base to work from and because conservation . . . is in a formative stage. However, it is clear that conservation is a major new element in energy policy. . . . It is therefore important to have some sort of measuring methodology to allow for future conservation efforts in the forecasts.'[25]

The problem arising here of the incompatibility of traditional econometric methods and the forecasting of new developments without historical antecedents was 'solved' by rather rough adjustments for energy conservation by some arbitrary allowances. The whole conservation model only covered two pages of the report.

Another interesting feature, at least politically, is the modelling of the iron and steel sector, which was treated apart from the rest of industrial demand in a distrinct subsector. Although the paper gives some very general methodological arguments for it, the main reasons were of a political nature. The iron and steel forecasts are the responsibility of the Department of Industry. The energy forecasts for this sector are based upon these official figures.

In 1978, the Department of Industry projected a crude steel output of 36 to 45.6 mt based on an annual output growth of 0.8 to 1.6 per cent per annum from 1970 to 2000. No mention was made of the fact that from 1970 to 1977 crude steel output had declined from 28.3 mt to 20.4 mt, a decline of 4.5 per cent per annum. The relevant point is that the government policy, as stated in the 'Plan for Steel', was based on these growth rates, and no other figures were politically acceptable. As a result, industrial energy demand was considerably overestimated, because telling the truth about the decline of the steel industry was politically impossible.[26]

The Leach scenarios

When Gerald Leach presented his low energy scenarios (G. Leach *et al.*, 1979; Leach, 1979)[27] in early 1979, he and his IIED team had been working for two years, gathering a vast amount of data on energy demand and energy conservation, which had never been available to the policy makers until then. Gerald Leach, formerly Science Correspondent for the *Observer*, a national Sunday newspaper, had worked at the International Institute for Environment and Development, a policy research institution, since 1974. For the OECD he had done a study on natural resources in relation to the motor-car industry, and became interested in the development of energy demand in the UK (P. Bunyard, 1979; Bunyard and Leach, 1979). When the idea of an energy study emerged at the IIED in 1976, the official policy documents at the time offered almost nothing. In the 1975 report of the Select Committee on Science and Technology, there were only four pages of energy forecasts and the forecasts were very high. This lack of basic information was a main reason why Gerald Leach asked for funds from the Ford Foundation.[28] The money was given and the study was carried out between 1977 and 1979.

At this time, the interest in more disaggregated forms of energy forecasting was growing. The Science Policy Research Unit (SPRU) of the University of Sussex published a sectoral approach to UK energy forecasts for the year 2000 in February 1978. The SPRU report contained a range of possible demand projections of between 335 and 577 mtoe for the year 2000. The most recent official forecast then was 560 mtoe. The SPRU work was based on a rather conventional view of the future and did not involve any far-reaching assumptions on changing lifestyles. The difference between SPRU and the government's view of the energy future was rather about the forecasting method. SPRU was an advocate of the 'bottom-up' approach in contrast to the more conventional 'top-down' forecasting procedure (J. Chesshire and A. J. Surrey, 1978).

The first results of the IIED study were made public in 1978, in a provisional report and in the form of evidence before the Windscale Inquiry, (Leach *et al.*, 1979: 150–6, 242).[29] Leach did not see any reason for expansion of nuclear capacity in the next ten to twenty years because of declining economic growth and the ample possibilities for considerable energy conservation.[30] He criticized the official forecasts of the Department of Energy, which were too high in his opinion.

The final report (Leach, 1979) in early 1979 was a fierce attack on both the level of the official energy forecasts and the hitherto prevailing forecasting methodology. He showed that even a decrease in energy demand would be entirely compatible with a strong economic recovery. The myth of a macro-economic coupling between GDP and energy demand was broken by using a highly disaggregated technological model instead of some aggregated econometric equations.

Leach and his colleagues based their study 'on a very detailed breakdown of the final use of energy by different fuels, types of appliance and end-use purpose in 1976, which ran to nearly 400 possible categories' (Leach, 1979: 81). The outcome of the study was a fall in energy demand of 7 per cent by the year 2000 and 20–5 per cent by the year 2025. Only well-known and cost-effective technologies were included. Leach's assessment of the role of Combined Heat and Power (CHP) and renewables was even more conservative than those contained in the Department of Energy forecasts. The contribution of coal would be well below the Plan for Coal target, and the nuclear power programme would be much more modest than the official plans. The IIED study did not make a choice against nuclear energy. It only stated that it 'could easily be abandoned if desired' (Leach, 1979: 83). The two IIED scenarios explicitly failed to present the most probable energy future, but indicated a possible future for which some policy action was required.

The IIED scenarios provoked a lively discussion on energy forecasting and its methodological basis. Immediately after the publication of Leach's alternative energy futures, there were a number of highly emotional reactions, but as the discussion went on, the IIED scenarios proved resistant to most of the critiques, though some weaker points also became clear. As the projected fuel demand in the IIED scenarios was so much lower than established views, it was only natural that the strongest objections were made by people from the fuel industries, who, however, more than once misquoted Leach's intentions by, for example, regarding him as an anti-nuclear opponent, which he was not.

Another misunderstanding was the 'Big Brother' type of government, which was allegedly required for the low energy scenario.[31] Curiously, the *Financial Times* indicated the results of the study as a 'bleak forecast for U.K. energy'.[32] Of course, it was a bleak forecast for investment in the fuel industry. In a highly intellectualized form, the UK Atomic Energy Authority reacted by an article in *Atom*. L. G. Brookes, in his review of the IIED scenario, focused very much on the methodological problems of disaggregated models, stressing the 'fallacy of decomposition'. According to Brookes, 'it is not possible to produce sensible forecasts of energy demand with the use of physical models only. In the end it is economic influences and social attitudes that prevail.'[33]

The position of the Department of Energy was somewhat different from that of the UKAEA. Although they had many reservations about the IIED scenarios, they could only confess that their own knowledge about energy conservation was still modest. At the press conference accompanying the publication of the IIED report, Frank Hutber, the chief forecaster for the Department of Energy, made an interesting revelation:

> When writing the energy conservation chapter in the Government Green Paper on Energy policy . . . , Hutber declared, 'he simply had no idea how much could be achieved, so he had in fact guessed a figure of 100 million tons of coal equivalents for "energy conservation" by the year 2000, published it and waited for the response.' (Patterson, 1979: 43).

Internal documents of the Department of Energy show that there was considerable doubt about many aspects of the Leach scenario, but the scenario as a whole was certainly not considered irrelevant.[34] The Department of Energy emphasized that the industrial problems arising from a heavily reduced investment programme in the energy-supply industry (in favour of investment in conservation) should be taken into account in assessing the costs and benefits of the strategy as a whole. The Department of Energy was less optimistic than Leach about both the scope for conservation (and the underlying economic assumptions) and the ease with which the proposed energy conservation measures could be implemented. The IIED scenario was thus seen by the Department of Energy as valuable, but far too optimistic. For example, Peter Jonas, Head of Energy Conservation Technology at the Department of Energy, said at an energy conference in London in April 1979, 'Leach has basically talked to a lot of knowledgeable and professional people, and he has selected opinions from those people which lead to minimum energy growth. . . . We would be very wrong if we based our plans . . . on the kind of optimism . . . which is incorporated in Leach's approach.'[35]

The other low energy scenarios and the ETSU study

The report on *Energy Technologies for the United Kingdom*,[36] published by the Energy Technology Support Unit (ETSU) of the Department of

Energy in 1979, followed up the earlier ACORD document of 1976. Central in this report was, again, the question of which technologies should be supported in order to ensure that energy demand could be met satisfactorily.

The still exclusive supply focus of the ACORD document had shifted a bit more in the direction of energy demand. The choice between investing in energy supply and investing in reducing demand, already put forward in the recommendation of the Select Committee on Science and Technology in 1975, was now seen as essential: 'There is a trade-off in the use of energy between expenditure on the purchase of a fuel and expenditure on either the equipment that uses the energy or on conservation measures . . . that may reduce the need for energy.'[37]

The scenario approach, already used in the ACORD paper, was developed further. It consisted of the use of a number of rather disaggregated models, making use of detailed technical data, an approach nearer to the work of Leach than to the older, official forecasts. (In fact, the members of the ETSU team maintained good contact with Leach while both were working on their respective studies.)

ETSU presented one scenario in particular in a very detailed manner. This 'exploratory scenario' reflected a conventional future with nuclear power and synthetic gas playing an important role. Installed nuclear capacity was foreseen to be 30 GWe in the year 2000 and 104 GWe in 2025.[38] The exploratory scenario was not the preferred scenario of ETSU. None the less, it came to play the role of the high-growth, high-nuclear 'establishment' scenario in the low energy debate.

The study on low energy futures, on 'unconventional' energy futures, in which ETSU played an important role, was already under way at the time, and some very preliminary results were included in the report. ETSU, in co-operation with the University of Aberdeen, did not see sufficient reasons for rejecting any of the proposed scenarios, but the report stated that 'the main areas of doubt seem likely to be the questions of social and political plausibility'.[39]

At the Windscale Inquiry in 1977, the only well documented low energy future had come from Gerald Leach. At this time (1980), there were already seven low energy scenarios; the two IIED scenarios, three scenarios prepared by the University of Bristol, representing the views of the Conservation Society and the National Centre for Alternative Technology (R. Lecomber and S. Price, 1981),[40] and the two scenarios by Earth Resources Research for the Friends of the Earth.[41] Later on, ERR would design two additional low energy scenarios. All scenarios being 'unconventional' in some respect, there was a fundamental distinction between the IIED scenarios and the other scenarios. Leach confined himself to presenting arguments which showed the possibility of a high conservation scenario, but he neither attacked economic growth, nor wanted to be identified with the anti-nuclear movement. He acted as a technical expert, not as a representative of a political or social movement. The other scenarios were quite the contrary: they were put forward by socially and politically motivated 'cause and interest groups',[42] whose concern was much wider than energy policy alone, questioning, for example, the benefits of economic growth. The energy policies proposed by these groups were also much more dramatic. The most extreme scenarios were designed by ERR for the Friends of the Earth. The ERR 'Conserver Scenario' opted without compromise for a Lovins-type soft energy path (A. Lovins, 1977).[43] Primary energy consumption in 2025 would be only 27 per cent of the 1976 level. In 2050 more than 60 per cent of primary energy demand would derive from renewables and more than 30 per cent from biomass alone.[44]

The plausibility of these scenarios was lengthily discussed by the ETSU/Aberdeen team, and the final assessment was published by ETSU in July 1982[45] (Pearce *et al.*, 1980). ETSU had reservations on many aspects of all scenarios, but was most negative towards the ERR/FOE scenarios. About the ERR/FOE 'Conserver Scenario' and 'technical fix scenarios' it says: 'They *overstep the limits of credibility* in several places on technical and economic grounds and, in general, *on the grounds of what could reasonably be achieved within the timescales assumed*.'[46]

The discussions at that time, taking place in the research and policy institutions, had two important consequences for the position of low energy forecasts. First, the Leach scenarios, not long before considered as 'purest heresy',[47] were suddenly regarded as representing a rather moderate middle position in the debate. Secondly, the debate about the ultra-low options worsened the position of the more exotic technologies, such as active solar and biomass. By having assigned unrealistically high contributions to, for example, biomass, the ultra-low energy forecasters did their cause no good. The scenarios were discarded as incredible and, unfortunately, so were the technologies. After this debate, only the more conventional options, such as straightforward conservation measures and city heating, were considered serious alternatives.[48]

The evolution of official forecasts

The 1979 revisions to the 1978 Green Paper were published in *Energy Projections 1979*[49] by the new Conservative Government. There was a significant reduction in expected electricity demand. A significant change was also made in the forecasts for the steel industry. The political difficulties in presenting pessimistic outlooks for the steel industry seemed to have disappeared when the Labour Government left office. The British Steel Corporation was now envisaging drastic reductions of domestic capacity (Pearce, 1982: 43). The 1979 forecasts were repeated in the first report of the Select Committee on Energy in July 1981.[50]

The apparent consensus, 1981–1984

The debate on low energy futures for the UK had practically come to an end by 1981. Things had changed considerably: the final ETSU report had been published in the aftermath of Tony Benn's energy discussion, but the new Conservative Government meant quite a different climate for British energy policy. Many ideas of the low energy forecasters in the seventies had already been absorbed into the more established institutions, such as the wide scope for conservation, the improbability of high growth scenarios and the more modest expectations about nuclear energy.[51] Thus, there were two circumstances by which the low energy debate gradually faded away: the gap between established and critical views had become narrower and the government was much less inclined to support the opposition.

An illustration of the growing consensus on long-term energy demand forecasts is the evidence given by the Central Electricity Generating Board[52] to the Sizewell-B Inquiry[53] in 1982. The CEGB had used a multi-scenario approach.

> It is estimated that UK primary energy demand in the year 2000 will be less than in 1979 in all scenarios except scenario B, in which it would rise by an average of about 1% p.a. over the intervening years. The energy ratio . . . has been declining since well before the 1973 oil crisis . . . the fastest rate of decline would be associated with a buoyant, successful, service-led economy.[54]

This could have been written by Gerald Leach instead of the CEGB. As was mentioned at the beginning of this chapter, the 1979 Leach scenarios fit very well into the scope presented by these CEGB scenarios (see Figure 6.1). But the band width of the CEGB scenarios is considerably larger. These scenarios represent an 'anything could happen' philosophy rather than a clear policy choice. This is consistent with the general ideological framework of the Conservative Government's policies and, more specifically, with the Conservative conception of energy policy.

The Conservative energy policy conception derives from a strong belief in the market place and, and a consequence, a general aversion to any kind of central planning. Nigel Lawson, the then Secretary of State for Energy, said in June 1982:

> I do *not* see the Government's task as being to try and plan the future shape of energy production and consumption. It is not even primarily to try and balance UK demand and supply for energy. Our task is rather to set a framework which will ensure that the market operates in the energy sector with a minimum of distortion, and that energy is produced and consumed efficiently.[55]

An important consequence of this revival of the rhetoric of the market place and flexibility is the decreasing importance of energy forecasts: 'By treating energy as a traded commodity, we greatly reduce the need for, and importance of, projections of UK demand and production' (P. Tempest (ed.), 1983).

The Secretary of State admitted, however, that there remains some room for forecasting in two areas. First, the logistic planning requirements of the electricity industry need some forecasting. Second, forecasting is needed in order to raise questions about policy coherence. For the latter reason, the Department of Energy has recently completely rebuilt its model, replacing the old model with one more in line with the current government's philosophy of reliance on the market.

Bearing these developments in the government's general philosophy in mind, the consensus which has been reached on long-term energy demand can be seen as a partial and also a vulnerable one. The low energy scenarios in the seventies all implied an important choice, quite contrary to the flexibility arguments prevailing in the eighties. The normative backgrounds of these two types of scenarios are thus very different, and only by coincidence do they produce the same range of demand forecasts. If, for example, the prospects for economic growth, and more specifically for growth in industrial activity, were to improve in the years to come, the current approach to energy policy and forecasting would produce high energy forecasts again.

In 1983, the last document of the British low energy discussion, which had virtually come to an end already, was published by Earth Resources Research (D. Olivier and H. Miall, 1983). It contained the final version of the two scenarios already presented to the ETSU low energy study alongside two scenarios accounting for a slower technical change than had been supposed in the earlier scenarios. The document failed to have any impact on energy policy and the energy discussion, although Sir Kelvin Spencer, a former Chief Scientist at the Ministry of Fuel and Power in the 1950s, said in the Foreword, 'This is one of the most important reports published in recent years on energy policy. It deserves careful study by all, and *action* by Parliament and the Executive.' (Olivier and Miall, 1983, page IX.) The report was hardly recognized in the British press. Clearly, it did not sufficiently break new ground,[56] and the institutional debate on conservation and renewables had already been ended.

THE POLITICS OF ENERGY FORECASTING IN THE UK

The historical account of the development of energy forecasting in the United Kingdom between 1965 and 1984 points to considerable change in many aspects during that period of time. The most obvious development is the systematic decrease in forecast energy demand in the long term. Attention shifted gradually from the supply side to the demand side; energy conservation was gradually taken more seriously; and growing attention was paid to energy forecasts which did not only consider the short, but also the long and very long term. The methodology, hardly developed in the first period, evolved continuously during the 1970s: the attention shifted from simple trend extrapolations and macro-economic models to highly disaggregated models on a technological basis. The political status of forecasting has been going up and down. In the seventies, there was a real boom in energy forecasting, but in the eighties the value attached to energy forecasts in the policy process declined.

No single causal explanation can be given of this complex historical process. The specific course of events was defined by coincidences, such as the fact that North Sea oil was being developed at the same time that international oil prices increased rapidly. It was also defined by the patterns of Labour and Tory Governments, for example. In the following part, however, no more attention will be given to this rather incidental course of events. Instead, the question will be asked what specific British condition, in terms of the UK political system, can be held responsible for the general characteristics of the role energy forecasts played in the UK energy discussion and in UK energy policy.

The political system and conflict resolution

In terms of the analytical distinction between technocratic and corporatist ways of legitimating decisions and resolving conflicts, the UK falls very much on the technocratic side of the scale. The underlying causes cannot be discussed here, but two examples from British energy policy show clearly that the dominant style is expert decision-making rather than seeking conflict resolution through extensive negotiation and consultation procedures.

The Windscale Inquiry shows a typical British solution to a societal problem. The inquiry procedure is similar to a legal procedure. The inspector decides on the basis of all the evidence and is helped by a number of experts to do so. There is no negotiation element in this type of procedure, for example between the 'establishment' and the 'critics'. This is the case in the corporatist type of solution, such as the Dutch have adopted. Tony Benn's energy discussion is only apparently a divergence from the British technocratic model. Tony Benn did not actually 'fund the opposition', but he obliged the opponents to defend their cause in an expert debate. No negotiation took place between the parties involved, the low energy forecasters and the establishment. Instead, the acceptability of the low energy forecasts was assessed by experts who had been chosen by government. In the government's low energy study, the technical organization ETSU played a central role. This is quite different from the Dutch situation, where this role was given to the corporatist Steering Group.

It is also an important feature of the British low energy debate that the energy question never became an important issue in party politics. Therefore, the critical groups were not able to form effective ties with members of parliament, as they did in the Netherlands and West Germany, and there was no need to form a unified, coherent opposition. This is reflected by the variety of low energy forecasts and their often contradictory assumptions.[57]

The use of models for energy forecasts in the UK

The variety of forecasting methodologies and models reported on here is proof of a vivid discussion in the UK on modelling methodology. The hot topics of macro-models versus disaggregation, of econometric versus technological models, have been mentioned more than once in this chapter. But serious attempts to unify the modelling approach seem to be absent in British energy policy. There was no pressure on the low energy forecasters in the ETSU study to use the same modelling approach in order to make the low energy scenarios comparable. The comparability issue, central in, for instance, the Dutch energy discussion, is thus virtually absent in the UK 'energy scenario game'.

It is crucial to observe that the importance of econometrics as a central modelling paradigm is absent in the UK: Britain does not have the historically based econometric planning tradition of the Netherlands and Norway. There is no governmental planning institution with exclusive econometric models which are used to legitimize all different kinds of policy decisions. There is rather a variety of models. The Treasury, for example, has a mid-term macro-economic forecasting model, which is now publicly accessible for anyone with the competence and the facilities to use it. There are also a substantial number of university-based models. But the only model with a detailed energy sector is the Cambridge econometric model.[58]

In Britain, there is no need for integration of different energy models, and low energy forecasters are not forced to use established econometric models. This may be related to the technocratic nature of the political system. In, for example, the Dutch corporatist tradition of consensus formation, the different scenarios had to be 'negotiable', hence constructed on a comparable basis. In the UK type of politics there was no question of negotiating between different scenarios. Their relevance was simply assessed by experts on the basis of the intrinsic quality of the arguments put forward.[59] Comparability and the unification of models and methodologies, therefore, did not play an important part in the UK. On the other hand, the trustworthiness of the scenario builders, in terms of their acceptability as research institutions, did prove a hot issue.[60]

The British style in the low energy debate caused a relatively fast institutional reassessment of energy forecasts. It actually started when Leach was preparing his low energy study between 1977 and 1979, and already in 1982 the new positions had been chosen: the Leach scenarios had become respectable, but the more exotic renewable-energy options had not. In this respect, the debate was highly effective, but it may seriously be questioned what its contribution has been to the resolution of the underlying societal conflicts over the production and use of energy.

Notes

Steve Thomas from SPRU at Sussex University did some preparatory work for this chapter, which he could not complete for practical reasons. Comments by Gerald Leach, David Pearce, David Collingridge, Peter James, and Simon Price were a great help in preparing this chapter.

1. Assuming no major change in overall energy intensity of the economy, i.e. no major change in energy efficiencies and, no major change in the economic structure, e.g. in the direction of a service economy.
2. *The National Plan*. Cmnd. 2764, London: HMSO 1965.
3. *The Second Nuclear Power Programme*. Ministry of Power, Cmnd. 2335, London: HMSO 1964.
4. *Nuclear Power and the Environment*. Royal Commission on Environmental Pollution, Sixth Report, Cmnd. 6618. London: HMSO, September 1976.
5. The rate of contraction was largely constrained by other than energy policy deliberations. It was defined so that the industry's morale would not collapse. The loss of 105,000 jobs by 1971 was seen as this maximum. Policy measures were to control the timing of the industry's contraction.
6. *Fuel Policy*. Ministry of Power, Cmnd. 2798. London: HMSO 1965.
7. *Fuel Policy*. Ministry of Power, Cmnd. 3438. London: HMSO 1967. See also Posner, 1973: 297–313; Cook & Surrey, 1977: 13–31. At that time, the Government's attitude towards indicative planning was very positive. *The National Plan* was published in 1965, and at the same time that the 1967 *Fuel Policy* paper appeared, a White Paper on *Economic and Financial Objectives of the Nationalised Industries* (Cmnd. 3437) was published. The importance of quantitative analyses and forecasts had just been accepted in government (Cook & Surrey, 1977: 14–15). There was, however, a discrepancy between this growing importance of centrally made forecasts and the guidelines for the nationalized industries which aimed at a considerable delegation of responsibilities (Cook & Surrey, 1977: 19).
8. The gas/coal price ratio on a cost per therm basis decreased from 3.02 in 1965 to 1.79 in 1970 and 0.77 in 1975. J. Chesshire & C. Buckley, 'Energy Use in UK Industry', *Energy Policy*, September 1976.
9. *Plan for Coal*. London: National Coal Board 1974.
10. *Coal Industry Examination*. Final Report, Department of Energy. London, August 1974.
11. *Energy Conservation*. First report from the Select Committee on Science and Technology. London: HMSO 1975.
12. *A Critique of the Electricity Industry*. Open University Energy Research Group, Research Report ERG 013, March 1976.
13. The author also played an important role in the debate on 'energy costs' of generating energy. P. F. Chapman, G. Leach and M. Slesser, 'The Energy Costs of Fuels', *Energy Policy*, 2 (3), 1974: 231–43.
14. *Energy R&D in the United Kingdom, a Discussion Document*. ACORD/Department of Energy, Energy Paper 11, London 1976.
15. There was discontent about the nuclear dominance in ACORD. In the 3rd report of the Select Committee on Science and Technology (1976–7, HC 534–1) we can read: 'We therefore recommend that ACORD should revert to its original role. Its advisory functions as regards alternative sources should be transferred to an Advisory Council on New Sources (ACNS) . . . ' The Government did not agree: *The Government's Response* to HC 534–1 and HC 564, Cmnd. 7236. London: HMSO 1978.
16. See R. Williams, 1980: 290: 'Tony Benn had become firmly identified with the concept of 'open government' while in opposition in the early seventies'.

17. *National Energy Conference*, Energy Paper 13/1, 2, Department of Energy, 1976. Note that at this time nuclear energy was not a particular issue in party politics. Only the Liberal Party opposed the Windscale decision taken in parliament. Williams, 1980: 306.
18. See note 4, above.
19. *Nuclear Power and the Environment: The Government's Response to the Sixth Report of the Royal Commission on Environmental Pollution*. Cmnd. 6820. London: HMSO, May 1977.
20. *Energy Policy Review*. Department of Energy, Energy Paper 22. London: HMSO 1977.
21. *Energy Policy, a Consultative Document*. Department of Energy, Cmnd. 7101. London: HMSO 1978. Alternatively called the Green Paper. See also Pearce *et al.*, 1979: 7–20.
22. *Energy Forecasting Methodology*. Department of Energy, Energy Paper 29. London: HMSO 1978.
23. In the *New Statesman*, 12 January 1979: 'The Department of Energy, despite Benn's reforms, still sees itself as a conduit pipe to deliver more and more energy. . . . So no one should be surprised that the Department is fighting one of Tony Benn's ideas: that the independent energy researchers, like the Leach team, should get some official funding. "Why the hell should we support the opposition?" one senior civil servant said about the idea.'
24. The author's interview with David Pearce, London, 6 July 1984.
25. *Energy Forecasting Methodology*, op. cit. (above, n. 22) p. 9.
26. From Steve Thomas, SPRU, University of Sussex (personal communication).
27. See also Pearce *et al.*, 1979: 16–18 for a discussion of the dramatic consequences for nuclear planning. In the year 2000: official 90 mtce, Chapman 60 mtce and Leach 30 mtce nuclear energy.
28. The author's interview with Gerald Leach, 11 June 1984. (Compare the timing to the developments in the United States described in Chapter 11.)
29. Leach was a witness for the FOE London, part of FOE International, together with P. Chapman, W. Patterson, and A. Wohlstetter.
30. The *Guardian*, 18 Feb. 1978: 'No need to expand, says energy expert'.
31. *Financial Times*, 12 Jan. 1979: 'Government intervention would be of the utmost importance'. See also the critique by Sir Francis Tombs, Chairman of the Electricity Council (lecture to the NW Fuel Luncheon Club, Manchester, 7 Feb. 1979, excerpted in *Atom*, April 1979: 99): 'Such an approach strikes me as showing a lack of respect for the market mechanism as well as for civil liberties . . . '.
32. *Finanial Times*, 11 Jan. 1979: 'Bleak forecast for UK energy'.
33. Book review by L. G. Brookes in *Atom* 269, Mar. 1979. In 1980, the UKAEA presented a better worked-out version of basically the same critique: G. V. Day, H. H. Inston, and F. K. Main, *An Analysis of the Low Energy Strategy for the United Kingdom as proposed by the International Institute for Environment and Development*, UKAEA Discussion Paper No. 1, Economics and Programmes Branch, UKAEA, May 1980.
34. *A Low Energy Strategy for the United Kingdom*. Comment by the Department of Energy, London 1979 and Internal memorandum by EC/S Branch, 12 March 1979.
35. In *Atom*, June 1979: 151: 'Energy Saving, an Urgent Need'.
36. *Energy Technologies for the United Kingdom—An Appraisal for RD, and D Planning*, Volumes i and ii (Annexes), Department of Energy, Energy Paper No. 39. London: HMSO 1979.
37. *Energy Technologies*, p. 1.
38. *Energy Technologies*, Annex C. The figures for future nuclear capacity concur with a UKAEA forecast reported by RCEP in 1976 (See Table 6.4).
39. *Energy Technologies*, p. 2.
40. Unfortunately, two of the three members of the Bristol Research team (R. Lecomber and P. Sears) died during and immediately after the research project. This also caused the loss of most of the organizational memory about the project.
41. The ERR report to ETSU was not available for writing this chapter. A conference paper has been used instead: Hugh Miall and David Olivier, *A Summary of the E.E.R. Energy Efficient Futures*, submitted to the 2nd International Conference on Soft Energy Paths, Rome, 15–19 Jan. 1981.

42. A terminology developed by the Aberdeen people; see *Energy Technologies*, Annex D, Section D2: 'Classification of Low Energy Groups and their Motivations'.
43. Lovins, a member of the Friends of the Earth and internationally well-known critic of nuclear energy, argued that a fundamental choice had to be made between 'soft' (renewables) and 'hard' (plutonium) energy paths.
44. Based upon the Conference Paper (note 41).
45. *Low Energy Futures, a Study Carried out by ETSU*, ETSU R11, Chief Scientist's Group, Energy Technology Support Unit, Harwell, 1982. The Aberdeen report to ETSU has never been made public, and was not available for writing this chapter.
46. ETSU's final report, p. 79.
47. The reviewer of *Energy Policy* (P. Coyne, 1979).
48. The author's interviews with Gerald Leach and David Pearce.
49. *Energy Projections 1979*. Department of Energy. London: HMSO, June 1979.
50. *The Government's Statement on the Nuclear Power Programme*. First report from the Select Committee on Energy, HC 114–1. London, Feb. 1981.
51. The feasibility of a number of conservation options, such as conservation in the building sector and the application of combined heat and power, were recognized, but there remained many barriers to the implementation of conservation policies. See Select Committee on Energy, 5th Report, London 1982 (on conservation in the building sector) and Select Committee on Energy, 3rd Report, Session 1982–1983: *Combined Heat and Power*, London, 1983.
52. *Statement of Case Sizewell 'B' Power Station Public Inquiry*. Central Electricity Generating Board, April 1982.
53. *Proof of Evidence, Sizewell 'B' Inquiry*. Department of Energy 1982. See also W. C. Patterson, 'A Report on Sizewell', *Bull. At. Sc.* June–July 984.
54. CEGB *Statement*, p. 52.
55. Nigel Lawson, 'The United Kingdom Energy Framework'. Speech on 28 June 1982, in P. Tempest (ed.), 1983.
56. The author's interview with Hugh Miall, London, 5 July 1984.
57. The opposite situation occurred in the Netherlands (see chapter 5). There the environmental movement presented only one common low energy scenario.
58. For a splendid review of energy models in the UK see S. C. Littlechild and K. G. Vaidya, 1982, chapter 2: 'A Survey of Energy Models'.
59. See the comments on the British style of decision-making on energy projects by M. Thompson: 'Postscript—A Cultural Basis for Comparison', in H. C. Kunreuther, J. Linnerooth *et al.*, *Risk Analysis and Decision Processes: The Siting of Liquefied Energy Gas Facilities in Four Countries*. Berlin: Springer Verlag, 1983.
60. Leach's academic style clearly fitted well into the British political system: it is quite the contrary to Potma's strategy in the Netherlands, who made a clear coalition with the environmental movement. The difference between these styles should not be explained in psychological terms, but rather in terms of the contextual differences arising from the contrasting political systems.

Bibliography

Bunyard, P. (1979), 'Fuelling Hope'. *New Ecologist*. Jan–Feb.

—— and G. Leach (1979), 'Man of Energetic Vision. *New Ecologist*. Jan–Feb.

Burn, D. (1978), *Nuclear Power and the Energy Crisis*. London: Macmillan.

Chapman, P. F. (1975), *A Fuel's Paradise*. Harmondsworth: Penguin Books.

—— G. Leach, and M. Slesser (1974), 'The Energy Costs of Fuels'. *Energy Policy*, 2 (3).

Chesshire, J. and C. Buckley (1976), 'Energy Use in UK Industry'. *Energy Policy*. Sept.

—— and A. J. Surrey (1978), *Estimating UK Energy Demand for the Year 2000: A Sectoral Approach*. SPRU Occasional Papers No. 5, University of Sussex, Feb.

Cook, P. L. and A. J. Surrey (1977), *Energy Policy: Strategies for Uncertainty*. London: Martin Robertson.
Coyne, P. (1979), Book review: 'A Low Energy Strategy for the United Kingdom'. *Energy Policy*, June.
Leach, G. (1979), 'A Future with Less Energy'. *New Scientist*, 11 Jan.
—— *et al*. (1979), *A Low Energy Strategy for the United Kingdom*. International Institute for Environment and Development, Science Reviews, London.
Lecomber, R. & S. Price (1981), *Report to the Energy Technology Support Unit on Behalf of the Conservation Society and the National Centre for Alternative Technology*. University of Bristol, Department of Economics.
Littlechild, S. C. and K. G. Vaidya (1982), *Energy Strategies for the UK*. London: Allen & Unwin.
Lovins, A. (1977), *Soft Energy Paths*. Harmondsworth: Penguin.
Olivier, D. and H. Miall (1983), *Energy-Efficient Futures—Opening up the Solar Option*. London: Earth Resources Research.
Patterson, W. C. (1978a), Energy Conservation, 'Not doing without but doing more'. *Bulletin of the Atomic Scientists* Dec. London.
—— (1978b), 'The Windscale Report, a Nuclear Apologia'. *Bull. At. Sc.*, June.
—— (1979), 'Conservation Cornucopia', *Bull. At. Sc.*, May.
Pearce, D. (1982), 'United Kingdom Energy Policy: A Historical Overview'. Discussion Paper (unpublished).
—— *et al*. (1979), *Decision Making for Energy Futures: A Case Study of the Windscale Inquiry*. London: Macmillan.
—— *et al*. (1980), 'Low Energy Scenarios for the United Kingdom: Their Social, Economic, Political and Environmental Implications'. A report to ETSU, Aberdeen, August (unpublished).
Posner, M. V. (1973), *Fuel Policy*. London: Macmillan.
Tempest, P. (ed.) (1983), *British Energy Economics*. London: Graham and Trotman.
Williams, R. (1980), *The Nuclear Power Decisions*. London: Martin.

Appendix

UK Energy Forecasts for the Years 2000 and 2025
a. Forecasts for 2000 (mtce)

Year of Publication	Organization	Total Primary Energy Demand	Coal	Nuclear + Hydro	Renewables
1974 (a)	National Coal Board	—	170	—	—
1976 (b)	Open University ERG	398–508	200		
1976 (c)	ACORD				
	Scen. 0	570–600	130–150	110–130	7–30
	Scen. 1	420–440	150–170	70–90	7–20
	Scen. 6	730–760	130–150	210–240	7–30
1977 (d)	Energy Com.	560	170	95	0
	Energy (low case)	450	—	—	—
1977 (e)	Dept. of Energy	500–650	100–165	27–102	0
1978 (f)	Dept. of Energy	560	170	95	—
1978 (g)	SPRU (Chesshire & Surrey)	335–577	83–162	40–102	—
1979 (h)	ETSU Exploratory scen.	470	155	72	2
1979 (i)	Dept. of Energy	445–515			
1979 (j)	IEED (Leach *et al.*)				
	High	364	122	34	9
	Low	330			
1979 (k)	Dept. of Energy	512	165	95	—
1979 (l)	Friend of the Earth				
	TF	244	65	2	33
	CS166	44	0	26	
1981 (m)	Conservation Soc. and NCAT/Bristol				
	High	377			
	Low 1	292			
	Low 2	264			
1982 (n)	CEGB, Sizewell-B	258–418			
1983 (o)	Friends of the Earth				
	A1	253	77	3	29
	A2	308	102	7	18
	B1	186	56	0	25
	B2	237	81	0	17

b. Forecasts for 2025 (mtce)

Year of Publication	Organization		Total Primary Energy Demand	Coal	Nuclear + Hydro	Renewables
1976 (b)	Open University ERG		460–680	200	—	—
1976 (c)	ACORD	Scen. 0	860–940	—	—	—
		Scen. 1	460–500	—	—	—
		Scen. 6	1270–1390	—	—	—
1979 (h)	ETSU Exploratory scen.		605	203	239	9
1979 (i)	IIED (Leach *et al.*)	High	356	148	30	36
1979 (l)	Friends of the Earth	TF	194	32	0	114
		CS	96	17	0	60
1981 (m)	Conserver Soc. & NCAT/Bristol	High	325	153	0	62
		Low 1	240	118	0	50
		Low 2	196	82	0	65
1983 (o)	Friends of the Earth	A1	185	81	0	110
		A2	238	84	0	63
		B1	92	17	0	58
		B2	140	46	0	44

Sources:

a. *Plan for Coal* (Note 14) and *Financial Times*, 11.1.1979.
b. *A Critique of the Electricity Industry*, 1976 (18).
c. *Energy R&D*, 1976 (20).
d. Energy Commission Paper 1 from Chesshire and Surrey, 1978 (41), p. 71.
e. *The Energy Policy Review* (27).
f. The 'Green Paper' (29).
g. Chesshire and Surrey, 1978 (41).
h. *Energy Technologies*, 1979 (52).
i. Belvoir Inquiry.
j. Leach *et al.* 1979 (1), (38).
k. DOE's comment on Leach (50).
l. Miall and Olivier, 1981 (57).
m. Lecomber and Price, 1981 (56).
n. CEGB, 1982 (69).
o. Olivier and Miall, 1983 (73).

PART IV

Élites and Professionals among Themselves

7

Electricity Forecasting in Norway: Administrative Centralism

Tormod Lunde and Atle Midttun

INTRODUCTION

ENERGY forecasting in Norway, has up to very recently been synonymous with electricity forecasting, as other forms of energy have been largely distributed without direct government involvement or control. In contrast to most other countries, the electricity supply system in Norway relies almost exclusively on hydropower. The abundant availability of suitable river sites, which have been developed at very low cost compared with thermal power, has ensured that a relatively large share of total energy consumption in Norway is covered by electricity.

The supply of cheap electrical energy from hydropower projects was an important factor behind the rapid industrial development in Norway at the beginning of this century. Foremost in this development was the establishment of electro-chemical and electro-metallurgical manufacturing plants in remote areas near the source of hydropower. Furthermore, relatively low electricity prices have induced consumers to use electricity for heating purposes to a larger extent than may be observed in other countries.

A brief description of the supply and demand pattern for the most important energy carriers in Norway is given in Table 7.1. Even though many decisions in the electricity system are decentralized, the overall responsibility for supply of electricity rests with the central government. Since World War II, new capacity for electricity production has in fact been built to a large extent by the central government through its utility, the State Power Board (SKV).

Plans for the electricity sector are prepared by the Norwegian Watercourse and Electricity Board (NVE), a directorate originally subordinate to the Industry Ministry, but moved over to the Ministry of Oil and Energy when the latter was formed in 1978. The board has traditionally handled questions on both the operation and the dimensioning of the supply system. Constructing a large-scale hydropower project takes from five to ten

Table 7.1 The Supply and Demand for Energy in Norway (in TWh). (From Bjerkholt *et al.*, 1983)

	Electricity	Oil products	Other fuels
Production	82.5	73.9	9.6
Imports	1.8	46.7*	13.1
Transmission losses**	7.4	—	—
Domestic net demand	74.7	95.0*	20.8
Energy intensive industries	28.1	15.6	10.8
Other manufacturing industries	11.1	17.2	5.0
Other industries and government	13.0	40.3	0.0
Households	22.5	21.9	5.0
Exports	2.2	25.6	1.9

* Exclusive of the use of fuels in ocean transport
** Losses in the electricity transmission and distribution network
Source: Energy accounts of Norway 1980.

years. Planning and processing the concession application also requires several years. Projections of future demand are therefore needed for properly planning the dimension of the electricity system.

Until a few years ago these demand forecasts were constructed by means of rather simple extrapolation methods based on official forecasts of overall economic development. Following the rise of the environmental movement in the 1970s, electricity forecasts became increasingly controversial. This controversy has given rise to debates on several levels:

(1) over the results of forecasting;
(2) over the assumptions made;
(3) over the models employed;
(4) over the organization of the forecasting activity.

The following discussion gives an outline of the short history of Norwegian electricity forecasting while relating it to the above mentioned debates.

TENDENCIES IN NORWEGIAN ELECTRICITY DEMAND FORECASTS

The large variation in electricity (and energy) demand predicted by different forecasts for the same future point of time has become a subject of debate in both professional and political circles in Norway. In this section we compare the most important predictions made in Norway and describe their differences with respect to central forecasting features.

The politically most important forecasts are published by the government in two series of documents: the so-called Long-Term Programmes (LTPs), which are prepared every four years;[1] and the bi-annual parliamentary reports on energy.[2] The forecasts published in these documents

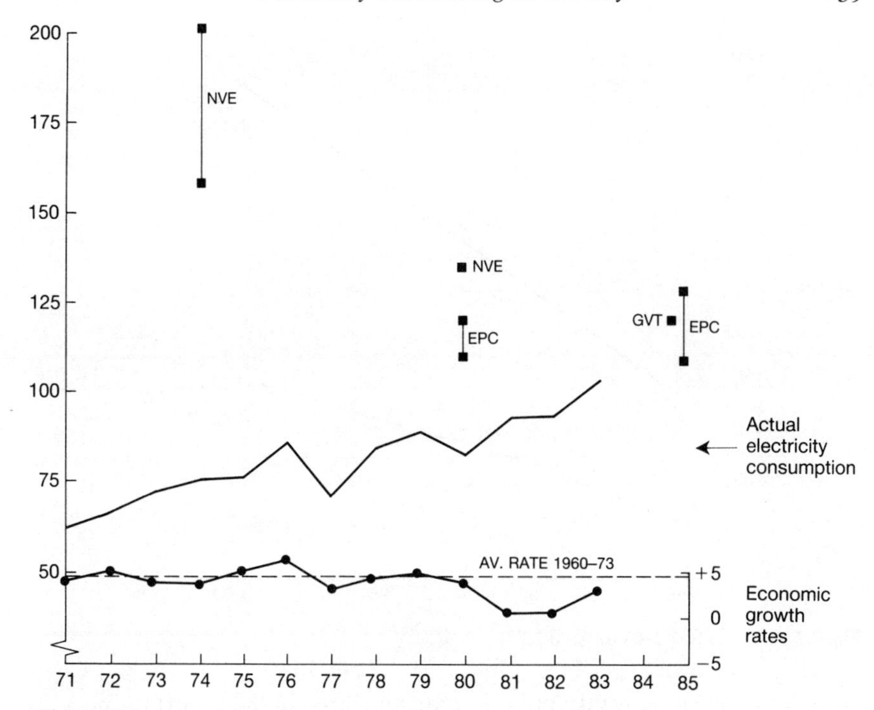

Fig. 7.1 Norway
Forecasts of primary electricity consumption for year 2000 (in TWh)

NVE: Norwegian Watercourse and Hydropower Authority
EPC: Energy Prognosis Committee
GVT: Government

Note: See also the table in the Appendix to this chapter.

have historically been prepared by the Watercourse and Electricity Board and more recently by a number of modelling teams and research bodies.[3] Figure 7.1 provides an overview of the forecasts made in the 1970s and early 1980s for the future growth of electricity demand in Norway. The first obvious impression is of the very large gap between the highest and lowest value of the same forecasts. Of course, the shorter the time span between the date at which the forecasts were made and the date for which demand is forecast, the smaller is the gap between highest and lowest forecast.

The second finding is that later forecasts are lower than earlier ones. This lowering of forecast demand goes hand in hand with a shift in the pattern of future electricity demand growth. The earlier forecasts all assume a linear trend. The latter ones, such as the one presented in Figure 7.2 from the Parliamentary Report on Energy of 1979–80, assume a break point in the future with the marginal electricity consumption growth becoming noticeably lower thereafter.

The first signs of a shift towards lower energy demand and a lower

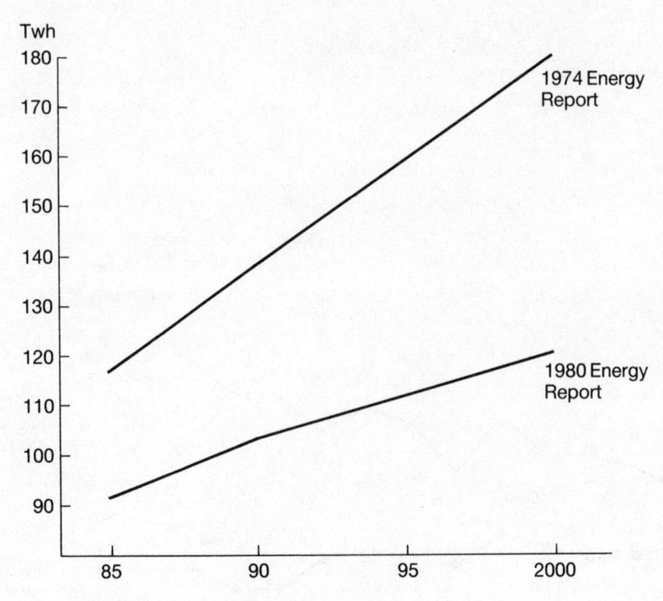

Fig. 7.2 Average forecasts for year 2000

growth rate at some future point in time appeared in the report of the Vatten Committee published in 1975.[4] The forecast presented by this committee differed dramatically from all previously published forecasts. In fact it is very close to most of the subsequent forecasts (at least for the years up to 1985, the forecasting horizon of the Vatten Committee)—(see Appendix to this chapter). The high forecast of the committee is about half-way between earlier forecasts and those made subsequently by the *ad hoc* Energy Prognosis Committee (EPC).

The forecasts published by the EPC in the 1980 Parliamentary Report on Energy (St. meld. no. 54 (1979–80)) follow more or less the path outlined by the Vatten Committee. However, they set the break point in the energy demand growth rate as late as 1990.

The forecast made by the Watercourse and Electricity Board included in the same report is also lower than the board's earlier forecasts and assumes an even lower growth rate. But it lies substantially above the high EPC forecast. And the energy demand growth rate is supposed to accelerate after 1995 in opposition to all the forecasts made by EPC.

Economic and political changes in Norway, similar to those in other industrialized countries during the 1970s, explain this downward revision in energy demand forecasts. On the economic side, growth assumptions were continuously lowered as the decade progressed and slow economic growth continued to persist. Stagnating population had a similar effect. The energy price consequences of the oil crisis led to the questioning of the future of energy-intensive industries in Norway. These industries remained internationally competitive only because they could use heavily subsidized electricity. The state subsidies, however, became increasingly questioned as the 1970s progressed. The decision was finally taken to set a maximum amount of cheap electricity which could be used by the energy-intensive industries,[5] a fact that obviously limits future electricity demand from this industry. Government also decided to make energy conservation an important policy goal, and this orientation too had an effect on forecasting.

Politically, the most important factor in generating lower energy demand forecasts was the emergence of an ecological movement. This movement managed to introduce into the debate an alternative model of societal development relying on reduced economic growth rates. The social and cultural values lying behind this movement were institutionally anchored when the Ministry for the Environment was set up in 1972. This change in administrative structure both reflected and accelerated the ongoing shift in political priorities in Norway. Power was redistributed among administrative bodies and among as well as within political parties. At times this led to open political conflict between environmentalists and more growth-orientated groups, supported by different ministries and professions. The reduction in energy demand forecasts was both a cause of this conflict as well as a result of these struggles.

THE USE OF MODELS IN ENERGY PLANNING

Changes in economic growth potentials, in societal values and attitudes, and in goals pursued by political and administrative systems have had an effect on energy demand forecasts. These factors, however, were not an integral and explicit part of the models used to forecast energy demand in Norway. They were either introduced through exogenously determined variables or they were embedded in structural parameters within the models. This means that there exists a 'grey zone' within established model-building and available data sources where ideological values and political interests play a substantial role. This problem has often been neglected and undercommunicated in current model-building and model-use in Norway as well as in many other countries (Koreisha and Stobaugh, 1979). As a consequence, political and scientific games around model-building and model-use have remained hidden. Technical language, mathematical formulations, and statistical judgements have formed an almost impenetrable shield of scientific objectivity which has masked the presence of political judgement, and prevented penetration from outsiders into Norwegian energy forecasting. An important task for institutional analysis is therefore to shed light on normative and political biases in model-building and use so as to allow a more qualified political and public debate. It is with this in mind that we here present the development of Norwegian energy modelling over the last years.

The first forecasts[6] were simple trend extrapolations. They were the result of the addition of independently made county-level forecasts ordered and collected by the Watercourse and Electricity Board from the regional electricity supply organizations. The generation and flow of information was therefore under the exclusive control of actors within the electricity supply system. Their interests in and evaluations of the energy demand situation were completely dominant in the forecasting process.

The first change in this set-up came already in 1969 with the 1970–1973 Long-Term Programme and later planning documents. Here a more detailed division of economic sectors was included in the models, and energy use in each sector was forecast on the basis of sectoral economic growth assumptions. Thus, knowledge of production and of the relation between production and energy use in each sector allowed a forecast of its energy consumption. Electricity consumption was then estimated on the basis of the division of the total energy consumption between different energy carriers. This was the first time that the relation between energy consumption and economic development had been modelled explicitly in the forecasting process. Energy price developments were however not yet included in the model (NVE, 1978).

One of the consequences of this development was that the Watercourse and Electricity Board (NVE) and the regional energy supply organizations lost their monopoly of information in energy forecasting. The new methods presupposed data on economic development within different sectors and thus laid down the basis for a link between electricity forecasting and general macro-economic planning in Norway. This link was not established practically before the oil crisis in 1974 when NVE, together with the Central Bureau of Statistics (SSB), started the development of so-called 'after-models' to the macro-economic models used by the Ministry of Finance. The 'after-models' contained energy consumption coefficients for each sector for which separate energy forecasts were made. This allowed the estimation of energy consumption once the gross product of the sector was itself forecast.

The determination of the coefficients for energy use already illustrate the difficult dividing line between science and politics. The size of these coefficients depends to a large extent on the time period which is used for their estimation. This opens up an opportunity to influence the forecasts by selecting an estimation period which gives the desired value.

The forecasting method used with the 'after models' did not basically differ from the one developed for the preceding model generation under the control of NVE. The main difference was that the explicitly defined sectors were increased and that the sector definitions were now closely matched with those in use in the governmental macro-economic planning models.

The energy models now became directly dependent on data generated by the macro-economic MODIS models. The latter were controlled by SSB and the Ministry of Finance.

This increasing dependence of energy models on macro-economic planning did not change with the multiplication of energy modelling that started emerging from the mid-1970s on. All energy models except one were closely linked to the existing economic ones. It was this virtually absolute dependence which in the late 1970s led almost naturally to a high degree of integration of the energy and economic planning system.

The deviating model was EFI-ENERGI, developed by the electricity supply industry's own research institute (EFI). The model represents the physical energy consumption system in Norway or a limited geographical area within the country. This sets it clearly apart from the more economic models. This model, however, did also, like NVE's 'after models', rely on input from the economic model apparatus (EFI, 1978a and 1978b).

Around the same time, three other institutes—EFI, the Christian Michelsens Institute, and the Institute for Energy Research—started the development of the ENOR model. This model may be characterized as a hybrid, with aspects from both EFI-ENERGI and the general macro-economic model built into it. This means that it combines physical and economic aspects (see IFA, 1978 and 1979).

Parallel to this development, an updated version of the macro-economic MSG planning model, which was also to include energy, was under preparation. This model, referred to as MSG-E, was a project undertaken by the Central Bureau of Statistics and the Institute of Economics at the University of Oslo.

The work on the model linked to the resource accounting system was also speeded up during the same period (Garnåsjordet and Longva, 1979; Hervik and Longva, 1978; NOU, 1977; SSB, 1981). This model was sponsored by the Ministry for the Environment and, as the name suggests, the energy accounting was included in a broader accounting system including other resources such as land use, fish, forests, and so on. Energy accounting was however given priority. This submodel was given the name EMOD (Ek *et al.*, 1979). It was an 'after model' to the macro-economic model MODIS.

The praxis of energy forecasting however lagged by several years behind the development of energy models. The NVE models were the only energy models used for elaborating official energy forecasts up to the preparations for the parliamentary report on energy of 1979–80. The other energy models were not operative before that time, or so the argument went. However, even during the preparation of the report, pressure from the public and interest organizations was necessary to institutionalize greater pluralism in energy forecasting. An interdepartmental committee, the Energy Prognosis Committee (EPC) was charged with the task of elaborating an official forecast.[7] The EPC contracted all existing energy modelling groups and let them make forecasts based on a set of common assumptions.[8] The committee's main difficulty was how to choose from among the forecasts supplied by the different teams. Lacking objective, scientific criteria, the choice had to be made partly on a normative basis. This the committee admitted.[9]

This choice had of course great political significance if one takes into consideration that the choice of model implied a selection among modelling groups with different professional perspectives and with different interests in hydropower construction. Furthermore, since models are based on fundamentally different concepts of reality, for instance physical or economic perspectives and data, they also dictate different measures in energy policy. Pricing is a typical measure suggested by an economic model, while changes in regulations for house building and production processes are more natural political tools if one works with a physical model.

The Energy Prognosis Committee in the end chose to use the EMOD model for all except the service sectors, for which the NVE 'after models' were preferred. The model-building teams of EMOD and of the NVE 'after models' were the only teams directly represented on the EPC. These modelling groups were also indirectly represented through the Central Bureau of Statistics (SSB), the Ministry for the Environment, and the Ministry of Oil and Energy (OED). It seems likely that the choice of the two models was part of a political compromise where representation on the committee was a dominant criterion.

The EPC also had to make controversial decisions with respect to some of the exogenous variables which had to be given to the modelling teams. Some of these variables were simply handed down from the ministries. This was the case for the assumption about future oil prices (made by OED) and about exchange rate and international trade developments (made by the Ministry of Finance (FD)). Others were supposed to be made on a more scientific basis. One of the most crucial decisions which EPC had to make concerned the choice of energy demand elasticities with respect to (energy) prices.[10] This choice belongs to a field were solid theory is lacking, and where normative assumptions therefore play a substantial role.

Existing energy statistics in Norway made it difficult to estimate a correct, current price elasticity. Using estimates based on earlier data did not really solve the problem because the elasticity could then be estimated only for a period with relatively stable energy prices and a relatively lower energy price level than was foreseeable. It was unlikely that elasticities from before the energy crisis would still be valid and would remain unchanged during the forecasting period. Nevertheless, the EPC argued that existing data pointed to a low price elasticity in the range of -0.1 or

−0.3. The final choice of −0.2 suggests that once more a compromise was reached by splitting the difference.[11]

Other economists have arrived at elasticities that deviate dramatically from this. According to Longva (1980) the elasticity was put too low. One of the reasons was that the methodological basis for estimating such elasticities in Norwegian energy models is weak. Based on the American economist Pindyck, Longva himself estimated an elasticity of −0.65.

The use of Pyndick's instead of the EPC's price elasticity assumption results in a further reduction of electricity demand by 10 THWh in 1990 or by about one eighth of today's consumption. The choice of elasticities is in other words of great consequence for energy policy. The uncertainty about estimation procedures gives room for considerable normative influence from modelling groups and administrative bodies in charge. It is interesting to note that the EPC report remained completely silent on this issue of price elasticity and failed to mentioned the high uncertainty involved in the choice of its price elasticity estimate. Nor did it present alternative forecasts based on other elasticity assumptions.

THE INSTITUTIONAL FRAMEWORK AROUND FORECASTING

It seems that changes in model structure often go together with a change in the organizational set-up of model users. A deeper understanding of forecasting must take this into consideration.

The general development of the energy sector in Norway over the last ten to fifteen years is marked by increasing complexity in the organization of public administration and an increase the number of actors who want to participate. This has led to an increased variety and complexity of perspectives and cognitive models. Thus, alliances and conflicts have been created and accentuated, which has put the decision-making system under considerable pressure. The electricity forecasts have played a central role in, and have, in turn, been affected by this development.

Partly, the conflicts have been between public authorities and interest organizations and grassroots movements. Partly they have taken place between different public organs. The first line of conflict has its root in the upsurge of new attitudes linked to ecology, environmental protection and the questions of life-quality and new lifestyles (see Inglehart, 1977; Knutsen, 1983).

These attitudes have increasingly come into conflict with growth-orientated ideologies of the organizations dominant in hydropower construction. The new ideologies have led to a strengthening of ecologically orientated interest organizations which consequently have pursued a more offensive policy.

The second line of conflict, between different parts of public administration, is connected to the incorporation and strengthening of a new environmental administration within a sector where industry and growth-orientated interests previously were completely dominant. We here specifically refer to the establishment of the Ministry for the Environment in 1972. There has, of course, been a close interplay between the conflict development within the administration and broader political conflicts outside it.

The increase in complexity and tensions within public administration in the hydropower field has had direct consequences for forecasting in several ways. Public organs with different perspectives and interests began to demand the introduction of new assumptions, elements, relations, and methods into forecasting. This again, as previously mentioned, has partly led to the building up of new modelling groups, which again has resulted in alliances between societal interests, ministries, and modelling groups that mutually support each other, but where each alliance pursues its own modelling strategy.

This will be exemplified in the following overview of the historical development of the organization of Norwegian energy forecasting. In order to structure the exposition, an introductory model of some key elements and relations is presented (see Figure 7.3). We are here making a gross division between the internal organization of public administration around forecasting and what we call external pressure groups and milieux, symbolized by circles outside the frame.

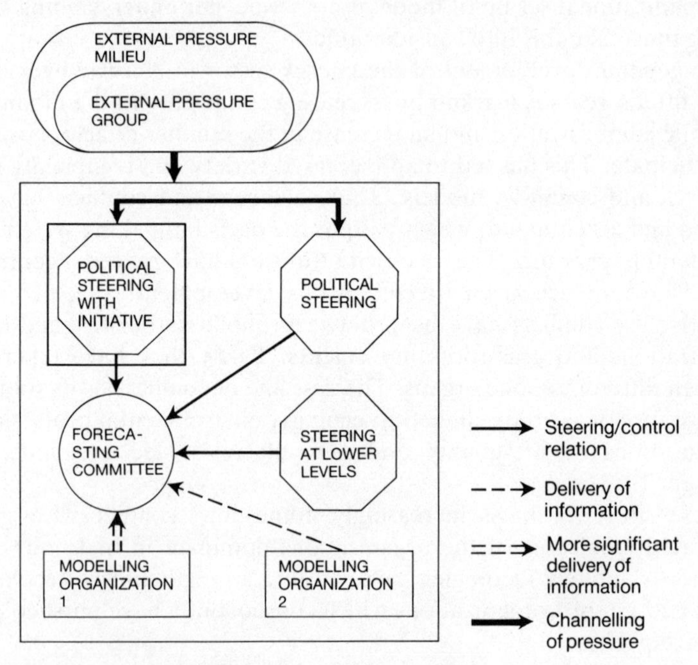

Fig. 7.3 Key elements of the institutional mapping

External pressure may be channelled into the public administrative apparatus in several ways: through party and parliamentary channels; through direct contact between interest organizations and administrations, or through lobbying; and through irregular politics in interplay with mass media and public opinion. Our model does not go into details about this, but only points to the existence of such pressure. The main weight, in this exposition, lies on the organization of public administration itself.

The early forecasts, up to the end of the 1960s, were undertaken within an organizational milieu with strong technical orientation. The Ministry of Industry was politically responsible and the forecasts were developed by the Directorate for Electricity, a subdivision of the Watercourse and Electricity Board (NVE), based on information from local electricity companies, electricity co-ordinating boards, and local electricity consultants. Assumptions about economic growth were provided by the Ministry of Finance (see Figure 7.4). At this early stage the environmental movement had not yet taken any interest in forecasting. The industrial sector in general, and especially energy-intensive industry, was therefore the dominant political pressure group. This is mirrored, for example, in the estimation of future energy use in the energy-intensive sector. The industry's own wishes, based on heavily subsidized tariffs, were included in the forecasts without modification.

In connection with the energy report in 1970 (St. meld. no. 97, 1969–70),

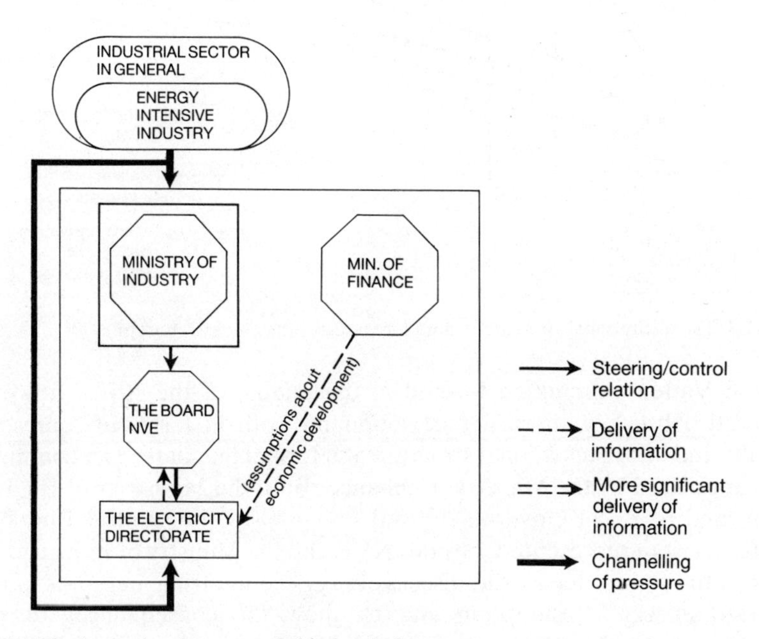

Fig. 7.4 The institutional structure behind early forecasts

a council was created by the Ministry of Industry to supervise the forecasting activity. The council initiated investigations and also developed new models in co-operation with the Central Bureau of Statistics (SSB). As mentioned earlier, forecasts were now linked more closely to economic models. The new organizational structure (see Figure 7.5), with a new council and a research group added to the already existing actors (NVE and the Ministry of Industry) did not imply a dramatic break with traditional power and influence structures. The energy council had a solid representation of industrial interests with strong links to hydropower construction and with a growth-orientated economic paradigm. Even though econometric methodology was newly introduced, and principles like pricing of electricity at long-term marginal cost were discussed, the forecasts that were presented did not deviate substantially from those made by the Watercourse and Electricity Board.

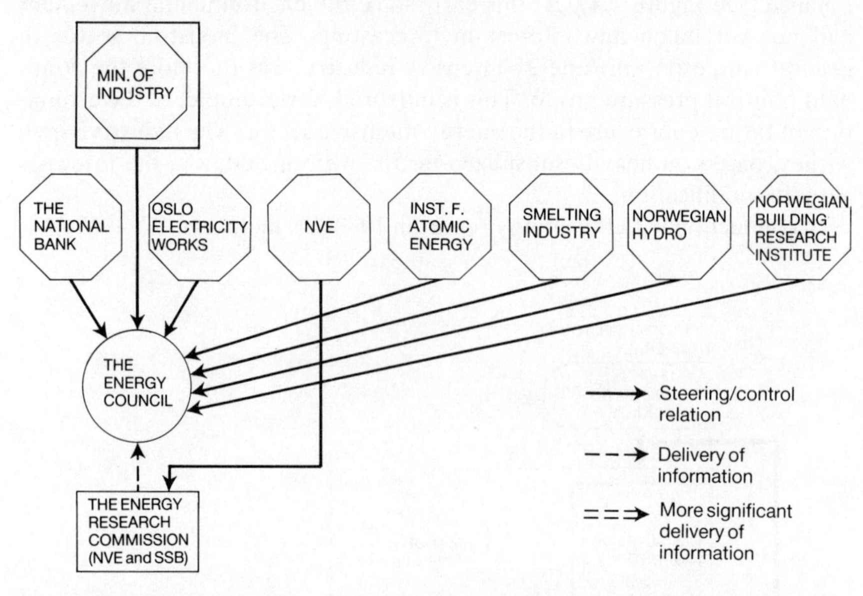

Fig. 7.5 The institutional structure behind forecasts for the energy report in 1970

The Vatten Committee formed in the middle of the 1970s, however, marked a break with earlier development both in terms of forecasting results and of organizational structure and formal mandate. The committee was appointed by the Ministry of Finance. Both the Ministries of Environment and of Local Government and Labour and the Central Bureau of Statistics were represented, besides NVE and the Ministry of Industry. The task of the committee was to discuss energy economizing measures in relation to energy consumption and to show the consequences of such measures for future electricity demand. The oil crisis in 1974 and the sub-

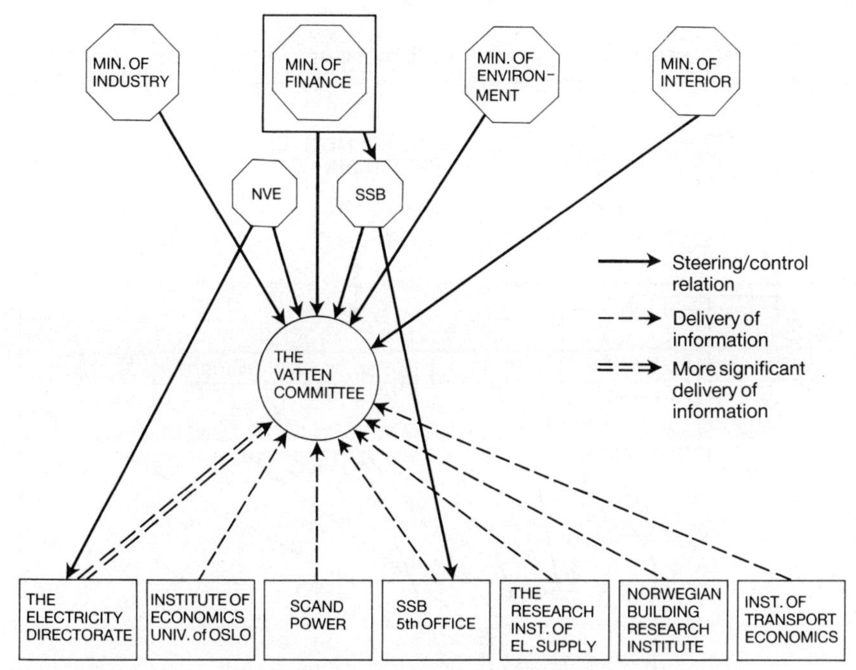

Fig. 7.6 The institutional structure behind the forecasts of the Vatten Committee

sequent energy debate had obviously also had its effect on forecasting. As previously pointed out, the committee arrived at forecasts that showed considerably lower electricity consumption than previously expected.

As Figure 7.6 indicates, the Vatten Committee made use of a number of modelling institutions for the preparation of its forecasts. Its work, in fact, started a learning process which has affected the development of competence in energy modelling in several of these institutions.

A continuation of the tendencies started with the creation of the Vatten Committee, led to the Energy Prognosis Committee, created in connection with the parliamentary energy report in 1980. Further complexity was added by the inclusion of the new Ministry of Oil and Energy and the Ministry of Transport in control positions.

The development of the EMOD-model by the new Resource Accounting Group of the Central Bureau of Statistics gave this group a prominent position in energy forecasting. This model was eventually used as the main basis for the official 1980 and 1985 forecasts, which implied a significant shift in control from the Watercourse and Electricity Board (NVE) and the energy-intensive industry to the Ministries of Environment and of Finance which had sponsored the EMOD model. As a consequence, the committee's forecasts of electricity consumption were considerably lower than those prepared by NVE. The organizational structure around the forecasts

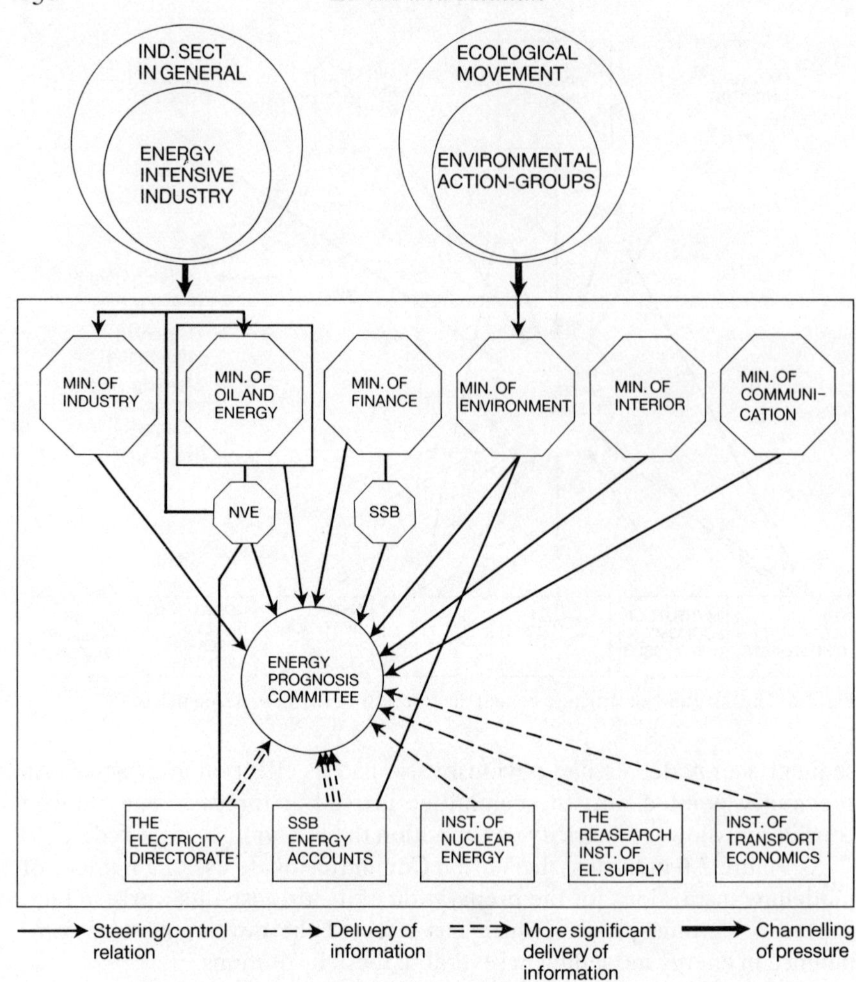

Fig. 7.7 The institutional structure behind forecasts for the 1980 energy report

presented in the 1980 energy report is of central importance, because it has now become permanent and therefore characterizes the latest (1985) energy report as well.

The shift of control away from the hydropower sector, and the new dominance of the Central Bureau of Statistics over the Watercourse and Electricity Board as the leading milieu for energy forecasting, has been closely linked to the introduction of economic paradigms and methodology. Both the environmental and financial authorities have had common interests in furthering this development, since they both have wished to gain better control over investment policy in the electricity sector and over expansion of hydropower plants.

The Ministry of Finance, with its overall responsibility for economic planning, has wished to impose economic criteria for investment and exploitation of resources. The establishment of such criteria would increase its control over the economy in general and over co-ordination of long-term planning in different sectors.

For the Ministry of Environment, lower energy consumption fits in with central goals in other parts of its field of responsibility. In a situation with heavily subsidized electricity prices, an application of economic investment criteria will lead to reduced expansion. The legitimacy of economic science and the possibility of an alliance with the Ministry of Finance made it easier for the Ministry of Environment to get through with an argumentation based on economics than with, for example, purely ecological arguments. In this context, the Ministry of Environment was also, to a large extent, a spokesman for the environmental movement. The movement, however would surely wish to go considerably further in establishing an energy policy along ecological guidelines.

The introduction of economic perspectives and models was seen to be against the interests of the utilities. This became obvious for all when the research institute owned by the utilities (EFI) introduced economic variables in its model and was subsequently heavily criticized for it.

The shift of bias in forecasting toward economic and resource-orientated milieux, however, has not affected the Hydropower Authority's dominant role when it comes to insight into physical and organizational structures of the hydropower supply system. It is interesting to note that these aspects have been given greater prominence in later years.

The controversy over electricity forecasts in the autumn of 1978 illustrates how control over forecasting models may be used strategically in political decision-making. New forecasts reflecting the decline in economic growth were made in the Watercourse and Electricity Board but were kept secret until the political decisions about some of the most controversial hydropower projects were made by the National Assembly. Even the Ministry of Finance was not able to gain access to these forecasts, and finally had to make its own, using the newly developed energy model provided by the Central Bureau of Statistics (see Lunde and Baumgartner, 1980; *Miljømagasinet*, 1 (1980)).

SUMMARY AND CONCLUSIONS

We have presented an outline of Norwegian energy forecasting showing that normative choices and political judgements enter into and affect the building and use of energy models on several levels. Our analysis points out many ways in which models can obtain a political bias in spite of the natural science ideal of objectivity. This, of course, also implies the possibility of using models and forecasts as a means in political struggles.

We have witnessed a development towards greater pluralism in modelling institutions and in the range of interests that had a say in energy forecasting matters in Norway in the 1970s. This has led to greater openness and to a broader professional and political debate. The competition on the modelling market and the development in computer technology may also eventually allow groups outside the public administration to elaborate forecasts based on alternative assumptions. This would give such groups the possibility to initiate public debates without having to rely on co-operation from the established élites.

Still, energy modelling in Norway today is primarily organized from a technical/economic point of view, although our analysis indicates that political and institutional factors play a major role in forecasting.

Notes

1. The first Norwegian electricity forecast was published in the second LTP in 1965 (for the period 1966–9). Each of the following LTPs (in 1969, 1973, and 1977) contained an energy forecast. This tradition was interrupted with the publication of the sixth LTP.
2. The first energy white paper was published in 1966 (St. meld. no. 19 (1966–7)).
3. Of course, new estimates are continually made with the models in the course of further developing and updating them, and in order to provide new data for policy-making. However, we concentrate our presentation and discussion here on the official, published prognoses, because only these provided inputs to the public debate.
4. The Vatten Committee was set up by the Ministry of Finance with the task of evaluating the consequences of several energy conservation measures. The final report was, however, kept confidential.
5. See St. meld. no. 54 (1979–80). The earlier policy is set out in LTP (1970–3). The EPC forecasts an electricity consumption of 34 TWh in 2000 for the energy-intensive industrial sectors, which, of course, would have wanted the amount to be substantially larger.
6. That is those presented in LTP 1966–9 and in St. meld. no. 19 (1966–7).
7. This Energy Prognosis Committee is to be distinguished from the previously mentioned *ad hoc* Energy Prognosis Committee.
8. Only the MSG-E model was left out because it was not then operative when EPC had to turn in its prognosis to the Energy Ministry.
9. From the committee's report: 'All of the models have strong and weak points. In addition, the models are based on somewhat different principles. These two factors make it impossible in practice to compare the models directly. The choice of one of the models for a given sector forecast has therefore necessarily to be based to some extent on judgment, at the same time as data base, sector definitions, and so on, are taken into consideration.'
10. This variable expresses the percentage change in energy demand for a one-percentage change in energy price.
11. Scientific arguments would rather have pointed to −0.3, the upper limit of the EPC range. It was likely that the historical elasticities underestimated the future elasticities because price fluctuations around a low price level are much more insignificant in terms of cost impact than changes around a high level. And this argument would not even have accounted for the long-term effect on elasticities of the large price increases in the immediately preceding years.

Bibliography

Bjerkholt, O., S. Longva, and Ø. Olsen (1985), *Analysis of Supply and Demand of Electricity in the Norwegian Economy*. SSB, Oslo.

EFI-TR no. 2248 (1978a), 'Energihusholdning: Dokumentasjon av energimodellen EFI-ENERGI'. Trondheim.

EFI-TR no. 24000 (1978b), 'Elektrisitetsetterspørselsanalyser—konsekvensanalyser for å kartlegge alternative utviklinger av elektrisitetsforburket'. Trondheim.

Ek, A., T. Kjølberg, and T. Sira, (1979), 'ENOR—An Energy for Norway', *IFA*.

Garnåsjordet, P. A. and S. Longva (1979), 'Outline of a System of Resource Accounts'. SSB, Oslo.

Hervik, A. and S. Longva (1978), 'Notat om energiregnskapet'. Oslo.

IFA Working Paper no. 1/78 (1978), 'Utvikling av energimodeller—en redegjørelse for samarbeidsprosjektet mellom CMI, IFA og EFI'.

IFA Working Paper no. 1/79 (1979), 'Statusrapport og presentasjon av nye sektormodeller i ENOR'.

Inglehart, R. (1977), *The Silent Revolution—Changing Values and Political Styles among Western Public*. Princeton University Press.

Knutsen, O. (1983), 'Post-materialisme, middelklasson og elitegrupper i Norge', *Tidsskrift for Samfunnsforskning*, 2.

Koreisha, S. and J. Stobaugh (1979), 'Limits to Models', in J. Stobaugh and D. Yergin (ed.), *Energy Future*. New York, Princeton University Press.

Longva, S. (1980), 'Prognosene i energimeldingen'. *Sosialøkonomen*, 1.

—— *et al.* (1979), 'Energy in a Multi-Sectoral Growth Model'. Paper presented at the *Nordic Economic Meeting in Helsingør*, 1–4 June.

Lunde, T. and T. Baumgartner (1980), 'Norske energimodeller'. Report No. 5 from the Energy and Society Project. *Institute of Sociology*, Working paper no. 155, University of Oslo.

Miljømagasinet, 1 (1980).

NOU 1977/31 (1977) 'Ressursregnskap' (Resource Accounts).

NVE Report 22/78 (1978), 'Forventet etterspørsel etter elektrisitet. Alminnelig forsyning. (Expected demand for electricity, general consumption) for 1985, 1990, 1995 og år 2000.

Statistik Sentralbyra (SSB) (1981), 'Statistical analysis no. 46'. Resource Accounts.

St. meld. (Parliamentary Report) no. 63 (1964–5), *Langtidsprogrammet* (*Long Term Programme*) *1966–9*.

—— no. 19 (1966–7), *Om elektrisitetsforsyningen i Norge* (On Norwegian Electricity Supply).

—— no. 55 (1968–9), *Langtidsprogrammet 1970–3*.

—— no. 71 (1972–3): *Langtidsprogrammet 1974–7*.

—— no. 100 (1973–4), *Energiforsyningen i Norge i fremtiden* (On Future Norwegian Energy Supply).

—— no. 54 (1979–80), *Norges framtidige energibruk og -produksjon* (On Future Norwegian Energy Use and Production).

—— no. 71 (1984–5), *Norges framtidige energibruk og -produksjon* (On Future Norwegian Energy Use and Production).

Appendix

Norwegian Forecasts of Electricity Consumption (in TWh)

		1975		1980	1985	1990	1995	2000
Parl. Rep.	MI	72.0	(1974)	94.0	112.0			
No. 100	NVE h			106.0	128.0	151.0		203.0
(1973–4)	NVE m			98.0	118.0	138.0		
	NVE l			90.0	107.0	125.0		160.0
The Vatten	h	70.0		87.0	102.0			
Committee	l			88.0	94.0			
1976 Rep.								
Parl. Rep.	NVE				100.0	110.0	120.0	131.0
No. 54	EPC1				90.0	105.9		124.6
(1979–80)	EPC2				86.3	98.9		110.7
	EPC3				90.0	103.8		118.6
	EPC4				88.3	102.0		116.4
	MOE				92.0	105.0		
Parl. Rep.	EPC h					106.0	116.0	128.0
No. 71	EPC l					102.0	107.0	112.0
(1984–5)	MOE					104.7	111.5	120.0

MI Ministry of Industry
NVE Norwegian Watercourse and Electricity Board
EPC Energy Prognosis Committee
MOE Ministry of Oil and Energy

h: high
m: medium
l: low

8

Electricity Forecasting in Denmark: Conflicts between Ministries and Utilities

Knud Lindholm Lau

INTRODUCTION

THE head of the Danish Association of Electric Utilities (Danske elværkers forening) commented in early January 1981 with some bitterness regarding his experiences with the preparation of the second Danish Energy Plan, to be completed later in the year:[1]

> From this, and from the experiences in 1980 in general, we conclude that the supply industry should in the future concentrate its attention on the work in those ministries that determine the basic assumptions that are going to be used in the long-term planning work of the Ministry of Energy. This might be our main task in 1981.

The utilities had originally preferred a highly aggregate macro approach to the forecasting of electricity consumption. But they had to accept the use of a sectoral approach, including the assumption of falling energy intensities due to technical progress in energy conservation. They now planned to concentrate their future lobbying on the economic assumptions underlying energy planning.

The resignation in the above quote should be compared to the utilities' rather triumphant comment on the first energy plan of the government, published in 1976: 'The statement made by the forecasting committee must be recognized as a very thorough one and it represents the most qualified analysis of future energy consumption patterns carried out so far'.[2] This chapter describes and analyses key events in the elaboration of the electricity forecast for the second energy plan and compares the methods used to estimate electricity demand growth for the first and second energy plan. It thereby explains why the electricity industry's enthusiasm about forecasting turned to bitterness within five years.[3]

The forecast picture in Denmark looks similar to that in many other industrialized countries. In 1973, just before the onset of the energy crisis, the electric utilities expected electricity consumption to grow to more than

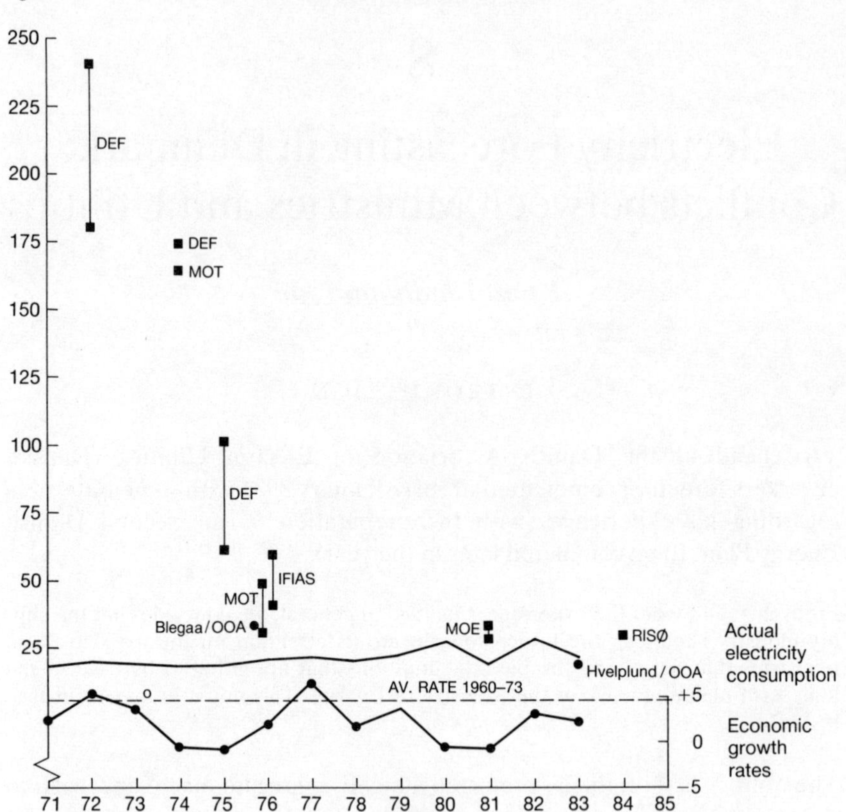

Fig. 8.1 Denmark
Forecasts of primary electricity consumption for year 2000 (in TW)

DEF: Danish Electricity Companies
MOT: Ministry of Trade
OOA: Anti-nuclear organization for Information on Nuclear Power
Blegaa: The Blegaa *et al.* report 1976
IFIAS: International Federation of Institutes of Advanced Studies
MOE: Ministry of Energy
Hvelplund: The Hvelplund *et al.* report 1983
RISØ: Risø Laboratory for Nuclear Research

80 TWh in 1990, about a fourfold increase in electricity consumption over the 1973 level. The first forecast after the oil crisis foresaw only slightly lower consumption in the year 1990. Both government and utilities were still basically in agreement about future energy developments.

Subsequent forecasts were substantially revised downward with the government projecting a consumption of only 32 TWh in 2000. This downward revision was accompanied by the emergence of non-establishment forecasts. They too were revised downward, with the latest one foreseeing a basically stagnant demand between now and the year 2000 (see Figure 8.1).

The second section discusses the reasons for the wish of the utilities and other actors linked to them to have the government come up with expansive electricity forecasts as part of the governmental energy planning exercises. The electricity industry saw its desire come true in 1976, hence the satisfaction reported above. It was disappointed by the 1981 forecasting exercise, as we have seen. The third section provides a factual description of forecasting methods used and assumptions made in the 1976 and the 1981 electricity forecasting exercises respectively. The comparison suggests that the methods have evolved technically and the assumptions have become more expert-based—at least for those about future energy efficiences, not necessarily those about socio-economic developments. But on the higher, conceptual level not much has changed: the (wished-for) socio-economic development sets the context for the energy development. Energy consumption is still not seen as a variable that could and should be consciously steered subject to the desirability of its socio-economic and other consequences.

In the fourth section, I describe a few key influence and decision processes around the choice of forecasting method and of the socio-economic scenario underlying the electricity forecast, around the determination of technical energy coefficients, and around the reconciliation of the initially incompatible views of the utilities and the Ministry of Energy on forecast electricity consumption. The conclusion places the forecasting episodes described and analysed here within the context of the larger question of state autonomy and power.

FORECASTS, ENERGY PRODUCERS, AND THE DANISH ENERGY SITUATION

The rise in oil prices in the 1970s hit the Danish economy and its energy system hard. Oil and gas satisfied close to ninety per cent of Denmark's primary energy consumption. Ninety-nine per cent of this oil and gas had to be imported, mostly from the volatile Middle East. The fuel import bill in the late 1970s equalled the deficit in the balance of payments, and this deficit was quite large. The utilities switched over quickly to the use of coal, and the development of Danish offshore oil and gas began to raise the prospect of a slightly diminishing import dependence. However, everybody was quite aware that dependence on increasingly expensive oil from the Middle East would sooner or later force the government to attempt to restructure the Danish energy system. The energy suppliers began to realize that markets would be at stake.

Denmark in the 1970s had not yet started to exploit nuclear power, despite the utilities declared interest in doing so.[4] The discussion on the viability of nuclear power concentrated on questions of safety and economic profitability. The latter was seen to depend on the future growth in electricity consumption. Hence the interest in electricity forecasts.

In 1979, the Economic Council, an advisory body to the government consisting of three distinguished economists, published a study indicating that nuclear power plants (NPP) were economically feasible.[5] Both the utilities and government were eager to claim that the report had proven that NPP would provide considerable economic benefits. The council, however, cautioned against this interpretation. Its cost–benefit analysis had shown a net present value of only US $600 million for a programme of seven NPP given an annual electricity consumption growth of 2 per cent. In case of stagnant consumption, this benefit would be halved.[6] It therefore became important that electricity forecasts should be able to show substantial future growth rates if the electricity industry were to get its NPPs.

Expansive electricity forecasts also became important in connection with the increasing 'politicization' of the heating market and the implication this had for the introduction of nuclear power. In the wake of the oil price crisis of 1973, the state had acquired the power, within certain legal and political limits, to allocate geographically prescribed heating markets to different energy sources. In 1979, the state acquired the right to force building owners to connect to collective energy supply systems, a right thought necessary to assure the economic viability of the planned natural gas distribution net in housing areas. This plan became important for the electricity industry because densely built-up areas were reserved for pipeline-based heating systems, thus shutting out the utilities from these potential heating markets. However, the plan excluded all pipeline-based systems from the so-called 'Area 4', the rural areas with low building densities. Here it was left up to individual house owners to choose from among other energy forms.

The utilities hoped that a situation of surplus generating capacity would tempt the government to transfer 'Area 4' to electric space and water heating (to the detriment of oil, coal, wood, straw, and so on). Furthermore, a possible orientation of generating capacity growth, especially if based on NPP, on forecast daily peak-load demands in winter would enlarge proportionally the so-called 'night-cap' of unused capacity.[7] This could tempt politicians to look at night-time electricity as a 'free good', ready to be used for the massive expansion of electric space and water heating without any additional investment apart from heat accumulators in buildings. Appropriate cost calculations of joint-supply electricity production could make this solution appear the most cost-efficient by far.

Low growth forecasts, on the other hand, threatened the autonomy of the utilities. Discrete, but large additions to generating capacity require careful matching of investment in new generating plants with anticipated demand growth. The standard power station built around the time of the preparation of the second energy plan had a capacity of 600 MW. Uncertain and low demand forecasts were likely to induce the government to tighten its control over power plant construction in order to prevent uncoordinated decisions taken by competitive utilities leading to uneconomic overcapacity.

The utilities were possibly hoping that expansive demand forecasts would forestall such government intervention, but could then later be used to induce government to shift markets in the utilities' direction in order to allow them to exploit the capacity that was going to be built in response to such optimistic forecasts. The forecasts would of course be vindicated in the end as the utilities would use the government to produce the demand increases necessary to make the forecast development come true. The utilities themselves could also use marketing policies to ensure that the forecast demand growth was actually going to occur. They could have engaged in price differentiation to exploit market segments with a high sensitivity to low electricity prices. There is, however, no empirical evidence that the Danish utilities have taken this course of action in the past.[8]

Industry too was interested in high electricity demand forecasts and the corresponding investment programmes. An input–output analysis suggests that the electric utilities' investment volume of DKr. 980 million had a domestic content of about half that sum. This was about equal to 25 per cent of the investment goods produced and sold in Denmark at the time by the construction and the metal fabrication sectors. A similarly sized investment programme for the years 1977 to 1989 would therefore have generated direct and indirect employment of about 80,000 man-years.[9] This compares very favourably with the estimated employment effect of 20,000 man-years from the building of the natural gas distribution system, the major Danish public investment project at the time.

This discussion suggests that not only the electric utilities themselves but all other energy suppliers had a substantial interest in the outcome of the forecasting exercise. So did industry in general and of course those groups that were against implementing the nuclear option.

THE ENERGY PLANS OF 1976 AND 1981

Neither the 1976 nor the 1981 energy plans stated explicitly the underlying socio-economic development goals. This fuzzy definiton of the societal future opened up opportunities for playing games around forecasting methods and assumptions, games that were played especially during the preparation of the 1981 plan. I will come back to these games in the next section. In this section, I describe the main features of the forecasting method and assumptions in the 1976 and 1981 plans.

The 1976 energy plan[10]

Only the assumptions about the major economic and industrial developments were fully spelt out in the plan. Economic growth was estimated to be between 3.5 and 4 per cent per year. Low and high growth cases for industrial production were defined with 3 and 5 per cent per annum respectively. However, important variables affecting energy consumption such as demographic growth, urbanization patterns, income distribution, or even energy prices to consumers remained unspecified. Only new housing construction was quantified and translated into increases in living space.

As a consequence, energy demand forecasts, for example, for electricity, were essentially based on trend extrapolations. Electricity usage for private and public lighting, in agriculture, and in the service sector was calculated assuming constant energy consumption growth rates. Only for lighting was a technical adjustment factor used to reflect advances in light bulb design. Electricity consumption in industry was calculated with the help of a constant marginal energy intensity of industrial GNP. In the household sector, electricity consumption in appliances was calculated by using penetration rates, use frequencies, and specific energy intensities; technical conservation gains in response to higher energy prices were not considered. In some sectors, future electricity consumption was simply guessed.

A number of calculations were made using different economic growth rates and energy coefficients in order to get an inkling of the robustness of the forecast thus produced. In short, the electricity consumption forecast in the 1976 energy plan rested on the projection of an almost unchanged present into the future of the next twenty years. Energy demand development was basically a function of time. No active energy policy measures were considered in the plan. Nor did the plan consider possible interaction effects between energy consumption, the evolving structure of the energy supply system, and the economy and polity.

The 1981 energy plan[11]

The model used for the elaboration of this energy forecast gave at least the appearance of greater complexity than in 1976. However, the elaboration of the underlying socio-economic scenario basically followed the same path as in 1976. The same was also true for the calculation of the electricity forecasts: some additional technical complexities modified somewhat the relationship between socio-economic developments and planned energy consumption developments. But *structurally* it still remained an exercise that projected the past into the future.

The socio-economic scenario[12]

As in the previous planning exercise, no explicit political goal setting preceded the elaboration of the socio-economic scenario. But the Ministry of Finance's econometric model ADAM was used this time to estimate expected economic growth, both in the aggregate as well as sectorally. This model has been described as being of a modified Keynesian type. Government demand, tax rates, exports, and productivity changes are important exogenous variables. The model also gives housing investments, the basis for the average number of inhabitants per dwelling (thought to fall towards close to two within the planning period).

The central growth scenario used for the elaboration of the energy forecasts assumed a GNP growth-rate of about 2 per cent annually at the beginning and at the end of the planning period, and of 4 per cent in the middle of it. A low growth case—maintained with difficulty in this planning exercise, as we will discuss later—assumed an annual growth rate of only 1.6 per cent.

The scenario assumed a structurally stable economy over the next twenty years, suggesting that infrastructure, production methods, urbanization patterns, and so on, would remain unchanged. This reflects on the one hand the built-in stability of any econometric model such as ADAM. On the other hand, such assumptions also reveal deeply held beliefs that market-driven developments would coincide with implicit political goals. Fuel prices were assumed to rise annually by 4 per cent. up to the year 2000.

The electricity demand forecast[13]

Electricity demand was derived sector-wise from the socio-economic scenario. In the *household sector*, electricity consumed by appliances was again calculated using penetration rates, use frequencies, and specific energy intensities. Some penetration rates this time were derived from a regression on private consumption levels. In addition, a vintage model was used to account for the spread of technical conservation improvements through the appliance stock: dishwashers were assumed to consume in 2000 only half the electricity they needed in 1980. Washing machines, tumblers, television sets and, refrigerators were thought to experience a 20 to 40 per cent improvement in energy efficiency. Electricity used for lighting was assumed to remain constant, improved lighting standards compensating the savings from improved light bulb design. Electric space and water heating was thought to remain restricted to 'Area 4', that is, the area where no pipeline-bound energy forms were allowed. A market share of 50 per cent for electricity was assumed. Electricity consumption in *industry* was estimated separately for five more or less homogeneous sectors and a sector of otherwise unclassified industries. The historical correlation between electricity consumption and sector output, controlled for investment levels,

Table 8.1 Electricity Forecasts in the 1976 and 1981 Energy Plans

	1976		1981					
Year	TWh	%	TWh	%	TWh	%	TWh	%
1975	18.05	4.5						
1980	23.60		21.9	2.3	21.8	1.9	21.8	1.8
1985	—	4.4	24.4	2.6	23.9	2.0	23.8	2.4
1990	36.20		27.7	1.9	26.4	1.8	26.8	1.7
1995	—	3.1	30.4	1.7	28.5	1.6	29.2	1.2
2000	49.30		33.1		30.8		31.7	
Scen.	expected		expected 'normal'		low growth		controlled	

hours worked and energy cost developments, was used to explain sectoral variations in energy intensities. On this basis, electricity consumption intensities were thought to fall by between 4 and 21 per cent to the year 2000, except in the case of mining and quarrying and the manufacturing of non-metallic products in which case increases of 13 per cent were assumed. Future electricity consumption was then calculated by combining these electricity intensities with the sectoral growth rates calculated by ADAM.

Similar, although less substantial calculations were made for agriculture, horticulture, construction, services, and the public sector. Table 8.1 compares the 1976 electricity forecast with the three alternative forecasts of the 1981 energy plan. The three alternatives were based on 'normal' and 'low' economic growth and on the assumption of normal growth combined with a slightly active energy conservation policy. Major elements here were a tax on electricity consumption, mandatory labelling of all electrical appliances with specific electricity consumption efficiencies, as well as a conservation information campaign. The resultant electricity consumption in 2000 was to be 45 per cent higher than in 1980 and only 5 per cent below the result of the normal growth scenario. The low growth forecast actually produces almost the same electricity consumption level in 2000 as the 'controlled energy consumption' scenario. In this case, lower electricity demand is almost compensated by a slower renewal of the stock of appliances with more energy efficient machines.

The calculation of the 1981 electricity forecast differed most clearly in its inclusion of technical energy conservation possibilities from that of 1976. The 1976 forecast basically overlooked such possibilities apart from the minor improvement of light bulb design. The 1976 forecast reduced electricity consumption to a function of economic activity and time (and the two were themselves colinear). The 1981 electricity forecast, in contrast, broke up this tight, almost autonomous relationship between economic growth and electricity consumption. This somewhat more technical perspective of the 1981 forecast was also highlighted by its sectoral basis, a point to which we will return in the next section.

However, this evolvement of forecasting methodology did not mean that the insertion of the electricity system into wider society had changed: the forecast demand was still to be matched by additional generating capacity. Energy supply was not going to constrain societal development. Market forces determine development. There is no deliberate policy action to change the link between energy and the economy. The conservation measures analysed in the preparation of the electricity forecast were of a traditional and limited kind. The use of the econometric model ADAM for elaborating the socio-economic scenario cemented the basic feature of this forecasting exercise: economic activity is stipulated first, energy intensities are estimated second, and the two then give forecast energy consumption.

However, the shifts in forecasting methodology, slight as they were, nevertheless went against the interests of the electric utilities. The latter were quite aware of this but proved unable to prevent them. On the other hand, the utilities did not have to fight against farther reaching changes; none of the central actors proposed any of them. The process leading to this middle-of-the-road forecasting result is highlighted in the next section where we describe some key events that took place during the preparation of the 1981 forecast.

EVENTS FROM THE ELECTRICITY FORECAST 'BATTLE' OF 1981

Several steps in the elaboration of the 1981 energy and electricity forecasts gave rise to discussions and conflicts between ministries and actors in the energy system. The forecasting method was attacked by the electric utilities and other pro-growth actors; a struggle opposed the Ministry of Energy to the Ministry of Finance (and some people within the former) over the type of socio-economic scenario that was to be used to calculate the energy forecast, and the estimates of technical improvements in energy efficiencies of machines and production processes were constantly challenged and doubted by the electricity supply industry. The losers of the methodological battle, however, did not give up pushing for an expansive forecast. They continued to challenge the low figures produced by the Ministry of Energy's sectoral analyses. The resolution of this forecast reconciliation process is described in the final part of this section.

The actors

We begin the story of the electricity forecasting 'battlee' of 1981 with a description of the central actors involved in this forecasting game and of their

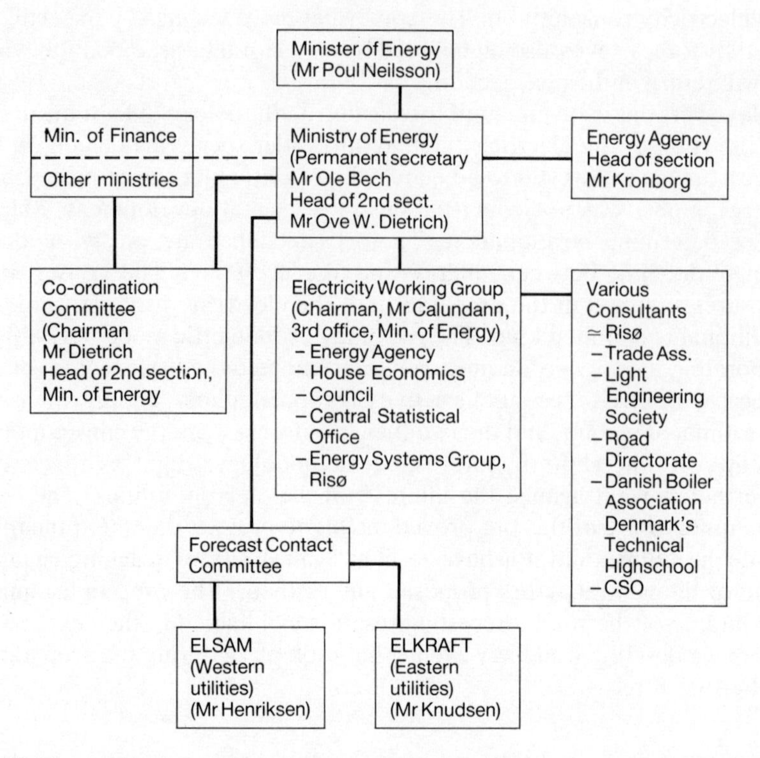

Fig. 8.2 The organization of electricity forecasting for the 1981 energy plan

relationships (see Figure 8.2). Energy forecasting was the responsibility of the Ministry of Energy. It was formed rather late, in 1979, taking over energy matters from the Ministry of Commerce. In Danish administrative practice, a ministry is split into two. The department handling all the general functions, including planning, has direct access to the minister. It is normally called the ministry. During the forecasting exercise of 1981, it was headed by Permanent Secretary Ole Bech, who therefore had direct access to the Minister of Energy, Poul Nielsson. The other part of the ministry is made up of the agency handling all the technical and administrative details. In our case this part is called the *Energistyrelsen*, or the Energy Agency. This agency is subordinated to the first part of the Ministry, the department.

The electricity forecasting work proper was the responsibility of the Electricity Working Group (*Arbejdsgruppen for Elforbruget*). It was headed by an official from the Ministry of Energy and had representatives from the ministry, the Energy Agency, the Home Economics Council, the Central Statistical Office, and the Energy Systems Group at Risø National Laboratory, the nuclear research centre. A number of other organizations were attached to this group on a consultancy basis. Other working groups were feeding the Electricity Working Group with detailed results.

The forecasting work of the different groups was supervised by a higher, more politically orientated Co-ordination Committee with representatives from all interested ministries and public organizations. It too was chaired by an official from the Ministry of Energy. However, the committee played only a minor role in the elaboration of the forecast, both because of its diverse membership and the infrequency of its meetings. The latter was apparently due to a decision of the committee's chairman.

The third important group, the Forecast Contact Committee, provided the meeting place for the officials from the Ministry of Energy and the representatives from the two electric utility associations, ELSAM (covering the west of Denmark) and ELKRAFT (organizing the east).

Give me the method and I shall give you the forecast

The interim report to the 1981 energy plan compared two forecasting methods, the so-called macro and sector methods. The report also indicated that the former would result in an unexpected annual electricity consumption growth rate of 4 per cent the latter of 2 per cent.[14] The final report settled on a figure close to 2 per cent and was based on the sector method. The writing of the interim report was the occasion definitely to eliminate the macro method as a serious contender to methodological respectability. The rather neutral presentation in the interim report of the two competing methods, and of their implications for forecast consumption levels, was the result of one week of intensive redrafting and struggle over the way to present the state of this methodological conflict. The supporters of expansive forecasts, mainly the utilities and a high civil servant in the Ministry of Energy, lost here their first big battle. As we will see later, this also meant that the utilities were eliminated as a central player from the electricity forecasting game. I describe and analyse this process in this section.

The macro method was championed by the utilities with the support of the head of section in the Department of Energy, Mr Ove Dietrich, responsible for writing the energy plan. It modelled total electricity consumption econometrically, as depending on a few, highly aggregate, macro variables such as growth in GNP and in private consumption, the oil price development, and so on. The sector approach, sometimes also called the micro approach, derived electricity demand as the sum of sectoral forecasts that were based both on socio-economic and on technical energy considerations.

It was widely believed at the time—and correctly so—that the macro method could produce more expansive forecasts than the sectoral one. Those actors interested in an expansive forecast intended to make sure that, at a minimum, the energy plan included a reference to the macro method and forecast derived from it, besides the results of the sectoral method. In this way they hoped to suggest to everyone concerned that the sectoral method with its two per cent growth forecast defined the lower limit of the likely development.

However, by that time the Ministry of Energy (as a whole) already championed the sectoral approach which, anyway, tied in neatly with the sectoral configuration of the econometric simulation model of the Ministry of Finance. The first draft of the relevant passage of the interim report (written on 11 December 1980) pointed out that the macro forecast was likely to be too high, the sectoral one too low.[15] The author of this passage clearly intended to suggest that none of the methods was fully satisfactory and that the appropriate forecast was likely to lie somewhere between 2 and 4 per cent. Given the situation, this formulation would have meant victory for the supporters of the expansive forecast.

This draft was an attempt to save in the larger committee what had already been lost in the Electricity Working Group. Its chairman wrote in a letter (dated 11 December 1980) that the forecasting group had decided early on in its work that it was correct to disregard the regression (or macro) method, one of the reasons being that this method could not take account of the likely energy technical developments.[16]

This technical decision of the Working Group had already been challenged earlier by the utilities. Their representatives on the Forecast Contact Committee had finally managed in a meeting on 5 November 1980 to get the civil servants on the Electricity Working Group to express a willingness to carry out a macro-based forecasting run despite the fact that the civil servants had been arguing against this method during the meeting.[17] But the civil servants obviously never really intended to follow suit; and they were in a position to procrastinate as they were actually the ones to do the modelling and calculating. Three weeks after the 5 November meeting, the utilities inquired in the Forecast Contact Committee about the state of the macro model run and they pushed once more for its quick execution.[18]

The reference in the interim report to a macro-method model run and a forecast of 4 per cent (calculated in fact by ELSAM) was obviously an attempt to force the hands of the forecasters. To no avail. Twice more the utilities tried to get things moving through the Forecast Contact Committee. Nothing happened. In a committee meeting on 31 March 1981, they were again promised a run of their model as soon as the 'low growth' and the 'controlled energy use' cases had been calculated with the sector model.[19] These cases were obviously designed to placate the 2 per cent sector-based forecast as the upper limit of future development, thus doing just the opposite of what the utilities wanted to achieve. Another promise was not kept, and was possibly never intended to be kept.

The Electricity Working Group had met three times during this period from 5 November 1980 to 31 March 1981. The proceedings of two of these meetings fail even to mention the topic of a macro-method forecasting run. It is true that a work schedule specified a time limit of 1 April 1981 for the execution of this run.[20] But fools' day indeed, as suggested by the renewed promise on 31 March 1981 and the continued failure to deliver before the publication of the final report.

The Electricity Working Group had obviously enough autonomy and backing to resist the demands of the utilities and to work with the method that the forecasters judged to be the correct one. The utilities obviously were quite prescient when they tried to outflank the unwilling forecasters, as mentioned above, by getting included in the interim report a reference to the macro method and its forecast thus providing it with an officially consecrated respectability.

Mr Dietrich was known at the time as anything but a low growth enthusiast. And it is true that the 2 per cent forecast of the forecasting group was surprisingly low and in opposition to official Danish views. But the manœuvre was too obvious and the reaction to the first draft of this part of the interim report was strong. In a letter of 11 December 1980, on the draft of the interim report, the chairman of the Electricity Working Group used strong language to express his dislike of the macro method.[21] He further argued that even the sector method was based on macro-economic assumptions. He concluded that there was anyway no reason to expect that the promise of a high forecast through the macro method could be fulfilled. He therefore proposed to delete any promise to use both methods for the preparation of the final forecasts. He did not succeed with this proposal despite another effort the next day. His effort deserves quotation for its bluntness: 'There is no reason to expect that the "request" of Mr Dietrich for a higher forecast, would be fulfilled by the method (macro, KLL) proposed in the paper'.[22]

Mr Kronborg from the Energy Agency also tried to eliminate the statement in question from the interim report. He suggested in a letter to Mr Dietrich, dated 12 December 1980, that no mention should be made of a connection between forecast method and forecast growth rate. All that should be said was that both methods would be applied in order to have some kind of a forecast sensitivity analysis. As he said, 'The work has to continue under all circumstances. It is therefore inappropriate to commit oneself that the macro method will produce a higher forecast than the sector method would do.'[23] This compromise position became more or less incorporated into the interim report with the important difference, however, that the actual forecast figures were given together with the description of the methods.

The utilities and high-growth supporters had gained the battle without realizing that they had lost the war. We have already seen that the Electricity Working Group continued its refusal to excute a forecasting run based on the macro-method. And no mention of it is made in the final report. The utilities did not give up however. The annual report of ELSAM for 1980 comments on the interim report pointing out that, besides the sector-based forecast of 2 per cent, there did also exist a forecast of 4 per cent based on a macro model. The report fails to mention that this was the model championed by the utilities, that this 4 per cent forecast was ELSAM's own and not the one from the Ministry of Energy.[24]

We will see later that the utilities continued to put forward this 4 per cent forecast whenever the sector-based forecast was discussed in the run up to the writing of the final report, and that they continually demanded that the sector-based forecast be revised in the light of this magic 4 per cent. The official consecration given to the macro method and its forecast in the interim report was maybe legitimation enough for the utilities to turn it into the standard against which any other forecast had to be compared.

The imposition of the socio-economic scenario

At its first meeting, the Co-ordination Committee (see Figure 8.2) decided to assume an annual 4 per cent real energy price increase as a basis for the elaboration of the socio-economic forecast that was to be calculated by the Ministry of Finance's econometric simulation model ADAM.[25] This economic forecast was delivered four days later but was dated with the day of the meeting that fixed the assumption.[26] The forecast had probably already been prepared beforehand by the Ministry of Finance, and the energy price assumption settled on by the group must have reflected the value usually used by the Ministry of Finance in its own economic simulation work.

This ability of the Ministry of Finance to constrain the forecasting work of the Ministry of Energy is also demonstrated by the way the Ministry of Finance tried, and sometimes even managed, to avoid having to elaborate alternative socio-economic scenarios. The Ministry of Finance's follow-up letter to the delivery of the mentioned scenario above bluntly pointed out that the ministry would be unable to calculate other scenarios until much later, after the annual report on the state of the economy had been completed.[27]

The Ministry of Energy originally had had the intention, even before the above mentioned first meeting of the Co-ordination Group, to explore the energy consequences of alternative economic assumptions. The views of the other ministries were to be taken into account. It was even planned to explore a 'heretical' economic scenario with less than full employment and capacity utilization after 1990.[28] (It has to be remembered in this context that government economic forecasts always predict full employment in the future after the 'temporary' trough of unemployment and unused capacity.) But with a busy Ministry of Finance in charge of the economic forecasting work for the energy planning exercise, the Ministry of Energy dropped the evaluation of the low growth scenario referring to delays in general.[29] It required the personal intervention of the energetic Minister of Energy, Mr Poul Nielsson, whose power had been strengthened during a public hearing on the progress of the energy planning exercise, to get this case again on the agenda of the Ministries of Finance and of Energy.[30]

The Ministry of Finance's control over the socio-economic scenarios underlying the energy forecasting exercise has the advantage of assuring uniformity among all the economic forecasts used in government policy documents. However, in exchange, all planning exercises are also subject to the same bias as the economic forecast. Discerning this bias is not easy as not much is known about economic forecasting procedures used by the Ministry of Finance and some of the assumptions behind a forecast are not even published. (This is for example the case with the adjustment terms that reflect political knowledge or intuition.) According to a central civil servant, exogenous variables are normally set according to an optimistic outlook on the future.[31] An academic observer has remarked that official policy, where it is explicitly formulated, will always be translated faithfully into the appropriate assumptions.[32] The problem with this procedure is that the short-term economic forecasts justify the strong policy measures politicians are apt to take after the elections. Long-range economic forecasts, in contrast, have to promise the light at the end of the tunnel of austerity, thus justifying the sacrifices the public is asked to make now. Just the opposite sequence is required in the energy area. The fundamental change will lie in the future, and we should be taking decisions now whose long-term effects will bring about the required scenario.

Constraining energy plans with a standard socio-economic scenario links energy demand forecasting to a political mechanism that has nothing in common with the nature of the energy problem. The energy forecasts in 1981 were based on a socio-economic forecast, enshrined in an official white paper of the time, which predicted, or better assumed, a less than realistic but politically desirable full employment and a healthy balance-of-payments as of 1990.[33] Tying the energy plan in this way to the traditional world of economists and economic forecasts precluded therefore turning the energy plan into a tool for system transformation.

Who is choosing parameter values this time round?

Technical energy coefficients played a more important role in the preparation of the forecast in 1981 than in 1976. Their determination requires detailed technical knowledge, in the absence of which they end up being guesswork, open to biased manipulation. In 1976, the civil servants preparing energy forecasts had to rely upon the advice proffered by the electric utilities. It was the utilities' research institutes which provided the few energy coefficients and intensities required. The situation had changed dramatically by 1980. The energy forecasters in the Ministry of Energy could and did draw from the beginning upon specialized knowledge available in utility-independent organizations.

Statens Husholdningsråd (the Home Economics Council) evaluated the likely energy technical developments of household appliances. The council proposed to assume improvements in the energy efficiency of appliances of up to 50 per cent between 1980 and 1990. The utilities suggested specific energy coefficients that were between 3 and 107 per cent above those proposed by the council.[34] *Lysteknisk Selskab* (the Lighting Engineering Society), a private-sector advisory organization, and *Vejdirektoratet* (the Road Directorate in the Ministry of Transport) provided the input on technical developments in building and street lighting. *Dansk Kedelforening* (the Danish Boiler Owner Association, the organization of users of large boilers) prepared the technical input on the energy savings potential in industry, itself carrying out two of the five underlying studies.[35] The Technical University of Denmark, the Central Statistical Office, and other organizations contributed their specific knowledge on the energy system. The Danish Centre for Nuclear Power Research in Risø, at the time already starting a move into research on energy systems and renewable energy resources, played a large part as well.

This introduction of technical energy expertise, which was independent from the utilities and had subsequently developed since 1976, is one of the main reasons for keeping the standard 1981 forecast below that of 1976.

The 'reconciliation' of different forecast perspectives

The reconciliation of the different forecasting perspectives supported by the Ministry of Energy (and other administrative actors) and the utilities respectively, was a constant topic at the meetings of the Forecast Coordination Committee. The utilities attempted at least twice to revise upward the ministerial draft forecast and to make them more compatible with their own views. The question was raised a first time during the preparation of the interim report,[36] and then again in March 1981, when the final report had to be prepared.[37]

From the beginning, the civil servants had hoped that a single forecast could be presented. They tended to value highly agreement on a single forecast. In this way, all the actors would be committed to the same future. The civil servants did not have direct responsibility for the electricity sector. They wanted therefore to be sure that future supply shortages could be blamed on the utilities and not on the administration or the government. Mr Kronborg, head of section in the Energy Agency, wrote for example in a letter dated 20 November 1981 and addressed to Mr Dietrich, the official in the Ministry of Energy responsible for the preparation of the 1981 energy plan that: 'The Agency could get into a very awkward position if the electricity forecast from the utilities in the interim report differs too much from the official one. The Agency does not wish to take over the obligation to maintain the security of supply.'[38] And he added: 'I would find it most unfortunate if the [Forecast Contact Committee] did not get the opportunity to study for some time—this means over some months—the new forecast and the assumptions underlying it.' This letter was written about one month before publication of the interim report. It thus represented a direct plea to eliminate the utilities' forecast from this report. Mr Kronborg did not manage to keep the utilities' forecast out of the interim report, but he at least saw the matter referred to the committee for further study.

The ninth meeting of the Forecast Contact Committee took up this issue on 25 November 1980. Committee members did agree on the electricity forecasts for industry and for household lighting. All other sector forecasts were contested.[39] The subcommittee formed to look into the differences of opinion reopened the issue of the industry forecast and found disagreement there too.[40] The utilities again requested the use of something other than the sectoral forecasting methodology.[41]

A third version of the ministerial forecast, showing slightly increased consumption levels, was discussed during the eleventh meeting of the Contact Committee in March 1981.[42] The difference of opinion about future requirements for generating capacity was now very small indeed, at least with respect to the eastern part of Denmark and if one factored into the administration's forecast a special provision of 5 per cent for 'cycle security', for a five-year period. This was a completely new concept that loosened the strict link between consumption growth and capacity construction that had been assumed before and was apparently intended to cover the risk that demand was going to grow somewhat faster than envisaged.[43] This safety margin would allow satisfaction of the higher than expected consumption until additional capacity could be brought on stream.

This skilful introduction of a new concept helped shape a meeting of minds. It served its purpose even though the concept was dropped from the final forecast. But its introduction helped reduce the tension between the utilities and the administrators.

Having helped to save the utilities' face with the temporary introduction of the concept of 'cycle security' the Energy Agency and the Ministry of Energy began in March 1981 to shape a final agreement to their liking, that is, to get the utilities' acquiescence to the administrators' forecast. The eastern utilities seemed to be adamantly against such a cave-in. ELKRAFT's representative on the Forecast Contact Committee, Mr Knudsen, proposed to refer the contended points to DEFU, the utilities' research organization, in order to get its comments on the content of and the documentation for the contended points. ELSAM, the association of western utilities, took a more conciliatory line, suggesting that 'the differences in the present forecast are now so small that, taking into account the uncertainties involved, it is very doubtful that further substantial work is needed on the short-run perspective'. But Mr Dietrich from the Ministry of Energy was unwilling to get an agreement on the basis of such a formulation. The reference to 'uncertainty' was not precise enough for the ministry's needs.[44]

However, Mr Henriksen from ELSAM could not compromise further as he was bound by the statement his board had made five days earlier: '[ELSAM's] extension plans are not based on the forecast in [the forthcoming] 1981 energy plan because it is inappropriate to base the enlargement of supply capacity on changes in consumer behaviour that so far have not been detectable in the actual development of consumption.'[45] The Forecast Contact Committee failed to reach a formal agreement on the question of the precise figures in the forecast. But the utilities did not continue to express their disagreement either. The DEFU report, ordered as a compromise strategy, noted opposition to the draft forecast only with respect to agriculture and household appliances.[46] The difference was now down to 2 TWh in 1990, about 5 per cent of forecast consumption (and half of what ten years of 'cycle security' would have given). But the difference was 25 per cent of the consumption increase of 7.8 TWh during the ten years between 1980 and 1990 that the utilities wanted to forecast. But the figures were of course not presented in this manner.

The reason for this persistent disagreement is hard to fathom. The only explicitly quantified difference added up to 0.34 TWh and was due to differences of opinion about the technical development of household appliances. But the utilities had great difficulty justifying their objection. As DEFU's report mentions, the utilities 'do not understand what the supposed technical solutions, on which the household consumption forecast rests' are all about. This was obviously not precise enough and the objection remained without effect.

Power leaves the utilities

A comparison of the three draft forecasts from the administration discussed in the meetings of the Forecast Co-ordination Committee—the first in November 1981 and the third and also final version late in March 1981—suggests that the utilities had lost the forecasting battle. The final electricity consumption forecast for 1992 in the final version was only 3 per cent higher than in the first draft forecast.[47] The administration basically managed to block all attacks on the forecasting elements originally proposed by it. One can assume that the socio-economic scenario remained unchanged throughout the discussions—it simply depended too much on the weighty economic planning work in the Ministry of Finance. The energy coefficients and intensities also remained almost unchanged.The utilities' inability to substantiate their criticism of the figures that came from non-utility sources proved their major weakness. At the same time this failure reveals the large dose of self-interest that must have gone into the 1976 electricity forecast. Although the utilities did not know it, they had lost the game already at the beginning when the Ministry of Finance used its control over econometric modelling to impose its view, and when the Ministry of Energy found itself in a position to draw on independent technical expertise.

The administration emerged as the apparent winner from this 1981 edition of the forecasting battle. Only the future will tell, though, if this shift, and the widened participation in forecasting that went with it, will really have an effect on electricity consumption growth. One is tempted to answer in the affirmative. The utilities have lost the central position that allowed them to determine forecast consumption, build production capacity accordingly, and then make sure that the electricity produced was also sold. It was this extended control that had made forecasting largely an exercise in self-fulfilling prophecy during the decades before 1980. The utilities' switch from smug self-satisfaction with the forecasting outcome in 1976 to dissatisfaction and disagreement with the outcome in 1981 suggests that they too believe that something has irremediably changed.

CONCLUSION

One of the necessary conditions for the change in the electricity forecast has been the emergence of centres of expertise outside and independent of the utilities. Two new actor complexes can be identified: technical and trade associations, public interest organizations, universities and research institutes that used the time after 1976 to acquire substantial amounts of technical knowledge and expertise on the energy aspects of their areas of interest and specialization, on the one hand; and, on the other hand, the administration and government in general.

This development of independent centres of energy expertise was closely intertwined with the emergence of grass-roots movements with a commitment to a low-growth future, and with the establishment of energy research groups at the universities in Ålborg, Odense, and Roskilde and at the Technical University in Copenhagen.[48] The researchers at the universities were in part centrally involved in the grass-roots movements, in part had close links with their animators. OOA, the anti-nuclear movement, got involved in the energy forecasting debate upon the publication of the government's 1976 energy plan,[49] when its implication for the development of nuclear power became obvious. Almost the same team elaborated an alternative forecast to the 1981 energy plan of the government[50] (see also Figure 8.1). However, there is no trace in the government documents from the preparation of the 1981 energy plan that the non-establishment researchers and movements had exerted or had tried to exert any direct influence.

Their presence made itself indirectly felt, essentially through informal contacts with the government officials and the experts directly involved in the preparation of the 1981 energy forecast. This type of influence process is typical for Denmark, a small and homogeneous society.

However, the alternative forecast to the one contained in the 1976 government plan was no more than a sketch, put together within a few months after publication of the energy plan by a small team that consisted mostly of physicists. Its estimate of electricity consumption in 1990 was only 10 per cent below the government forecast. This was sufficient to cast doubt on the official document, to mobilize public interest in and research on alternative possibilities, and to draw political attention to the topic of energy forecasting.

The more carefully elaborated forecast alternative to the 1981 energy plan, published in 1983, differed more substantially than the earlier exercise from the government forecast. Its forecast energy consumption was 20 per cent below the government figure for 1990 and 80 per cent below the one for 2000. It presented a stabilization of the electricity consumption of 1980 and envisaged a consumption decline in the years after 1990 (see Figure 8.1).

Particularly the first alternative forecast was of great help to the politicians in parliament who were basically dissatisfied with official policy.[51] It gave them arguments with which to question and criticize government policy. This changed the political climate and, together with the public pressure maintained by OVE and OOA through their technical feasibility studies, forced the government to become concerned about its own energy expertise.

The realization that energy demand growth was not simply an act of God—to formulate it crudely—but was, among other things, dependent on socio-economic development, suffered or chosen, meant that the Ministry of Finance, with its dominance over national macro-econometric modelling, would also move into a key position.

The utilities, in contrast, seem to have rested on their laurels after the 1976 planning exercise, satisfactory as it was for their interests. When confronted with the changed forecasting climate of 1980–1, they proved unable to develop reasoned, quantified, and sophisticated arguments that they could oppose to those put forward by the new actors in the game. Realizing this, the utilities simply gave up once their initial defence of trying to suppress the sectoral and technical approach in favour of their old macro method had failed. The prolonged rearguard struggle to bump up the administration's electricity forecast was probably not much more than an attempt to save face, although it probably also was a process of learning about and getting to accept the weaknesses inherent in the primitive forecasting capability on the part of the utilities. ELSAM, at least, is now claiming that it uses both the macro and the micro methods in its planning, and that it even uses surveys of its major customers to determine likely development.[52] It remains therefore to be seen whether the utilities can realize their stated intention to influence the ministries' setting of the key assumptions in future energy planning exercises, among them the one currently in progress in the summer of 1985.

The state has asserted its authority in a new area of crucial importance for societal well-being. But not all state actors are equal. The Ministry of Energy has ably controlled the energy side of the energy forecasting and planning exercise of 1980–1. But energy forecasts are basically still dominated by the assumed socio-economic development—at least as long as the energy forecast does not consider strong energy conservation policies. But forecasting and determining socio-economic developments are the prerogatives of the Ministry of Finance, which is at the top of the ministerial hierarchy. This means that the socio-economic base of energy forecasting will be determined by the Ministry of Finance, and not by the Ministry of Energy or an environmental ministry or agency with an interest in the energy future. The Ministry of Finance has every reason to make sure that the same socio-economic scenario underlies all government planning. It has the power to ensure this. It obviously intends to act accordingly because it knows that the emergence of an alternative centre of expertise on socio-economic modelling would be the beginning of the end for its own dominance and the refutation of its ideology.

This hierarchical control over energy planning is not without dangers. I have pointed out that socio-economic planning is definitely subject to political considerations. Forecasting socio-economic endpoints means, more often than not, pretending that political wishes and goals are objectively reachable. It is therefore quite possible that energy forecasting in Denmark has been freed from the imposition of one interest perspective, that of the electric utilities, only to fall under the spell of another one, that of economics, of business interests, and of politicians who desire to assure their re-election. But it is exactly this perspective that has led us into the present energy (and ecological) crisis. The question may therefore be raised whether, despite the positive evolution of electricity and energy forecasting between 1976 and 1981, anything has really changed. A few power plants will not be built thanks to a more realistic electricity forecast. This is a worthwhile achievement. But it seems that the next crisis is nevertheless pre-programmed.

Notes

This report is based on my Danish thesis 'Political Processes and Prognoses: Political Functions and Origins of the Electricity Forecasts' for the Institute of Political Science, Aarhus

University, Denmark. I am most grateful to my thesis supervisor, Associate Professor Ole P. Kristensen and to the members of the Energy Group at SPRU, Sussex University, for their patience with their visiting fellow. Henrik Andersen and Arne Rolighed made indispensable comments.

1. The Association of Danish Utilities, *Briefing of Members.* Jan 1981.
2. *Commentary to 'Electricity Consumption up to 2000'.* Forecast developed by the Ministry of Trade, 1976.
3. I have had free access to the documents related to the 1981 energy plan. It therefore became possible to reproduce the decision and influence processes within and between ministries, and between them and the external lobbies. I am most grateful to Permanent Secretary Ole Bech, who made this possible. I also would like to thank the staff of the 2nd Section of the Ministry of Energy, who have helped me in my research effort.
4. The Danish parliament did definitely exclude, in late 1984, nuclear energy from future energy plans.
5. *Danish Economy and the Energy Problems.* The Economic Council, June 1980.
6. According to the author's own calculations.
7. *Danish Electricity Supply.* Association of Danish Utilities, Copenhagen.
8. Only about 10 per cent of changes in relative consumer prices can be attributed to changes in sales (estimated on 80 utilities). The French case also shows that price differentiation is not just a question of maximizing electricity consumption. Such a policy involves questions of fairness and equal treatment. These are political considerations and therefore not under the control solely of the utilities (*editors' comment*).
9. Author's calculations on the basis of the *Input–Output Table of 1973*, Tables A.1 and B.1, sector 105; *Statistical Ten Years Survey*, 1980: 28; *Danish Electricity Supply*, 1976.
10. The plan was outlined in *Danish Energy Policy 1976* and the electricity forecast described in detail in *Electricity Consumption through the Year 2000.*
11. *Energy Plan 81.*
12. This section is based on *Forecast for Heated Buildings in Denmark, 1980–2000: A Basis for the Preparation of Energy Plan 81*, Ministry of Buildings, Dep., 4. kt., 28 Oct. 1980, Lb. No. 369 and ibid. 14 Nov. 1980, Lb. No. 377. *Calculations based on Prognosis by Ministry of Buildings from Heated Buildings to Number of Homes 1980–2000*, Danish Bureau of Statistics, 1. kt., 11 Dec. 1980. *Future Development of Prices for Coal, Oil, Gas and Nuclear.* Working Group for Imported Fuels, EM, PL/TO1, 15 June 1981. *Draft: Future Development of Prices for Coal, Oil, Gas and Nuclear.* Working Group for Imported Fuels, Ministry of Energy, PL 101, 18 Sept. 1980: 16. *Long-Range Economic Planning 1981–92—Selected Planning Assumptions.* Print No. 16, Oct. 1980. Copenhagen: Ministry of Finance, Budget Department. *Numerical Material for Energy Plan 81.* Department of the Budget, 12 June 1981, J. No. Ø3487 njl/jk.
13. This section is based more specifically on Main results of *The Lighting Engineering Society's Views on the Consumption of Electricity for Lighting.* The Lighting Engineering Society 1980, The Home Economics Council, Copenhagen, 16 Mar. 1981. *Electrical Heating*, Ministry of Energy, 2. kt., 3g-Sbk/JD, 13 Feb. 1981. *ELPROG—A Model for Forecasting of Consumption of Electricity for Home Machines* by Poul Erik Morthorst, ESG, Testing Station, Risø, 18 Mar. 1981. Memo on *Working Method in the Production of the Neutral Forecast of the Development of Industrial Electricity Consumption through the Year 2000.* Energy Ministry, 2. kt., 2j-Sbk/Pf, 16 Mar. 1981. *The Development of Energy Consumption 1970–78. Food Industry.* Danish Boiler Association, 1980–10–02, EP/IS. *The Development of Energy Consumption in the Food Industry through the year 2000.* Danish Boiler Association, 1980–11–20, EP/ENB. *Energy Consumption in Farming 1970–78.* Frits Møller Andersen, Risø, ESG. 27 Oct. 1980. *Documentation of Forecast Development of Electricity Consumption in Farming, Market Gardening and Construction Industries.* Risø, FMA/sn, 3 Mar. 1981. *Energy Consumption in Market Gardening 1970–78.* Fritz Møller Andersen, Risø, ESG, 27 Oct. 1980. *Construction Industry 1966–73*, Risø, PEG/JL, 1980–09–03. *Forecast of Electricity Consumption in Sales and Service Sectors and Public Companies*, Risø, ESG. PEM/sn, 9 Mar. 1981.
14. *Energy Plan 81. Interim Report on Premisses for a Revised Energy Plan to be Presented to Parliament 1980/81.* Energy Ministry, 18 Dec. 1980: 11.

15. *Remarks to the Section on Electricity in the Interim Report*. Ministry of Energy, 3. kt., JC/lkh3e, 11 Dec. 1980.
16. Ibid.
17. *Response from The Contact Committee on Electricity Forecasts, Memo*. The Directorate of Energy, JA/bbl, 16 Dec. 1980.
18. *Report from the 9th Meeting of the Contact Committee on Electricity Forecasts 28 November 1980*. Energy Directorate 2. kt., JA/bbl. 3 Dec. 1980.
19. *Report from the 11th Meeting of the Contact Committee on Electricity Forecasts 31 March 1981*. Energy Directorate, 2, kt., JA/gms, 5 May 1981.
20. *Time Plan for the Work Group on Electricity Consumption*. Energy Ministry, 3 Feb. 1981.
21. An expert on the Danish Administrative system commented later that this language, unfit for public disclosure, reflected the relatively coarse ways of a young ministry, such as the Ministry of Energy.
22. *Remarks to the Section on Electricity in the Interim Report*, op. cit. n. 15 above.
23. Letter dated 12 December 1980 from Kronborg to Dietrich, Energy Directorate, 2. kt., THK/bbl.
24. *ELSAM. Annual Report and Accounts 1980*, pp. 23 ff.
25. *First Meeting of Co-ordination Group for Energy Plan 81. Decision Report*, EM, 2. kt., J. No. 2105–80, 17 June 1980.
26. Letter from Tom Togsverd, Ministry of Finance, to Ove W. Dietrich, Office Manager, 20 June 1980.
27. Ibid.
28. *Notes to the Co-ordination Group for Energy Plan 81. Some Economic Aspects of Prime Importance to the Energy Planning*, EM, 2. kt., 9 June 1980. (See also Note 33 below.)
29. Letter to Members of the Co-ordinating Group for Energy Plan 81 from Ove Dietrich, dated 4 Feb. 1981. Energy Ministry, 2. kt., 6a–Sbk/Pf.
30. Ibid.
31. *Informationen*, 19 Nov. 1980.
32. Rolf Nordstrand, Budget Department, *Economy and Politics*, 55 (1), 1981.
33. See Table 8.2.

Table 8.2 Unemployment and Current Account Deficit 1970–1983

	Unemployment Rate*	Deficit on Current Balance
1970	1.0	0.55
1971	1.3	0.42
1972	1.4	0.06
1973	0.9	0.47
1974	2.1	0.95
1975	5.1	0.55
1976	5.3	1.99
1977	6.4	1.78
1978	7.3	1.49
1979	6.1	2.91
1980	7.0	2.45
1981	9.2	1.80
1982	9.9	2.25
1983	10.5	1.32

* Registered Unemployment (excluding early retirement and employment schemes)

Source: OECD Economic Surveys *Comparisons between Estimates of Specific Energy Consumption 1970–1990*.

34. The Home Economics Council, 17 Mar. 1981. J. No. 1981 0222/29–1.
35. These estimates are printed in *Development of the Stone, Clay and Glass Industries' Energy Consumption through the Year 2000*. Danish Boiler Association, 20 Nov. 1980, EP/ENB, and *Development of the Food Industry's Energy Consumption through the Year 2000*, Danish Boiler Association, 20 Nov. 1980, EP/ENB.
36. *Report from 9th Meeting of the Contact Committee on Electricity Forecasts*, op. cit.
37. *Report from 11th Meeting of the Contact Committee on Electricity Forecasts*, op. cit.
38. Letter dated 20 Nov. 1980, J. No. 1980–3523, from Kronborg to Dietrich. THK/bbl, Energy Directorate.
39. *Report from 9th Meeting of the Contact Committee on Electricity Forecasts*, op. cit.
40. *Response from the Work Group of the Contact Committee on Electricity Forecasts*, Energy Directorate, JA/bbl, 16 Dec. 1981.
41. *Suggested Response from the Work Group of the Contact Committee on Electricity Forecasts*, JN–JM/bn.
42. *Report from 11th Meeting of the Contact Committee on Electricity Forecasts*, op. cit.
43. *Report on the Status of Electricity Forecasting*. Energy Directorate, 2. kt., THK/bbl. Memo dated 10 Feb. 1981, J. No. 3523.
44. This paragraph is based on *Report from 11th Meeting*, op. cit.
45. *Extension Plan 1981, Proposal and Detailed Statement*: ELSAM, Board Meeting 26 Mar. 1981: 25.
46. *Analysis of Differences Between Electricity Forecasts by the Ministry of Energy and by DEFU*. DEFU, JM/bin, 29 April 1981.
47. Letter from Dietrich to the members of the Contact Committee on Electricity Forecasts, EM, 2. kt., 25 Nov. 1980.
48. Especially important here were the anti-nuclear Organization for Information on Nuclear Power, the OOA, formed in 1973, and the Organization for Renewable Energy, the OVE, formed in 1975, and for the role played by these organizations in promoting wind energy and forcing the electricity utilities to permit the coupling of small windmills to the electricity grid).
49. Blegaa *et al.*, 1976: *Skise til alternativ energiplan for Danmark*, Copenhagen, Borgens forlag.
50. Hvelplund *et al.*, 1983: *Energi for fremtiden: Alternativ energiplan 1983*, Copenhagen, Borgens forlag.
51. Politicians answered the question 'What has OOA meant for the nuclear power policy of your party' in the following ways. Margrethe Auken from the Democratic Socialist Party: 'A lot. Close to everything. The shift in our policy would not have been easy to obtain without OOA. Because of OOA, information and public pressure, the work was easy.' Lone Dybkjær from the Social Liberal Party: 'The material from OOA has been very useful. OOA has also continuously suggested alternatives to the official energy policy. This is especially useful as we are very seldom presented with policy alternatives.' Birthe Weiss from the Social Democratic Party: 'Public pressure was decisive for the change of opinion within the Social Democratic Party' (*Atomkraft*, Jan. 1984).
52. *El & Energi*, June 1985. The Association of Danish Electricity Companies.

APPENDIX

Denmark
Forecasts of Primary Electricity Consumption for Year 2000 in TW

Year	ident	status	high	medium	low	average/forecast
1972	a	est	243.0	0.0	182.0	212.5
1974	b	est	0.0	174.0	0.0	174.0
1974	c	est	0.0	162.0	0.0	162.0
1975	d	est	101.1	0.0	60.9	81.0
1976	e	est	49.3	0.0	36.2	42.75
1976	f	n.est	0.0	38.3	0.0	38.3
1976	g	est	58.2	0.0	41.5	49.85
1981	h	est	33.1	31.7	30.8	47.8
1983	i	n.est	0.0	20.0	0.0	20.0
1984	j	est	0.0	32.0	0.0	32.0

Key
a. DEF: Danish Electricity Companies
b. DEF
c. MOT: Ministry of Trade
d. DEF
e. MOT: Ministry of Trade, 1976 Energy Plan
f. Blegaa/00A: The Blegaa *et al.* study used by the Organization for Information on Nuclear Power.
g. IFIAS: International Federation of Institutes of Advanced Studies.
h. MOE: Ministry of Energy, 1981 Energy Plan
i. Hvelplund/00A: The Hvelplund *et al.* study used by the organization for Information on Nuclear Power.
j. RISOE: Risoe National Laboratory for Nuclear Research (app).

9

The Ups and Downs of Electricity Forecasting in France: Technocratic Élitism

Louis Puiseux

J'aime l'ignorance de l'avenir.
Maurice Blanchot

INTRODUCTION

In France, the forecasting of electricity demand occupies a privileged position in the numerous debates which have been stimulated by the onset of the energy crisis over the last ten years. First, forecasting is the prime intellectual exercise to reveal ambiguous dialectics between necessity and freedom in the games played by interested actors. Secondly, the discussion about electricity forms the core of the whole energy debate. Electricity is energy of high quality. It can be produced by other forms of energy. Theoretically, it can satisfy almost all energy consumption needs. And it can claim, thanks to nuclear power, to become in time and for ever the dominant energy form.[1]

Finally, France is among all western countries the one which has the most centralized and most rationalized electricity supply system. It is also the only one which ten years ago committed itself exclusively to nuclear energy and which has succeeded in getting its population to accept this large-scale nuclear power programme. However, I hasten to add that the lessons to be drawn from this development remain for me a source of perplexity and discovery. Yet I have devoted twenty-five years of my life to the development of electricity. I spent fifteen years from 1960 to 1975 at Electricité de France (EDF) within its Department for General Economic Studies. By this I mean to say that I do not pretend to know the truth. I only want to share insights gained through my experience.

I begin this report with a historical overview of the developments over the last thirty years in the electricity sector in France. This provides the background for the subsequent presentation of the forecasts and the forecasting developments during the different phases which characterized the activities of EDF. I conclude with an interpretation of these developments.

FRENCH ELECTRICITY DEVELOPMENTS SINCE 1945

The sources

The passion of the Enlightenment for the rational, organizing approach to the world has been pushed nowhere further than in France. The desire to gain permanent control over the world through a reduction of its complexity and the definition of society in terms of a few evident properties is very strong. This intellectual disposition led very early to a social structure that made the state the pre-eminent, supreme organizer of life. Colbert's way of managing the economy in the seventeenth century set France on its way towards its modern form of planning. The class struggle contributed to the formation of the nation-state at a time when the rest of Europe was still made up of a jumble of free towns and rivalling principalities. The Revolution was nothing more than the pursuit of this ambition for national unification by other means (Furet, 1979). The educational system begun by Napoleon makes up another cornerstone of this edifice which is France. The *Grandes écoles* guarantee the reproduction of a 'managerial élite' which links service to the state with service to capital (Suleiman, 1978, 1981). Operating on the lines of a para-military organization—in contrast to the liberalism of the universities—they have inculcated generations of students with the spirit of technical–scientific rationality.

It was only to be expected that such an inheritance and such traditions would ensure that French energy policy was conducted in quite forceful ways.[2]

Nationalization and the period of reconstruction, 1946–1958

At the end of World War II, France found itself with a left majority and a desire for national independence, the latter personified by General de Gaulle. The nationalization in 1946 of electricity, gas, coal, and oil exploration was not only the result of the political dominance of the working class; it also resulted from the recognition that the state alone could finance the enormous investment required for new hydropower dams, the interconnection of the electricity grids, and the electrification of the countryside.

Two new organizations were created alongside those just nationalized: The Atomic Energy Commissariat (*Commissariat à l'Énergie Atomique*—CEA) which at first had a purely military vocation, but was later also entrusted with the development of research on civilian uses of nuclear power; the General Planning Bureau (*Commissariat Général au Plan*—CGP) as the co-ordinating body for the economic development of France. The bureau was also supposed to be the defender of economic rationality. Its energy commission was to play an important role in the development of the energy sector. The senior managers in these new public-sector enterprises (EDF, GDF, CDF, BRP, CEA, and so on)[3] and the high civil servants in the Planning Bureau and the Ministry of Industry and Research were—and still are—in most cases graduates of the same *grandes écoles* (Polytechnique, Centrale, Supelec, and so on). They are informally organized into two engineering clans, the Corps des Ponts—dominant at EDF—and the Corps des Mines, which dominates the CEA and the oil companies. There exists a traditional rivalry between these corps, even affecting policy formulation. But they are quite united against the graduates from the École Nationale d'Administration (ENA, another of the *grandes écoles*), who dominate the Ministry of Finance, which has to approve the budgets, and therefore the investment plans, of the public enterprises.

The power of these engineers does not derive from their wealth, but from their élite education with its strong scientific and technical orientation. They all believe that scientific rationality, so successful in appropriating and transforming matter, can also be applied to transforming the economic and social spheres. Some of them will become government ministers. But more important is the fact that they frequently switch from industry to the civil service and back, thus forming a network of like-minded people in all central and large institutions and organizations in France. For example, three successive chairmen of the Energy Commission within the Planning Bureau were later to be found together at EDF headquarters[4] whereas the new planning commissioner, Pierre Massé, a former EDF director, returned later to EDF as its president.

The post-war years were times of rapid and steady economic growth. Electricity consumption increased at around 7 per cent a year, mostly as a consequence of industrial growth, rural and railway electrification, and the spreading of household equipment. The EDF engineers were fully occupied developing the required new supply capacities, keeping a balance between hydropower and coal-fired capacity, minimizing costs by completing the interconnection of transmission and distribution nets, and orientating demand towards off-peak hours through, for example, the introduction of marginal cost pricing, and so on. The EDF managers did not have to think commercially, as the 'natural' growth in consumption was all they could handle. In addition, the electricity market was rather insulated from other energy markets. Electricity had its niches in industry for the driving of industrial motors, for lighting purposes, and for household electrical appliances. Electricity was rarely used for household heating purposes except for small-scale water heating during off-peak hours.

Political conflicts around decolonialization dominated the politics of that time. This blurred the fact that there existed a social consensus between left and right, the working class, the bourgeoisie, and the state not only on the absolute necessity for rapid development of the forces of production (that is, GNP), but also about the type of growth this would involve. This consensus was taken as self-evident. The goal was to catch up with the USA, imitating American methods and behaviours. It meant motorizing the population and forcing the development of the energy sector.

In 1957, three 'wise' men, Armand, Etzel, and Giordani, were asked by Euratom to write a report on the future of nuclear energy in the (then six-country) EEC, which was in the process of being created. They saw it as grandiose, a point of view which I will re-examine later.

Preparations for the nuclear age, 1958–1974

The collapse of the Fourth Republic and de Gaulle's return to power hardly worried the technocrats. On the contrary, they were rather elated. For the CEA, this guaranteed the continued development of indigenous nuclear power technology based on the gas-cooled, graphite-moderated reactor. The CEA had just started this programme as a spin-off of its military work. EDF too had great hopes. Its hydropower construction programme was coming to an end. Domestic coal clearly was becoming too expensive as a fuel for power stations. But cheap oil was being produced in increasing quantities in North Africa and the Middle East, making it an increasingly attractive source of primary energy for electricity production.[5] The creation of the EEC had a positive effect on French economic growth; but cheap energy was necessary to guarantee international competitiveness.

The engineers at that time were unbounded optimists. Coal had gradually replaced hydropower in the production of electricity. Fuel was now being substituted for coal. Nuclear power would be ready to take the place of oil when it in turn became rarer later in the century. Everybody seemed to concur with this picture of the energy future. Relatively falling prices for energy were the pre-condition for economic progress, reducing the effort of work in factory and household and thus increasing productivity and well-being.[6] Nobody doubted at the time, at least not until the end of the 1960s, that the prolific oil fields in the Middle East would continue to be available to the West, and this despite the convulsions of decolonialization at the time.

There was enough time to prepare the nuclear age. CEA and EDF even engaged in the luxury of a struggle over the choice of nuclear technology during the 1960s. It ended with a victory for EDF and American technology shortly after the resignation of de Gaulle in February 1969.[7] The stake in this battle was not so much the better technology as the future exercise of strategic control over French electricity, if not energy developments. The CEA would have acquired the key position if EDF had been forced to order reactors from it. EDF however would have been in control if French reactor producers had been working with American licences.

While these public battles went on, French energy independence was taking a rapid but unnoticed turn for the worse during the 1960s. This development, at least after 1965, was independent of the outcome of the struggle over nuclear technology. Demand for oil was allowed to go beyond all forecast quantities under the pressure of falling oil prices and the desire to profit from this situation to the utmost. In 1970, imported oil satisfied three quarters of energy needs, thus making France extremely vulnerable to international events. Oil accounted for 40 per cent of primary energy in the electricity sector.

This situation provided the 'electricians' with the opportunity to realize a strategic breakthrough that was to change the French energy game within a few years. EDF argued that energy independence at a low price was possible only by going nuclear. It was therefore imperative to let electricity drive out oil wherever possible. EDF should be allowed for the first time to advertise its services and the advantages of electricity for both industrial and household uses. Going nuclear would have the additional benefit of turning France into an exporter of nuclear technology. French PWR and breeder reactors would dominate the world in the twenty-first century as Boeing aeroplanes and IBM computers do today.

The Ministry of Finance was reluctant to accept this argument. It seemed to fear that EDF would grow into a state within the state. But this language was rather well received by politicians and the administration. In this way, rapid growth in the production and consumption of electricity became a kind of national priority. It even got into the sixth National Plan (CGP, 1971: 95). The Energy Commission proposed that there should be inscribed in the Sixth Plan the following objectives or programmes:

(1) A programme to order nuclear light water reactors of a minimum of 8,000 MW during the plan period. This volume should be increased if the construction capacity of French industry permits it.
(2) A *progressive acceleration of electricity consumption* in such a way that an annual growth of 8 per cent is achieved at the end of the plan period. It is important that this objective—*more ambitious than what would result given the actual outlook*—should be supported by appropriate actions.[8]
(3) A substantial investigation into the possibilities of using nuclear power for other than electricity generating purposes.

This turning towards a *commercial strategy* received its official seal of approval with the signing of the first contract between EDF and the state in

December 1970. Marcel Boiteux, EDF's president, concluded triumphantly during a press conference on 27 March 1972 that 'the future will be all-electric and all-nuclear'. He had to engage in self-criticism for this slogan only two years later.

The oil conflict and the economic crisis, 1974–1983

The 'electricians' were not completely surprised by the oil conflict which erupted in earnest at the end of 1973. The price increases were much larger than anybody had dared to predict, but this only further increased the competitive advantage of nuclear power. In EDF's view, the solution to the crisis was the acceleration of the planned nuclear programme. The Messmer government gave the go-ahead at the beginning of 1974. Nuclear electricity was from now on to cover the total increase in energy demand and even a little bit more in order progressively to substitute electricity for oil.[9] Between 6,000 and 7,000 MW of nuclear power capacity was to be ordered yearly between 1974 and 1980. This massive programme went its (almost) undisturbed way [10] while similar programmes in other western countries, above all in the USA and the FRG, began to collapse or became bogged down.

Those Frenchmen who were against this future were called stupid, obscurantist, nostalgic for the stone age, selling out to the Arab sheiks, and, in general, bad French citizens. The communists were completely in favour of this nuclear option. They had already supported EDF's commercial strategy. Right and Left were engaged in the same battle together, proof enough that this was the correct strategy.

This consensus started to break down even in France, not over nuclear power as such, but over EDF's role in bringing it about. Public opinion started to accuse EDF of high-handedness. Protests began to be voiced even within EDF through the channels provided by the socialist trade union, the CFDT (1975).[11] The administration too, which so far had limited its overseer role to financial aspects of EDF's management,[12] started to become much more distrustful. The government began for the first time to form teams of economists and engineers in the different ministries that were to monitor and control the strategic choices so far made exclusively by EDF. This included the making of the electricity demand forecasts. These teams could discuss from a position of equality with the famous and already well established Department of General Economic Studies, founded by Marcel Boiteux at EDF in 1958.

Doubts about the planned nuclear future were beginning to arise because it was becoming too incredible that only the French should be pushing the development of nuclear electricity while the rest of the world was left behind. Too many people in the advanced countries were speaking too seriously about new energy sources and about the conservation of energy for it to be possible that the pro-nuclear French could simply continue to treat this alternative solution with absolute disdain.[13] The 1974 commitment to an additional 7,000 MW every year was already weakened in 1975.

The growth in the demand for electricity began finally to slow down as a consequence of the economic crisis. While in 1974 electricity consumers had been threatened with supply interruptions as of 1980 unless the ambitious nuclear building programme was accepted, we discover in 1983 that we will have too much nuclear electricity before the end of the 1980s. It is now becoming imperative to find electricity export opportunities in order to exploit the power stations coming on stream. In addition EDF is groaning under the burden of the debt—in part because of the rising dollar—it had to take on to finance the building programme. In short, at the time of writing the main body of this report in June 1983, the climate had turned extremely morose in the circles involved in the making of energy policy. Nobody is sure when demand will begin to take off again, if it ever does.

Before ending this short overview it should be noted that the change in political power in 1981 has not really affected the basic orientations of French energy policy. They have at least changed much less than we were led to believe during the presidential election campaign in early 1981. The large, democratic debate about the energy future then promised was disposed of during a two-hour meeting of the managing committee of the Socialist Party. The breeder option was kept open. The expansion of the nuclear waste retreatment facility at La Hague was decided against all protests. It is true that new and not unimportant institutional and financial means were found to support the development of alternative energy technologies. The creation of the French Agency for the Control of Energy (*Agence Française pour la Maîtrise de l'Énergie*—AFME) opens up at least the possibility for the diversification of knowledge and expertise, something which had been demanded in vain for a long time by those contesting the official, pro-nuclear choices. France thereby acquired in 1981 an institution which most other western countries had already acquired around 1975.[14]

FORECASTING METHODS AND RESULTS

The most important producer of electricity demand forecasts is of course EDF, especially its head office (Direction Générale). The primary addressee of these forecasts is the Ministry of Finance. The important decision that has to be made—and for which the forecasts provide one input—concerns the financial means to be allocated to the expansion of the electricity system. These can come from new capital given by the state to EDF, from authorization to borrow domestically or abroad, or from letting EDF increase its cash flow through higher rates.

EDF's position in the forecasting game

EDF's electricity forecasts have rarely been challenged on forecasting grounds. In the beginning, nobody else had the forecasting capability to challenge EDF; and today, forecasts have little political clout.

Two government agencies mediate between energy interests, or between them and more general economic interest groups. They could, therefore, challenge EDF in addition to the Ministry of Finance, which dominates economic policy-making and most policy choices that involve the allocation of public funds. The Ministry of Industry, the tutelary organization of all public energy enterprises, has an energy directorate made up of a few high-level civil servants. But this directorate lacks study funds, and is therefore unable to undertake independent forecasting exercises. The Planning Bureau (CGP) defines development perspectives every five years. Its energy commission, now called the Groupe Long Terme Énergie (GLTE), has a permanent staff and a mandate to study long-term energy developments. It controls limited funds, however, and its five-year exercise is therefore mostly a collation of plans elaborated by working groups dominated by the different energy interests (Saumon and Puiseux, 1977: 136).

The Direction de la Prévision of the Ministry of Finance has substantial funds at its disposal. It also has the economic and political weight that would give importance to its own forecasts, especially since the loss of influence of the Planning Bureau. However, the Ministry of Finance is preoccupied with the short to medium term and the containment of the economic forces that always threaten economic stability. In addition, it has to take a political view. Energy forecasting has, therefore, never been of primary concern.

This situation may explain the predominance of five and ten-year energy forecasts, especially in the electricity sector.[15] These time frames encompass the planning horizons of conventional and nuclear power plants, and hence of capital allocation decisions that the Ministry of Finance has to make. It is also a time frame that meshes with the horizon of economic cost-benefit analysis and with the seven-year mandate of the president under the constitution of the Fifth Republic. It is, however, noteworthy that the energy crisis has not led to a lengthening of the official forecast horizon. Quite the contrary has occurred. No long-term forecasts to the horizon of the year 2000 were prepared between 1973 and 1983, at least not officially (Chateau, 1985: 2).[16]

The first long-term forecast published in 1983 in preparation for the Ninth Plan (1984–8) reflects both the pressure from the anti-nuclear and ecological minority within the Socialist Party as well as the socialist attempt to strengthen the position of the Planning Bureau and of the plan. But when this forecasting exercise suggested that no new orders for nuclear plants would be needed, because those already under construction were, anyway, going to produce substantial electricity surpluses by 1990, economic considerations, that is, preservation of the industry and of employment, prevailed once more. At least one plant, if not more, has been or is going to be ordered every year under the socialist presidency.

Non-establishment forecasts are extremely rare, for reasons that will become clear in the last section. Where they have been made, for example within the Socialist Party before the 1981 election, they have remained without political consequence apart from the fact that the Planning Bureau might not have been forced to make a forecast for the horizon 2000. They have not provided a new methodological impulse either. Energy decisions, and especially electricity ones, have remained the preserve of the closed, technical–administrative élite, as they have always been. The ecological movement has remained weak. The hoped-for breakthrough after 1981 did not happen. The coming to power of the socialists in 1981 instead forced silence on the anti-nuclear, low-energy-growth wing of the party. Any outside forecasting challenge was therefore denied political support from the beginning.

Forecasting methods before 1968

EDF used three forecasting methods before it switched to a commercial strategy in 1970:

1. *Simple exponential extrapolation*. Here average growth rates and their standard deviations during a certain period are projected forward. The base period should be at least as long as the period for which a forecast is to be made. This method is at the origin of the famous law of a doubling of electricity consumption over a ten-year period.[17] This method considers growth in electricity demand as an autonomous variable. Its development can be likened to the trajectory of an interstellar probe where it is sufficient to know its past path and its present speed to forecast its position at any future date.

2. *Models with independent, explanatory variables*. One assumes stable coefficients for a relationship (of the type $y = a + bx$) observed in the past and which is supposed to also apply in the future. Typical relationships studied link electricity consumption to GNP or GDP, an index of industrial production, of household income, or of any other economic activity deemed to be influential. Electricity demand is here considered as an input into the production and consumption system in order to obtain an output. The input–output link is very similar in its nature to a technical coefficient which, for example, specifies performances of a machine.

3. *Analogue models*. These refer to the experience of a country suppos-

edly 'in advance' of our own. This country was of course the USA, its development foreshadowing our own future. American electricity consumption figures, standardized for population, number of households, GNP, and physical production measures for a number of industries, were used to predict the energy consequences of French economic development. Economic development in this view follows a single, compulsory path. The only freedom left is to catch up more or less quickly (or possibly never) with the leader of the pack.

Each of these methods can of course be applied to estimate electricity demand as a whole or in different sectors (households, services, industry, and so on), in regions, and in individual industries. Any method can be called sector based which proceeds to estimate total demand by adding up a number of such sector forecasts.[18]

Simple extrapolation methods (of the first type) were dominant in the preparation of energy forecasts for the First, Second and Third Plan, that is until about 1962. The forecasting results were more or less checked against previous American developments with the analogue method.

Methods using explanatory variables began to be substituted for the extrapolation method with the preparation of the Fourth Plan (1962–6).[19] This methodological change was considered an undeniable intellectual progress. It was thought that a technical approach had been replaced by a more economic one. It was also thought that the new approach to forecasting was less 'schizophrenic' and more open to world developments. I remember however that the quarrel between the old and the modern methods still occupied all our attention at the congress of Unipede (the international Union of Producers and Distributors of Electrical Energy) in the summer of 1973 in The Hague. The following, condensed dialogue gives the gist of this confrontation:

'What good are these explanatory variables? It is again through extrapolation that you find them!'
' . . . maybe, but as long as we play this risky game of forecasting we might as well combine the forecasts of all the different sectors. One feels less lonely this way.'

With hindsight one could add that this answer carries little weight if the club of forecasters is made up of men formed in the same mould, sharing the same beliefs. If is also astonishing that in this discussion almost nobody questioned the total absence of price variables in the energy forecasting models.

Of course, econometrics would not have been of great help in this situation. Estimates of past price-quantity relationships made later failed to find coefficients which were significantly different from zero.[20] But in any case, the belief in an indefinite further fall in the relative price of energy, and especially of electricity, was still almost universal.[21] We were all marching into the void without knowing it.

Forecasting methods used after 1968

The adoption of a commercial strategy by EDF in 1968 and the signing of the first contract with the state in 1970 changed the nature of the forecasting problem. French electricity forecasters stopped being neutral observers—or at least believing themselves to be such—in search of a likely estimate of the future. They instead acquired the function of active combatants in the battle for national energy independence. The ambitious goals of the commercial services with respect to the gain of market share for electricity in the so-called competitive areas—where substitution of oil was possible as in the case of household heat—became virtually assumed in any forecasting exercise.

The forecasting methods remained the same, at least in appearance, but they had now acquired a *normative* content. I believe I realized that this epistemological and deontological revolution had occurred one day in 1967 or 1968. On that day the chairman of the Energy Commission of the Planning Bureau suggested privately to me that if only EDF would decide to engage in somewhat more vigorous commercial activities, it would be possible substantially to increase the value of the GNP elasticity of electricity consumption. In this way the numbers which resulted from my regression calculations stopped being natural constants and became instead political action variables. This was quite a shattering discovery for a naïve soul.

This development changed the traditional annual dialogue between EDF and the state.[22] Before 1970 it went like this:

EDF. We simply require the capacity to satisfy projected demand.
MINISTRY OF FINANCE. The demand for electricity won't be as high as you assume.

Now it sounded like this:

MINISTRY OF FINANCE. We do not have the financial means to realize such a heavy investment programme.
EDF. You have to decide once and for all if you are determined to reduce oil consumption and re-establish energy independence.

The oil price developments since 1973 have certainly lent additional weight to this last argument. But the oil price increases also made possible the alternative path of energy conservation. This implied controlling the demand for energy, not satisfying it through capacity expansion in the electricity sector. Energy conservation is the path most other western countries have finally preferred over the accelerated development of nuclear power.

In France, the first consequence of the oil price shock for energy forecasting was an extraordinary multiplication of forecasting methods. This was the time of futurologists and scenario analyses. Everybody got involved: political parties, trade unions, interest organizations, university researchers, and national newspapers. The technocrats responsible for forecasting engaged in a forward defence. They made their models more sophisticated, multiplying the number of explanatory variables, both quantifiable and non-quantifiable ones. Numerous expert surveys were made in an attempt to weigh the influence of a great number of factors and to determine interaction effects.[23]

The resultant model complexity had its hour of glory, but is likely to generate a smile today. We discovered quickly that all this effort produced results that were only as good as the experts consulted. The operational value of these methods was not much better than that of a good brainstorming session. The outcome of expert surveys depended on the people invited to sit around the table. 'The presence of an eloquent charlatan was possibly enough to switch the consensus opinion of the group similar to the litmus test in school chemistry where one additional drop of acid changes the colour of the paper.'[24]

The energy producers were quite willing to play the game of scenario analyses. They became quite adept at it too, to the extent that this method also lost its innocence. The energy companies would present scenarios covering a wide spectrum of possible futures. They assured their audience that they had studied all possible hypotheses. However, scenarios with low energy consumption growth were presented in such negative tones that they met with general rejection.[25] Instead, the most optimistic one with respect to electricity consumption was in general adopted for policy-making purposes.

Almost all the methods in use today, whatever their name, even if they are called 'scenario analysis', are sectoral in character. This is certainly true for the electricity sector in the forecasting exercises of the producer (EDF), the government, and even their opponents. The attempt is to be as concrete as possible, differentiating at least between types of energy and types of users. Nobody dares to present models that are aggregated. Nobody even dares to talk about extrapolation although this method is—shamefacedly—used for filling in the coefficients in the matrix of users and uses. Models with explanatory variables continue to be used, both with fixed or variable elasticities. But this modification does not really change the basic nature of the approach.

However a new consensus has formed behind the smoke-screen of methodological variety. Energy and certainly electricity consumption are not something to be estimated through forecasting. They can be steered through the use of appropriate policies. Everybody basically agrees on this point, from the energy and administrative technocrats to their opponents, the ecologists. The myth of scientific objectivity and neutrality of forecasting has been definitely shattered. Both sides, those in favour of accelerating non-oil energy production and those desiring to stabilize it, are ready to take the measures necessary to achieve their goals, out of free will and in a normative manner. This view is even openly expressed in EDF documents (Bergougnoux, 1982):

Under today's conditions—and they will certainly continue to exist during the next decades—the relationship between the demand for electricity and its exogenous determinants such as economic growth (and its distribution among sectors) on the one hand, and the price developments of substitutable energy resources on the other are definitely susceptible to deliberate intervention and substantial modification. These interventions can of course go in both directions. They can favour the substitution of electricity for imported energy resources. Or they can be used to encourage a more economic and rational use of electricity.

Forecasting results

Table 9.1 and Figure 9.1 present a comparison of EDF's electricity forecasts made for the different national economic plans. The plans incorporated these forecasts unchanged.

From 1945 to 1960, EDF had difficulty satisfying a growing demand for electricity. The forecasters generally underestimated this growth; the overestimation of demand only becomes systematic for the years after 1960. The relative size of the overestimation is rather stable. Ten-year forecasts go wrong by almost 20 per cent. Five-year forecasts err by between 5 and 10 per cent (Table 9.1). This error consistency had some wags comment that EDF could have done better by using the five-year forecasts as six-year ones instead.

The Planning Bureau—luckily one might add—underestimated economic growth for the periods covered by the Fourth (1961–5) and the Fifth Plan (1965–70). A rerun of the forecasting model used with actual growth rates results generally in predictions that are even around 50 per cent higher than the forecasts actually made (Table 9.2). The models (with explanatory variables) were using GNP elasticities of electricity demand that were far too high. This is not surprising as they had been based on the experience of the preceding years when GNP growth was 'electricity-intensive' due to reconstruction, development of basic industries, and rural elec-

Table 9.1 Forecast and Actual Electricity Consumption, 1961–1980, in TWh

	1965	1970	1975	1980
Actual Consumption	102.2	139.9	180.7	237.9*
1961 Forecast (4th Plan)	109.5	165.0	n.a.	n.a.
deviation in %	(7.1)	(17.9)		
1965 Forecast (5th Plan)	—	149.0	214.0	n.a.
deviation in %		(6.5)	(18.4)	
1970 Forecast (6th Plan)	—	—	200.0	280.0
deviation in %			(10.7)	(17.7)
1975 Forecast (7th Plan)	—	—	—	250.0
deviation in %				(5.1)

* Excluding consumption of Eurodif (uranium enrichment)

Table 9.2 Forecast Electricity Consumption Increases Using Predicted and Actual GNP Growth Rates

Forecast Year	Year Forecast	Forecast Consumption	Forecast if Actual GNP Growth
1961	1965	+7.1%	+11.0%
(4th Plan)	1970	+17.9%	+26.0%
1965	1970	+6.5%	+11.0%
(5th Plan)	1975	+18.4%	+16.0%

trification. The switch to a commercial strategy tended to correct for this error as the result for 1975 of the 1965 forecast seems to suggest.

The electricity forecasts for the Fifth (1970–5) and Sixth Plan (1975–80) in contrast were based on over-generous GNP growth estimates provided by the Planning Bureau to EDF. But this error was also lucky. The retrospective electricity forecasts using actual GDP growth rates give consumption figures for 1980 of about 6 per cent above actually realized ones. However, this error was in part compensated by commercial activities in favour of the all-electric household and office that were more successful than planned. GDP elasticity of electricity demand had obviously recovered from the low level of the 1960s because of deliberate policy choices. The forecasting error, however, remains serious as the commercial policy was decided—at least in principle—before the forecasts were made.[26]

The future seen from the present

It is not yet possible definitely to judge the quality of the forecasts made for the years 1985 to 1990. However, one cannot fail to be impressed by the spectacular decrease of the official forecasts made for these dates (Table 9.3 and Figure 9.1). As a consequence, nuclear power plants coming on stream from 1985 on, which were ordered on the basis of these forecasts, will not be needed except during periods of peak demand. But this makes nuclear electricity into an expensive proposition.

The report of the Long-Term Energy Group to the Planning Bureau in preparation for the Ninth Plan concluded therefore that no new nuclear power plant is needed before 1987 or even 1991 if the economic crisis continues and if the availability of sufficient electricity is the decision criterion applied (CGP, 1983: 21 (vol. 1), 51–5 (vol. 2); Dethomas, 1983). The number of nuclear power plants in operation in 1990 is probably going to be too high by 25 to 30 per cent, depending on economic developments in the meantime. The 1990 electricity supply surplus will probably reach 75 TWh, that is, 20 per cent of the consumption that was forecast in 1983 for that year.

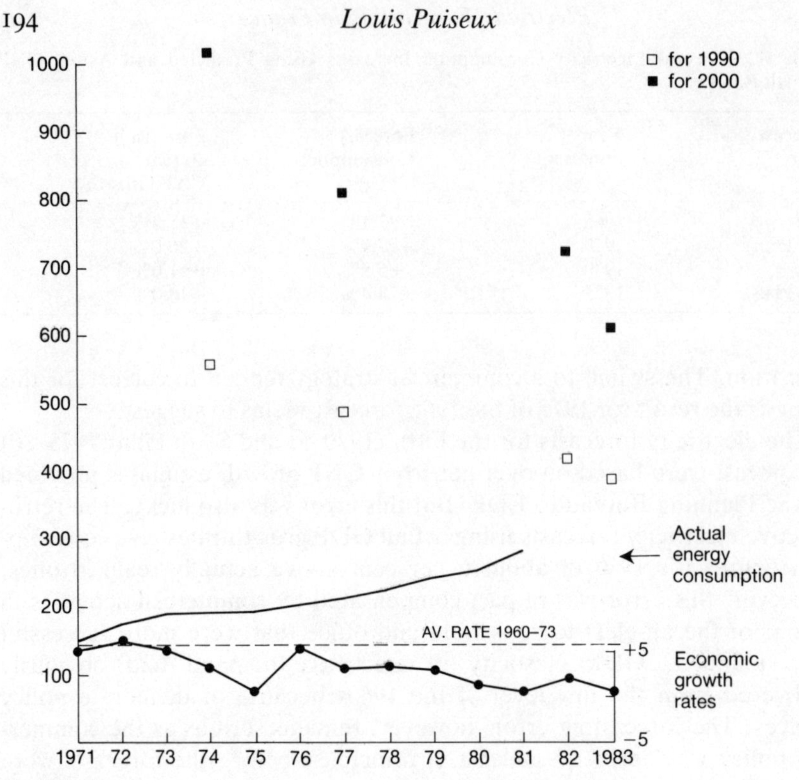

Fig. 9.1 France
Forecasts of electricity consumption for year 1990 and 2000 (in TWh)

All forecasts are made by EDF Electricité de France.

Table 9.3 Government Forecast of Electricity Consumption in 1985 and 1990, in TWh

Forecast Year	Electricity Forecast for 1985	Electricity Forecast for 1990
1962	500	n.a.
1966	430	n.a.
1974	400	560
1977	365	475
1982	315	415
1983	290	370

EDF's commercial strategy will therefore have to cover all aspects of its operation if the gap between the forecast and actual developments is going to be kept within reasonable limits. But the Achilles' heel of the French nuclear energy strategy, its rigidity—discovered already only a decade after it was launched in earnest—cannot be affected. None of the nuclear power plants ordered in the years from 1976 to 1981 can or will be cancelled. And at least one new plant will have to be ordered every year during the 1980s (instead of the minimally planned three full-sized ones) in order to keep the nuclear industry viable and to protect employment. As a consequence, 'nobody doubts that the major goal during the next few years will be to reduce the (conventional) surplus generating capacities' (Dupuy, 1983). The nuclear share of 63 per cent of electricity production in 1985 will therefore increase further, turning the nuclear plants—designed as base-load plants—into medium load capacity.

On the sales side, EDF has received the green light to go all out in exporting electricity. The Ninth Plan (1985–90) hopes that exports of about 15 to 20 TWh annually can be achieved. This seems to be realizable. Net-exports in 1984 reached 25 TWh, 40 per cent of this to Switzerland alone. A further expansion would require the construction of new transmission lines and this might run into socio-political resistance. Domestically, EDF is pushing electric space-heating, and is making active efforts to substitute electricity for other energy forms in industry by designing and selling the necessary electric equipment. This however would lead to a peak-load curve that is just the opposite of the base-load profile of a nuclear electricity system. EDF is therefore thinking of introducing seasonally differentiated rates in the household sector too. Plans toy with a ratio of 1:6, with the highest rate applicable for twenty-two days in winter at the choice of EDF. This would make bivalent space-heating systems competitive, and EDF is propagating them as the solution for the future. More generally, the government's commitment to energy saving seems to be waning. AFME, the energy-saving agency at least is constantly complaining about further reductions in its budget, endangering the realization of its mission.

The authors of the forecast made in 1974 for the period 1980–90, which served as the basis for engagement in the massive nuclear power programme, defend themselves with the argument that they had only used official GDP growth forecasts as the basis of their projection. They add that the projected electricity consumption was going to be necessary to make planned economic growth possible. Sufficient and cheap electricity makes economic development possible in this view, not the other way around. This is a rather disingenuous and also self-serving explanation; it however correctly points to the passive role that forecasts played in France in the elaboration of energy and socio-economic policies during the 1970s.

Chateau (1985) concludes from his review of official energy and electricity forecasts during the period from 1973 to 1983 that it was energy policy decisions with respect to the nuclear option, national independence, and mastery of energy consumption that shaped energy, and specifically the electricity forecasts, and not the other way around as is usually assumed and pretended in forecasting circles. This would explain the absence of the preparation of official long-term forecasts for the horizon of 2000 between

Table 9.4 Growth Assumptions to 1985 used in Official Energy Forecasts, 1970–1983

Year Forecast made	Av. annual Growth Rate GDP	Av. annual Growth Rate Energy	Elasticity E/GDP
1970	6.0%	4.8%	0.80
1973	5.0%	4.1%	0.82
1974	5.0%	2.9%	0.58
1975	5–5.5%	3.5%	0.64–0.70
1978	4.5%	3.2%	0.71
1979	3.0%	2.2%	0.73
1980	3.5%	3.1%	0.89
1981	2.5%	1.7%	0.68
1983	1.3–3.0%*	−0.8%−+1.0%	−0.62–0.33*

* Based on the hypothesis of CGP (1983) for 1980–90.
Source: Chateau, 1985: 3.

1973 and 1981. The contradictions between policy and likely system development would have become too obvious otherwise. It also concords with the fact that the official forecast revisions during this period were simply rebased to reflect slower-than-anticipated economic and energy growth and the delays in the nuclear plant building programme. Hence the relative stability in the energy growth rate as shown in Table 9.4.

Chateau (1985: 3–4) also points out that the downward revisions of the economic growth asumptions did not automatically lead to lower energy growth predictions. Above all, the decline of GDP elasticity of energy demand was reversed in 1975 and steadily increased to a value higher than even the one assumed in the pre-crisis 1970. These revisions went counter to received wisdom, disregarding the effects of rising energy prices and the government's own energy conservation policies. They also failed to take into account the consequences of the crisis for the economic structure (Criqui, 1982). The former effects are in retrospect sufficient to explain the gap between the forecasted and the actual developments (Chateau, 1985: 6). The slow adaptation of the forecasts to reality and the counter-intuitive revisions of the energy demand elasticity point to the weakness of non-establishment views on the energy future and to the manipulation of forecast revisions during the 1970s in the interest of political and industrial concerns. It was important to show that there did not exist energy demand barriers that would require a revision of the nuclear building programme decided by the conservative Messmer government in 1974.

It required the coming to power of the socialists, and with them of the ecological tendencies that had earlier decided to work through the socialists for a realization of their views, in order to revise drastically the assumptions going into the official long-term forecast. However, the ideological commitment of communists and the majority of the socialists prevailed again as pointed up in the decision to continue the (slowed-down) nuclear programme despite the forecast that the resultant electricity was not needed.

Another classic argument in favour of high electricity consumption forecasts concerns the so-called risk asymmetry inherent in forecasting (Bergougnoux, 1982). Aiming too high leads to over-investment in production capacity. This wastes scarce capital which could have been used more profitably somewhere else. Aiming too low entails the risk of future supply breakdowns, especially during winter when such supply interruptions are the most serious and politically the most dangerous. For the electricity producers, erring in the direction of over-supply is the preferred solution. Capital profitability is somewhat lower than in the other case. But they are also avoiding the possibility of a supply collapse with all its attendant economic costs. One can also imagine the sweat breaking out on the forehead of politicians when the possibility of a shut-down of factories and shivering householders is evoked, and all this as punishment for their stinginess seven or eight years earlier when they refused to allocate the required financial means. In short, it is preferable to err too high than to estimate too low, thus setting up a self-fulfilling prophecy.

The struggle over the allocation of capital is part of the dialogue between producers and financiers, or, if one prefers, between technocrats and political power, in all branches and all countries. The producer tries to hide his real reserves and capacities to produce, stressing his constraints instead. However, a large, integrated electricity production and supply system has considerable reserves and adaptability, not only in the short but even in the medium run. Four years is sufficient to build a conventional power station using coal or fuel. A gas turbine can be installed within two years. Electricity import contracts can be negotiated or renegotiated within weeks if not days. The same is true for contracts with large electricity consumers permitting interruptability or disconnection after suitable warning. All these possibilities should be sufficient to re-establish system equilibrium in case demand development estimates of seven or eight years earlier prove to be inaccurate. The cost of such adjustment methods would not be higher in total than the costs implied by the surplus nuclear power stations presently being built. Naturally, this course of action would force the electricity producers to live somewhat more dangerously than in the past.

It should be noted that the quantitative forecasting errors described above—based on wrong GDP elasticities of electricity demand—have gone hand in hand with equally large errors in estimating electricity production costs. The French nevertheless proceed according to plan, expecting in the medium to long run a fall in the price of nuclear electricity compared to all other energy forms (except for the breeder, where EDF has now recognized the commercial un-competitiveness of this technology). But the truth is that the price of nuclear energy remains unknowable as long as major uncertainties exist with respect to the cost of reprocessing, the disposal of nuclear waste, and the social cost of a serious accident or of an illegal diversion of nuclear material.

FORECASTING AND ITS MORAL IMPLICATIONS

It is the irony of our time that the development of the forces of production made engineers and technicians gain responsibilities for which they were not prepared. Ilya Prigogine (in Prigogine and Stengers, 1979: 294) argued that 'the problem of the interaction of human populations with populations of machines has nothing in common with the problem of constructing this or that machine, a problem which is relatively simple and which can be mastered'.

The technical vision of the world

Engineers and technicians typically have a 'technical' vision of the world, and not only because the daily exercise of their profession confirms it. This vision has since childhood been at the origin of their educational and career choices. It is a vision of objectivism. The engineer is placed in a universe where utilitarian considerations reign supreme. He subjects nature and raw materials to a rigorous determinism in order to realize precise goals. He loses himself in an interminable series of tasks and sees in houses only the quality of the wooden frame and the concrete walls, in waterfalls only the energy potential that could be harnessed, in humanity the labour power, the technical potential, and the ability to consume. The pleasure in this job comes from its dry rigour and its effectiveness, the hold it provides over the material world, but also from the dependence on the world it creates in him to the extent that being an engineer is a *serious* job. Gorz (1977: 70–1) expresses this in the following way:

> On the one hand I am, as an engineer, a little demiurge because I command, calculate, foresee, and invent. I take an existing world and make it into something new which exists only because of me. My thinking has become materialized. But on the other hand, the laws I observe are not thought up by me . . . The world takes my orders but only to the extent that I observe its rules. The ransom I pay for my effectiveness is my acceptance of the world's objectivity, to think only what is, to subject my thinking and my activity to the world's laws and structures, which I make come true. There is only room in this world for technical freedom, a freedom which never questions the universal laws and structures, which never asks 'what is it all good for?' . . . There are no doubts, no goals, and no ends, but only means-ends, that is, technical problems. Here I am, protected from myself. The urgency of tasks to be done keeps me busy. I am constrained on all sides in my activity. I am delivered from my terrifying subjectivity. Isn't this the last and most original aspect of my choice, this giving up of myself to the world, this drowning in the objectivity which makes me be and which dispenses me for all times from making myself exist?

This technical vision of the world is legitimated by the prestige which is conferred upon the builders of machines with undeniable utility. This vision with all its strengths and power can be transferred quite naturally to the task of forecasting energy demand. The electricity needs of intermediary and final consumers, of enterprises and households, all form an input into the 'machine' of society whose output is the standard of living. The link between input and output is analogous to the technical coefficient of a machine. To give up growth in electricity consumption would mean giving up progress.

One can oppose this logic only with a historical view of world development, so much older, so much more modest and ironic. Societal development from the stone age to the nineteenth century occurred with a very small increase in energy consumption. Only one energy revolution comparable to a biological mutation has ever occurred. It began with the switch to coal in the nineteenth century and is ending now with the transition to nuclear power. We all know that there exists an absolute, impassable limit to the growth in energy consumption. It is set by the warming of the oceans.[27] We will certainly one day have to decouple energy consumption from general development whether we like it or not, thereby reinventing a civilization which can live with the stable consumption of energy (Puiseux, 1981).

What is a 'good' forecasting method?

A 'good' forecasting method does not exist. It cannot exist because science always applies to the present. Futurology and prognosis-making can only discern the promises and threats for the future which exist today. In short there is only history, that is a sequence of events which sometimes generates something new, something which cannot be reduced to preceding events and which retrospectively modifies the way we understand the past.

The misunderstanding results from the experience in physics and more generally the exact sciences where there is no history. There are only apparently stable laws which can be taken as definitely established. The increase in entropy and the expansion of the universe certainly do give a meaning to the vector of time, but they do not bring something *new*. In other words and following Sartre (1960), nature is not dialectic or if she is we do not know it. Only the history of societies is dialectic. Energy use is part of this history. It is not a part of the history of nature. It is therefore only to a small extent a scientific problem.

I am therefore tempted to answer the question about a good forecasting method in the manner Paul Feyerabend (1979) answered the question about scientific discoveries: everything is good! That is, all methods, all the models are equal in their poverty and their lack of relevance.

If 'good' is however understood here as characterizing the method which has in practice provided the best approximation of actual developments, I would be forced to answer that a retrospective comparison of forecast with actual electricity consumption developments forbids any selection of this or that method as the best one. The 'law' about the doubling of electricity consumption every ten years has 'worked' very well for quite a while. The most sophisticated models have performed miserably during the last years.

If however 'good' designates the method which allows us to ask in the clearest possible way what the social and political values underlying planning are, that is, the method that permits the democratization of planning, then there is a clear answer. The sector-based models are superior to the aggregate or semi-aggregate models because they get closer to considering concrete needs and activities. These models are more pedagogical and more transparent than those using other methods. The non-specialist is better able to orientate himself and to participate in the debate. But I hear the objection that there is no need to involve non-specialists in this forecasting business. They are unable to understand anything of the problems involved, and anyway they do not care about them in the first place.

Marcel Boiteux (1975), the Director General of EDF, has revealed the gist of this hostile view to non-specialist participation in an article written for *Le Monde* in 1975. Boiteux answered a professor who expressed doubts about the economic justification of the French nuclear power programme and demanded a public debate on the subject. Boiteux countered that accepting a debate on this topic on television would be as dangerous as permitting a debate 'between those who pretend that the earth is round and those who know that it is flat'. Everybody would simply become confused. This argument is the more interesting as it comes from a man who went to university and not to one of the *grandes écoles*, a man known for his openness to discussion and his liberalism.

A crucial trick in this 'non-debate' consists in presenting the assumed low production costs of nuclear electricity as if they were a scientific fact, a natural phenomenon similar to the physical performance of a nuclear reactor. Yet the figures come straight out of the books of EDF. From this way of looking at costs it is concluded in good economic manner that the success of nuclear power is therefore both desirable and probable. It is therefore important that the financial, commercial, and publicity measures are taken, which will assure nuclear success. At the same time profit calculations are made using the most optimistic assumptions, which can be justified with the presumed rapid expansion of nuclear electricity. Ronald Reagan said once appropriately that 'my right hand is ignoring what my extreme right hand is doing'.

Everybody today knows that in the western world as a whole the choice is between more energy supplies and more security, and that this choice is a political and not a technical matter. It cannot be resolved with the help of a scientific algorithm exactly because it is not an 'objective' choice. This is of course not to say that it is easy to make this choice with the usual political procedure of national elections and referenda.

The inevitability of expert knowledge

No information campaign, no long-term education programme will ever transform the man in the street into a competent expert in all of the scientific domains and areas of modern technology. *L'honnête homme* of the seventeenth century will never return. Expert knowledge is therefore an unavoidable fact.

In France the means and the prestige of expert knowledge are unfortunately concentrated in the hands of a powerful technocracy. Its members are extremely unified. They form a decision centre which is highly autonomous in relation to the spheres of politics and private capital. In no other western country and in no other sector except energy does Habermas's dictum apply so well: 'The dependence relation between the specialist and the political sphere seems to have become reversed. Politics has become the executor for a scientific intelligentsia which determines optimal strategies within objective constraints and technical possibilities . . . Politics has become some kind of guided programming' (Habermas, 1973).

French energy policy until 1974 was never anything other than the policy of the energy producers (Puiseux, 1982). Only the shock of the oil crisis made the political class become aware that energy technology is something too serious to be left to the decisions of technocrats. Their know-how was too greatly implicated in the decision-outcome and could therefore no longer be considered as objective.

There is no other solution to this problem than to create counter know-how. Provided it has the same investigative means, the same access to computing power and to the mass media as the established experts, its holders will be able to discuss with the specialized bodies of the administration in order to evaluate the risks of the different technical possibilities, to discover the areas where modifications and change are possible, and to unearth the real goals behind and the implications of technological projects. This presupposes however that the French stop both on the national and the regional level to plan in the way they did until 1980. Instead of thinking in terms of optimizing a given situation they have to start to think in terms of variant solutions. The real role of the expert planner should be, as Michael Kalecki has said, to propose possible alternatives from which one can choose through a political process.[28] Each alternative should be internally coherent but reflect a different set of feelings and preferences. A future house-owner does not hire an architect to design him a 'good' house. He wants him to study the advantages and disadvantages of different technical solutions, and to determine their costs.

CONCLUSION

The demand for transparency in energy forecasting is part of the more general problem of the social control of technological development. Alain Touraine (1980) has shown how the dynamics of historical change are being displaced from the confrontation of private capital and the working class to the new conflict line between the large, modern technocratic empires and the population of the consumer-users.[29] Yesterday, the struggle was over the division of the productivity gains. What is at stake today is nothing less than control over the content of modernity.

Notes

Translator's note: translation and editing by Thomas Baumgartner. All quotations have been translated from French. Thanks are due to Claude Baumgartner for her help in finding the right words and expressions.

1. 'There is no doubt left about the very long run. Electricity based on nuclear energy will win out against its competitors given that oil and natural gas will inevitably become rarified.' (Boiteux, 1972) Marcel Boiteux was at that time director of EDF's General Economic Studies Department and is today President of EDF.
2. Saumon and Puiseux (1977) and Lucas (1979) provide succinct summaries of French energy policy. Mendershausen (1976) makes many of the same observations with respect to French oil policy.
3. See the list of acronyms below.
4. They were Albert Robin, Lucien Gouni, and Claude Destival.
5. Saumon and Puiseux (1977: 142) point out that the government's insistence on a domestic refining industry, coupled with the rapid expansion of demand for gasoline in the wake of motorization, made large quantities of cheap heavy fuel oil available.
6. Leyral and Monnier (1983: 2) mention that the massive expansion of low-priced public housing in the 1960s was made possible, in part, because of the absence of sound and heat insulation.
7. See Saumon and Puiseux (1977: 144–50). Finon (1979) tells the related story of the decision to develop the French breeder technology.
8. Author's emphasis. A note to this passage even stated 'that these appropriate actions should first of all involve rates and commercial and public aspects. These goals justify ending certain constraints, such as the prohibition to sell household equipment.'
9. The 1974 Messmer Plan reduced the 1985 energy forecast by almost 16 per cent to 240 mtoe, cut planned oil consumption by 46 per cent and expanded the contribution of nuclear electricity by 50 per cent to 60 mtoe (Chateau, 1985: 3).
10. There was however an appeal of 400 scientists against the accelerated nuclear programme and demonstrations occurred against the breeder reactor in Creys-Malville. After these resulted in deaths, the anti-nuclear movement decided to seek change through participation in the forthcoming elections (Nelkin and Pollack, 1982).
11. In the light of the conclusion I will be drawing, it should be mentioned that CFDT officials were appointed to many high-level positions within the AFME when it was formed by the socialist government in 1982.
12. This control included the electric rates because of their impact on the price level, salary settlements because of their impact on wage negotiations in the private sector, and the volume of investments because of its incidence on financial markets due to the need to borrow.
13. Alexis Dejou, former director of Études et Recherches at EDF, wrote in *Le Monde* (18 December 1973, p. 15): 'It has to be clearly said that it is not only illusory but obviously dangerous to rely on these alternative energy resources apart from certain very excep-

tional cases. . . . To let ourselves be diverted from (the acceleration of the nuclear power programme) by those who promise us the heavens with the alternative energy resources is the surest way to make the serious energy problem completely unsolvable.'

14. See on all these points Saumon and Puiseux (1977), Simmonnot (1978), Péan and Serini (1982), and Puiseux (1981, 1982).
15. Longer-term forecasts are quite necessary to decide on research orientations, to plan the evolution of the production structure (with respect to location and mix) and of the energy transportation network. But if they were made in the 1970s they have not been published.
16. Thompson (1983) identifies France as a hierarchical, highly secretive system that produces little in terms of written documents on the basis of which an outsider could reconstruct a governmental decision process.
17. A variation of the same law predicts a ten-fold increase in consumption every 33 years. André Decelle, Director General of EDF at the time said in a speech at the beginning of 1966: 'We have just crossed last year the threshold of 100 billion kWh, and we will without doubt reach a consumption level of 1,000 billion kWh at the end of the century.'
18. These methods reflected international forecasting standards in the early 1970s, at least as perceived by the electric utilities. See the manual published by Unipede (1972), the International Union of Producers and Distributors of Electrical Energy, edited by the author and presented at the Unipede Congress in The Hague in 1973. See also Finon (1974) and Frisch (1975) on these methods.

    The final annex to the Unipede manual, assessing the methodological future, issued a note of caution. It suggested that 'the concepts . . . will doubtless be to some extent obsolete by the time this manual is published' (Unipede, 1972: 159). The likelihood of future oil crises is mentioned. The continuation of economic growth is doubted by making reference to the limits-to-growth debate just getting under way at that time. And the interdependence of economic and social phenomena was touched upon while referring to systems analysis as a possible future method of some merit (Unipede, 1972: 159, 160). This prescient part seems to have had little influence on the French utilities' forecasting methodology.
19. The most frequently used model had the form $\Delta CE/CE = a + b\ \Delta GDP/GDP$, i.e. the annual growth rate in electricity consumption is a linear function of the GDP growth rate. The coefficients *a* and *b* were estimated on past developments, assumed to remain stable in the future. One can also write $\Delta CE/CE = \varrho\ \Delta GDP/GDP$ where ϱ is the famous elasticity coefficient.
20. A note from the working group 'Energy Prices' of the Planning Bureau dated 21 December 1970 proposed a few values for the price elasticity of energy demand. However, the group failed to go as far as recommending that elasticities be included in the forecasting models.
21. The relative price of electricity fell by about 2 per cent annually between 1940 and 1970. An internal EDF note of 8 December 1969 concludes with respect to the price elasticity of the demand for low-tension electricity: 'One objective which the electricity distributor is close to realizing is to bring the level of knowledge the individual consumer has about prices and advantages of electricity up to the level reached by architects and real estate promoters. In this way high elasticities will become more common as is already the case for electric household machines and for house builders. This will help to push up the consumption of electricity in the household sector given the likelihood *that electricity prices will continue to fall*'. (My emphasis.)
22. The representatives of the Planning Bureau and of the Ministry of Industry were almost always faithful spokesmen for the EDF interests against the treasurers and inspectors of the Ministry of Finance when it came to this type of argument.
23. Making use of Delphi Methods and cross-impact studies. Duperrin *et al.* (1975), for example, developed a complex model of the nuclear energy system that included technological as well as social conflict dimensions.
24. Comment made by Marcel Boiteux at the colloquium Université—EDF, Les Renardières, 18 November 1976.
25. For example in 1976 EDF studied discreetly an 'ecological scenario', projecting a consumption of 350 TWh for 1990 instead of the officially assumed 450 TWh. Consumption in 1990 is likely to be below 350 TWh. (This ecological scenario was presented at the

Colloquium Université—EDF, Les Renardiéres, 17 November 1976, whose proceedings were classified.)

26. See also on this subject Cassett-Carry (1980, 1981).
27. Robert Gibrat, a member of the French Academy of Sciences, reproached me in 1980 during a speech at the École Polytechnique with lack of faith in the creativity of technicians and engineers who would certainly be able to find ways to overcome the problem of waste heat disposal by using the cooling capacity of the oceans. I dared to answer him that my faith was certainly larger than his. Only I was sure that the researchers of the 21st century would find better things to do with their time than to stuff the bottoms of the oceans with waste calories.
28. Kalecki often repeated this definition during his seminars at Warsaw University, according to Ignacy Sachs.
29. Chomsky (1969) wrote: 'Nothing allows us to suppose that the men who claim power in the name of knowledge and technology will be more benign in exercising it than those whose pretensions to the exercise of power are based on wealth and aristocratic origin.'

Acronyms

AFME	Agence Française pour la Maîtrise de l'Énergie
BRP	Bureau de Recherche de Pétrole
CDF	Charbonnage de France
CEA	Commissariat à l'Énergie Atomique
CFDT	Confédération Fédérale des Travailleurs
CGP	Commissariat Général au Plan, the Planning Bureau
EDF	Électricité de France, now EDF-GDF
ENA	École Normale d'Administration
GDF	Gaz de France
GDP/GNP	Gross Domestic (National) Product
GLTE	Groupe Long Terme Énergie at the CGP, the Planning Bureau
PWR	Pressurized Water Reactor

Bibliography

Bergougnoux, J. (1982), 'La Prévision de la demande d'électricité. Paper presented at the Colloquium Futuribles, Avignon, 30 September.

Blanchot, M. (1969), *L'entretien infini*. Paris: Gallimard.

Boiteux, M. (1972), 'Rétrospective et perspectives de l'économie énergétique en France, 1950–2000'. *Revue française de l'énergie*, 23 (January).

—— (1975), 'E.D.F. a longuement étudié le programme électronucléaire'. *Le Monde*, 11 June, pp. 21 and 25.

Cassette-Carry, M. (1980), 'Prospective et énergie: examen critique des analyses prévisionelles et prospectives dans le domaine énergétique'. Thesis under the direction of Pierre Maillet, Université de Lille I, November.

—— (1981), 'Peut-on croire aux prévisions énergétiques?' *Revue de l'énergie*, 32 (November): 561–72.

CFDT (1975), *Nucléaire, énergie: nos conditions*. Paris: CFDT-information.

CGP—Commissariat Général du Plan (1971) *Rapport de la Commission de l'Énergie du VI Plan*. Paris: La Documentation Française.

—— (1983), *Énergie*. Rapport du Groupe Long-Terme. Paris: La Documentation Française, 2 vols.

Chateau, B. (1985), 'La Prévision énergétique en mutation?' *Revue de l'énergie*, 36 (January): 1–11.

Chomsky, N. (1969), *Les Nouveaux Mandarins*. Paris: Seuil.

Criqui, P. (1982), 'Impacts du premier choc pétrolier sur la consommation d'énergie finale'. *Économie prospective internationale*, 11: 39–55.

Dethomas, B. (1983) 'Pléthore d'énergie'. *Le Monde*, 14 May, pp. 1 and 27.

Duperrin, Ch., M. Godet, and L. Puiseux (1975), *Les Scénarios de développement de l'énergie nucléaire à l'horizon 2000: application de la méthode SMIC 74*. Paris: Commissariat à l'Énergie Atomique, Rapport CEA–R 4684.

Dupuy, G. (1983), 'Planification: les arbitres en première ligne'. *Énergie Magazine*, 31 Jan.

Feyerabend, P. (1979), *Contre la méthode, esquisse d'une théorie anarchiste de la connaissance*. Paris: Seuil.

Finon, D. (1974), 'Prévisions de consommation d'énergie et d'électricité, les méthodes utilisées en France'. Paper presented at the Workshop 'French–American Energy Systems Forecasting and Pricing', University of Wisconsin, Madison. Université de Grenoble: IEJE.

—— (1979), 'Décision publique et surgénérateur française', in P. Kahn (ed.), *De l'énergie nucléaire aux nouvelles sources d'énergie: vers un nouvel ordre énergétique international?* Paris: Librairies Techniques, pp. 215–67.

Frich, J. R. (1975), 'Évolution des méthodes de prévision de consommation d'électricité à moyen terme'. *Revue de l'énergie*, 26 (July–Aug.).

Furet, F. (1979), *Penser la révolution française*. Paris: Gallimard.

Gorz, A. (1977), *Fondements pour une morale*. Paris: Galilée.

Habermas, J. (1973), *La Technique et la science comme idéologie*. Paris: Gallimard.

Leyral, R. and E. Monnier (1983), *Consumer Energy Conservation Policies and Programmes in France*. CECP Technical Report vol. 5. Berlin: International Institute for Environment and Society.

Lucas, J. D. N. (1979), *Energy in France: Planning, Politics and Policy*. London: Europa Publications.

Mendershausen, H. (1976), *Coping with the Oil Crisis: French and German Experiences*. Baltimore: The Johns Hopkins University Press.

Nelkin, D. and M. Pollack (1982), *The Atom Besieged: Anti-Nuclear Movements in France and Germany*. Cambridge, Mass: The MIT Press.

Péan, P. et J.-P. Serini (1982), *Les Émirs de la République*. Paris: Seuil.

Prigogine, I. and I. Stengers (1979), *La Nouvelle Alliance*. Paris: Gallimard.

Puiseux, L. (1981), *La Babel nucléaire*. Paris: Galilée.

—— (1982), 'Les Bifurcations de la politique énergétique'. *Les Annales*, July–Aug.

Sartre, J.-P. (1960), *Critique de la raison dialectique*. Paris: Gallimard.

Saumon, D. and L. Puiseux (1977), 'Actors and Decisions in French Energy Policy', in L. Lindberg (ed.), *The Energy Syndrome*. Lexington, Mass.: Lexington Books, pp. 119–72.

Simmonot, P. (1978), *Les Nucléocrates*. Grenoble: Presses universitaires de Grenobles.

Suleiman, E. N. (1978), *Élites in French Society*. Princeton: Princeton University Press.

—— (1981), 'Élites et Grandes Écoles', in J.-D. Reynaud and Y. Grafmeyer (eds.), *Français qui êtes-vous?* Paris: La Documentation Française, pp. 101–7.

Thompson, M. (1983), 'Postscript: A Cultural Basis for Comparison', in H. C. Kunreuther, J. Linnerooth *et al.* (eds.), *Risk Analysis and Decision Processes:*

The Siting of Liquified Energy Gas Facilities in Four Countries. Berlin: Springer Verlag, pp. 288–325.
Touraine, A. (1980), *La Prophétie antinucléaire*. Paris: Seuil.
Unipede (1971), *International Manual on Medium and Long-Term Electricity Consumption Forecasting Methods*. Paris: Unipede.

Appendix

Table 1. Forecasts of Electricity Consumption in France for 1985, 1990, and extr. for 2000 (in TWh)

Year	Agency	1985	1990	% incr. 1985–90	1995*	2000*
1974	EDF	400	560	40.00	784.00	1097.60
1977	EDF	365	475	30.14	618.15	804.44
1982	EDF	315	415	31.75	546.75	720.32
1983	EDF	290	370	27.59	472.07	602.29

* Extrapolated assuming growth rates 1985 to 1990.

Table 2 Forecasts of Primary Energy Consumption in France (in mtoe)

Forecasts for 1985

year	ident	actual consump-tion	high	medium	low	average/ forecast	Forecast index 1973=100
1973	a	175.7		284		284.0	100.0
1974	b	176.4		240		240.0	84.5
1975	c	165.3		232		232.0	81.7
1978	d	184.5	230.0		215.0	222.5	78.3
1980	e	188.5		219		219.0	77.1
1981	f	184.3		197		197.0	69.4
1982	g	182.7	185.5		180.5	183.0	64.4

[Table 2 continued on next page]

Forecasts for 2000

year	ident	status	high	medium	low	average/ forecast	Forecast index 1973=100
1973	a	est		419		419	100.0
1974	b	est		327		327	78.0
1975	c	est		332		332	79.2
1978	d	est	328.0		280.0	304	72.6
1980	e	est		311		311	74.2
1981	f	est		245		245	58.5
1982	g	est	227.5		188.5	208	49.6

a. MI: Ministry of Industry
b. MP: Messmer Plan
c. CGP: Commissariat Général du Plan, 7th plan
d. MI: Ministry of Industry
e. CGP, 8th plan
f. RH: Rapport Hugon
g. CGP, Rapport du Groupe Longe Terme, Scenario a and c

Numbers for 2000 from 1973 to 1981 are extrapolated assuming stable growth rates.

PART V

Market Competition

10

Future Imperfect: Energy Policy and Modelling in Canada Institutional Mandates and Constitutional Conflict

John B. Robinson and
Clifford A. Hooker

INTRODUCTION: THE HISTORICAL CONTEXT

ENERGY development has always been an important part of economic and political development in Canada. (See Appendix E for information on current Canadian energy supply and demand.) For this reason, Canada developed national energy policies and energy data collection systems earlier in its post-war development than did many countries. Like most industrialized countries, however, Canada did not develop comprehensive policy analysis and modelling capabilities until these were required by the emergence of energy supply-and-demand problems.

In Canada the twin focuses of post-war energy policies were the geographic distribution of fossil fuel resources (mainly coal and later oil and natural gas) and the availability of US supplies and markets to Canada. Because of the long and narrow east–west distribution of both Canadian society and the energy resources it depended on, Canadian energy developments occurred mainly along a north–south axis and involved considerable energy trade with the US. The result was a strong export orientation to policy and a heavy reliance upon American markets, capital, and technology in the development of energy resources.[1]

The evolution of energy policy and development in Canada was also strongly influenced by the constitutional division of powers between the federal and provincial governments. Briefly, provincial governments have full ownership rights over energy resources within their boundaries. Except with respect to taxation, such resources only become subject to federal jurisdiction when they cross provincial or national boundaries.[2] This has had two major effects on Canadian energy policy. In the first place, because electrical power grids are primarily provincial or subprovincial in scale, electrical power is essentially within provincial rather than federal jurisdiction. With the exception of export policy, electrical power policy is set separately by each province. As will be seen in more detail below, this means that federal energy policy is essentially oil and natural gas policy.

Second, just because provincial governments have ownership rights, the evolution of oil and gas policy in Canada is a delicate balancing act, with major tensions and confrontations between federal and provincial governments. In essence, an agreement between the federal government and the oil and gas producing provinces (Alberta, and to a much lesser extent, Saskatchewan and British Columbia) is necessary for the successful development of national policy.

By the early 1970s it was becoming clear that the western world, including Canada, was moving out of the conveniently simple policy circumstances of the post-war period into more turbulent times. The 1973 OPEC announcement of massive, sudden increases in the price of oil and constraints on supply simply crystallized the growing perception.

Significant changes in energy and economic conditions combined to render both traditional fuel demand and supply projections nearly useless. The results of the forecasts prepared in the decade after 1973 varied widely as conditions changed. Ironically, the same changes that made forecasts so difficult also increased their importance, for these changes made it crucial to have some basis for the energy policy-making that was by 1976 perceived as urgent.

The purpose of this paper is to examine the relationship between energy forecasting and energy policy-making in Canada over the period from 1969 to 1984. The main body of the paper will consist of an examination of the forecasting activities of the two main federal energy agencies: the National Energy Board, a regulatory and advisory agency, and Energy, Mines and Resources Canada, the federal government department with responsibility for energy policy. The forecasts made by these agencies and their relationship to federal energy policy decisions will be outlined.

The National Energy Board and Energy, Mines and Resources Canada were not the only organizations to produce energy forecasts in Canada during the period under review. From the mid-1970s on, there emerged a substantive and methodological critique of Canadian energy policy and forecasting. A brief description of these alternative approaches is provided.

Finally, the paper concludes with an attempt to draw some general lessons about the forecasting–policy relationship.

The National Energy Board: the failure of forecasting

The National Energy Board (NEB) Act was passed in July, 1959, creating the NEB. The NEB presently consists of eleven members, all of whom are appointed by the federal government, supported by a staff of about 450 people. The historical context of the board's formation had a significant impact upon the provisions of the Act and these in turn served to concentrate the board's attention on energy forecasting methods.[3]

According to the NEB Act, the NEB is primarily intended to fulfil two major roles: that of a regulatory agency with authority in the federal sphere of energy jurisdiction, and that of an advisory board to the federal government on energy policy.

The NEB was given regulatory authority over the construction of major pipeline systems and over oil and gas exports. Since most major pipeline proposals have had export components, an important part of the NEB's activities in assessing pipeline applications has been their statutory requirement to determine whether the gas or oil to be exported was surplus to Canadian needs. The board's understanding of this requirement was in terms of projected demand measured against projected supply capability. This has meant that supply and demand evaluation and forecasting have become explicit components, and often the determinants, of the NEB's export decisions. Moreover, because the NEB was the only federal energy agency until 1966, and the pre-eminent one until the early 1970s, its focus on forecasting techniques coloured the entire federal perception of the proper approach to energy modelling and energy policy formation over the sixties and much of the seventies.

The export orientation of federal oil and gas policy in the early seventies was based upon the geographical proximity of US markets to western Canadian sources of supply and the overwhelming foreign ownership of the Canadian oil and gas industry. Significantly, however, it also depended upon the perceptions of abundance of supply so widely held at that time.

Because of this prevailing export orientation, the discovery in 1968 of oil and gas in Prudhoe Bay, Alaska, was perceived as both a threat and an opportunity by the government. The threat was that the delivery of American oil from Prudhoe Bay would reduce Canada's exports of oil to the US. The opportunity was that the Prudhoe Bay oil might be delivered by means of a new overland pipeline up the Mackenzie Valley through Canada, thus providing the chance of developing the oil and gas reserves of the Canadian north and of increasing Canadian exports.

In 1969, one year after the Prudhoe Bay discoveries, the NEB published a report entitled *Energy Supply and Demand in Canada and the Export*

Demand for Canadian Energy, 1966 to 1990 (NEB, 1969). The report's title itself reveals the close connection felt to exist between energy supply and demand forecasting and the question of exports.

The forecasts published in the 1969 report (see Figure 10.1) consisted of extrapolations of past trends and judgements concerning future market shares for fuels and economic growth rates. They were also closely tied to policy questions. On the one hand, the forecasts buttressed official optimism regarding future supplies by projecting a continuing excess of domestic supply capability over domestic requirements through the projection period. On the other hand, the forecasts demonstrated a continuing American need for Canadian exports. For example, the report contained a forecast of US oil supply and demand which showed a future US deficit (labelled 'Canadian opportunity'), even allowing for the production and delivery of Alaskan oil to American markets. Appearing one year after the Prudhoe Bay discoveries and reinforcing official perceptions of abundance, the 1969 forecast must be seen as both a cause and a consequence of the policy context of the time.

During 1969 and again in 1971, applications were submitted by various members of the gas industry for export permits. These applications were supported by industry forecasts showing the existence of large 'exportable surpluses' of natural gas in Canada. The NEB reviewed these and produced their own forecasts. The first set of applications was partially approved by the NEB, and subsequently the federal Cabinet in 1970 (NEB, 1970); the second was refused by the Board in 1971 (NEB, 1971).[4]

The favourable 1970 export decision, officially based upon NEB forecasts, was used by the federal government as a lever to influence the US to permit continued or increased imports of Canadian resources. The NEB's report itself was quite explicit on this point:

> The Board must judge natural gas export applications on their merits according to the law. The law requires the Board to have regard to the trends in the discovery of gas in Canada. In assessing those trends, the Board cannot ignore the present and future marketing prospects for oil, as well as natural gas, because of the joint nature of exploration for hydrocarbons. (NEB, 1970: 10–11)

This quotation clearly illustrates the intimate inter-relationships both between supply and demand forecasts and between those forecasts and policy goals. Decisions are explicitly based on 'most likely' forecasts, yet those decisions are intended to influence the supply and demand behaviour being forecast. Moreover the level of future demand is expected to influence the level of future reserves. Finally, the techniques used are based either on extrapolation of past trends or on judgemental alteration of those trends, although the underlying behaviour on which the trends are tacitly based will be influenced by these decisions.

In the context of these optimistic expectations, the 1971 NEB decision refusing further gas exports came as something of a shock to the oil and gas industry. This refusal was based upon the NEB's finding that 'Requirements for natural gas in Canada are increasing much more quickly than was previously foreseen, even as recently as August, 1970' (NEB 1971: 6–4). The refusal also apparently served a policy function:

> The Board would also hope that rejection of these export applications would be read by the producers as an indication of the need to find much greater reserves of gas in Canada, reserves in amounts that would keep pace with the recent upswing in requirements and, it is to be hoped, make available surpluses which would enable producers and Canada as a whole to benefit further from export opportunities. (NEB, 1971: 6–11.)

The 1971 turn-around in expectations regarding future natural gas supply and demand was quickly followed by a similar message with respect to oil. In 1972, the NEB released a gloomy report entitled *Potential Limitations of Canadian Petroleum Supplies* (1972) which projected oil shortfalls by 1986. One year later, in March 1973, the government imposed quotas on the export of Canadian oil to the United States, reversing the historical thrust of Canadian oil policy.

The NEB forecasts of 1971 (for gas) and 1972 (for oil) (see Figures 10.1 and 10.2) both of which predated the 'energy crisis' of 1973, set the stage for the apparently sudden reversal of policy perspective that occurred in 1973 and 1974. The new policy perspective influenced the results of subsequent forecasts, which for the next three to four years were to paint a universally gloomy picture of Canada's energy prospects. In 1974 and 1975 the NEB released one report on gas and two on oil, all of which indicated that shortfalls in supply were expected by the early 1980s (NEB, 1974, 1975a, 1975b).

The changed expectations regarding natural gas supply and demand were not unconnected to the emergence of frontier gas pipeline proposals. The new perspective made possible the promotion of such a pipeline on the basis of domestic need (that is, anticipated shortfalls in conventional supplies) rather than on the increasingly politically unpopular basis of export potential. However, by the time the NEB hearings on frontier gas pipelines reached their supply–demand phase in early 1977,[5] it was not so clear that the situation was as gloomy as had been thought only two years earlier. Though most industry spokesmen continued to argue that frontier gas was needed as soon as possible to avert expected shortfalls, signs of a current gas *surplus* began to emerge at the hearings. In fact, before the hearings were to end several new gas export proposals had been filed with the board.

Despite the changed supply–demand situations, the NEB devoted only two weeks out of an eight month hearing to supply and demand matters. In July of 1977 the board approved a compromise pipeline proposal, partly on the basis of new forecasts which, while considerably more optimistic than those published two years before, still indicated a domestic need for frontier gas in the mid-eighties. A major reason was its gas supply forecast which, as Helliwell notes, 'assumed a set of production weights on existing reserves and production from new discoveries that would only produce 90 per cent of the marketable reserves even with an infinitely long production life' (Helliwell, 1979: 204).

Part of the reasoning lying behind the NEB's choice of assumptions can perhaps be found in statements about forecasting made in the northern pipelines report.

> In the assessment of whether a pipeline is and will be required by the present and future public convenience and necessity to carry Mackenzie Delta gas to Canadian markets, the question is bedevilled by the inability to accurately predict demand and supply; this [and other uncertainties] make it prudent for Canada to provide for itself a safety margin in connecting new sources of supply earlier than an uncertain forecast may indicate a need. (NEB, 1977b: 1.66.)

In this quotation the basically conservative orientation of the NEB, its concern with the requirements of its statutory mandate, the importance of the judgements made by the board and the impact of current events on those judgements are all evident. Further evidence of the NEB's attitude towards forecasting comes two pages later.

> . . . the Board under the National Energy Board Act clearly has no mandate to force changes in the manner sought by certain public interest groups. Rather the Board has sought to perceive the rate of change which will probably occur in the complex milieu of Canadian society and to reflect this perception in its forecast of demand. (NEB, 1977: 1.68.)

Here an emphasis on forecasts as estimates of the most likely future demand comes through very clearly without any apparent recognition of the contradiction implied in denying responsibility for the outcome of forecasts while producing forecasts that are critically dependent upon assumptions concerning, for example, government policy or industry behaviour.

The NEB had not been ignoring oil policy issues during the course of the gas pipeline hearings and decision. During 1976 the board held the second in a periodical series of public hearings on oil supply and demand that had been recommended in a 1974 oil-export hearing. The report of that hearing was released in 1977 (NEB, 1977a) and contained both lower demand and lower supply forecasts than in the previous report (Figure 10.1).

Part of the reason for the changed demand forecasts in both the oil and the gas reports of 1977 was the adoption of an econometric model for demand forecasting developed by the federal department of Energy, Mines and Resources (and discussed further below). This model allowed explicit treatment of interfuel substitutability and price and income effects in the analysis.

In 1978, the board released yet another report on oil supply and demand

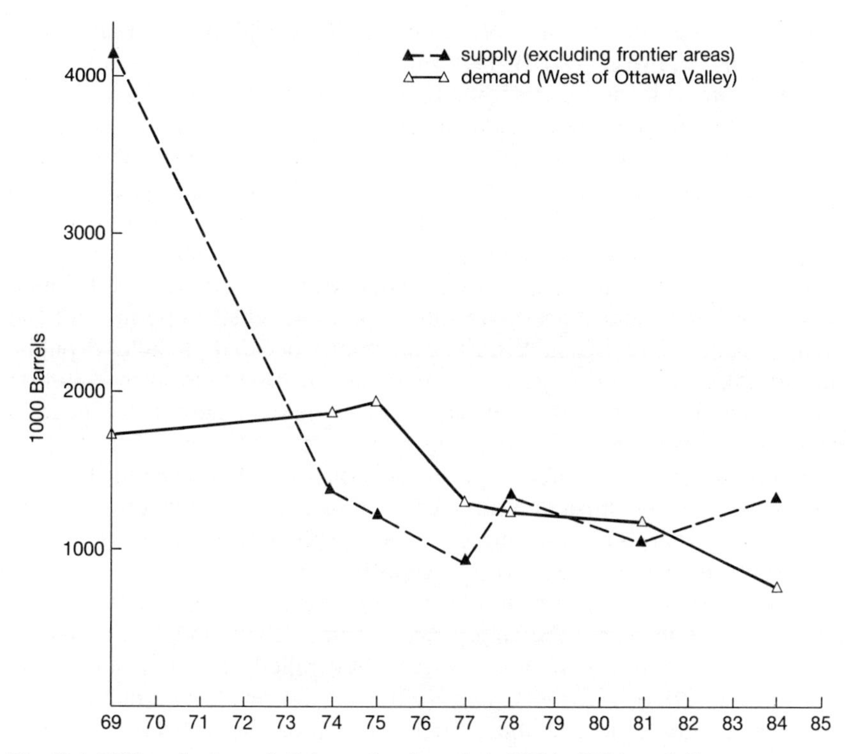

Fig. 10.1 NEB projections of oil demand and supply in 1990 in 1000 barrels/day

in Canada (NEB, 1978). In contrast to the 1977 report, which had projected oil supply deficits throughout the projection period, the 1978 report projected surpluses over the same period (see Figure 10.1). The main reasons for these differences were more optimistic board assumptions concerning prices, government incentive programmes, and the like. The report showed more sensitivity to the question of alternative forecasts and rather plaintively remarked on the connection between forecasts and behaviour:

Although the Board is pleased that its February 1977 report may have focused attention on the need for accelerated development of new supplies, its current forecasts must not be allowed to have the reverse effect. There is no reason to pause in efforts to increase the availability of new indigenous oil supplies from new sources; on the contrary, the Board has assumed that these efforts will increase. (NEB, 1978: 73.)

During 1978 and 1979, the gas 'bubble' disparagingly dismissed by the NEB in its 1977 northern pipelines report grew larger. In February 1979 the board released a new gas report which contained much higher supply and lower demand projections than in 1977 (NEB, 1979a). This paved the way for an export decision in November 1979, which approved new long-term exports to the US (NEB, 1979b).

With their twin gas reports, the NEB had come virtually full circle since 1969. Within one decade Canada had apparently gone from predicted continuing surplus to imminent shortfall and back to expected surplus again. The middle period of expected deficits had barely lasted long enough to allow approval of a frontier pipeline based on a predicted urgent need for frontier gas in domestic markets.

Several months after the 1979 gas export decision the NEB 'concluded that changing circumstances warranted a comprehensive appraisal of the energy outlook' (NEB, 1981: 2). The board decided to hold a public inquiry into total energy supply and demand, the first time since 1969 that the NEB had examined all forms of energy in the same appraisal. The total primary energy demand forecasts in the new report, issued in June 1981, were not significantly different from those in 1979. However in the 1981 report, the NEB published a number of different 'cases' of future supply and demand for both oil and gas. This explicit recognition of choice represents an improvement over previous reports.

In their discussion of demand forecasting techniques the NEB also discussed for the first time the emergence in some submissions of end-use or bottom-up forecasting methods, which they called 'deterministic model forecasting' (NEB, 1981: 19). The NEB itself, however, continued to rely on their econometric demand model and a judgemental choice of fuel-market shares.

During 1981 and 1982, the combination of increased availability of and reduced demand for domestic gas supplies led to a large number of new gas export applications. In their 1983 report on those applications, the NEB 'discovered' significantly more gas than had been projected in their 1981 forecasts. The result of such changes was that the supply capability forecast was 17 per cent higher throughout the forecast period than in the 1981 forecast (see Figure 10.2). Correlatively, the NEB's total energy demand forecast was significantly lower than that for 1981, leading to lower gas demand forecasts. The combination of lower demand and higher reserves doubled the exportable surplus calculated in 1981, resulting in large new export approvals (NEB, 1983).

The favourable supply–demand trend of the previous few years was continued in 1984 with the release of the NEB's second overall energy supply and demand forecast since 1981 (NEB, 1984). In that report, which contained forecasts for all forms of energy, the board projected significantly lower demand and higher supply forecasts for both oil and gas and total energy demand (Figures 10.1, 10.2 and 10.3).[6]

The NEB's 1984 report represented a kind of breakthrough in attitudes to energy conservation. After citing the conclusions of the Friends of the Earth soft energy path study (discussed below), the board went on to

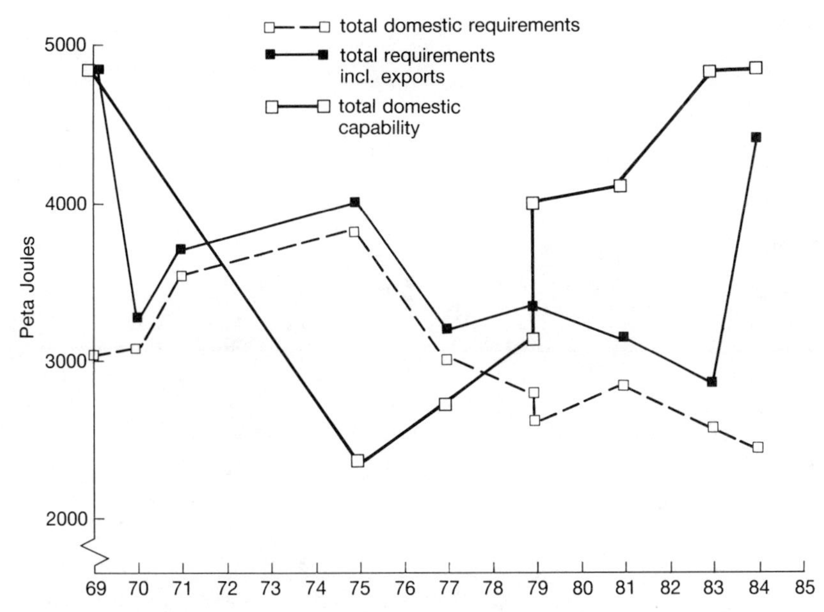

Fig. 10.2 NEB projections of natural gas
Requirements and production capabilities in 1990 in petajoules

argue, directly contrary to its conclusion in 1981, that conservation would continue to be substantial throughout the projection period, and that this would occur despite declining real prices.

In summary, over the fifteen-year period from 1969 to 1984, the NEB had successively refined its analytical capabilities. The adoption of a total energy econometric demand model and the subsequent in-house development of computerized models of oil and gas deliverability represented attempts to improve the NEB's quantitative forecasting capability. Despite these improvements, however, the NEB's record with regard to energy forecasting over this period was not such as to inspire confidence. In the first place, a veritable pendulum swing of forecasts and decisions had occurred. In addition, the new techniques were not able to prevent large and sudden changes in estimates, changes which had important implications for specific export and pipeline applications (see especially Figure 10.2 for the gas estimates). Finally it seems clear that, far from being neutral analyses, the forecasts prepared by the NEB often served clear political functions and closely reflected the general policy context prevailing at the time the forecast was prepared.

Given the great variations in NEB forecasts during the 1970s and early 1980s, it is interesting to examine the forecasting activities of the federal Department of Energy, Mines and Resources (EMR) over roughly the same period. The activities of EMR and the NEB have overlapped, both in

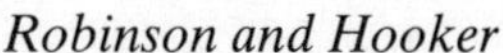

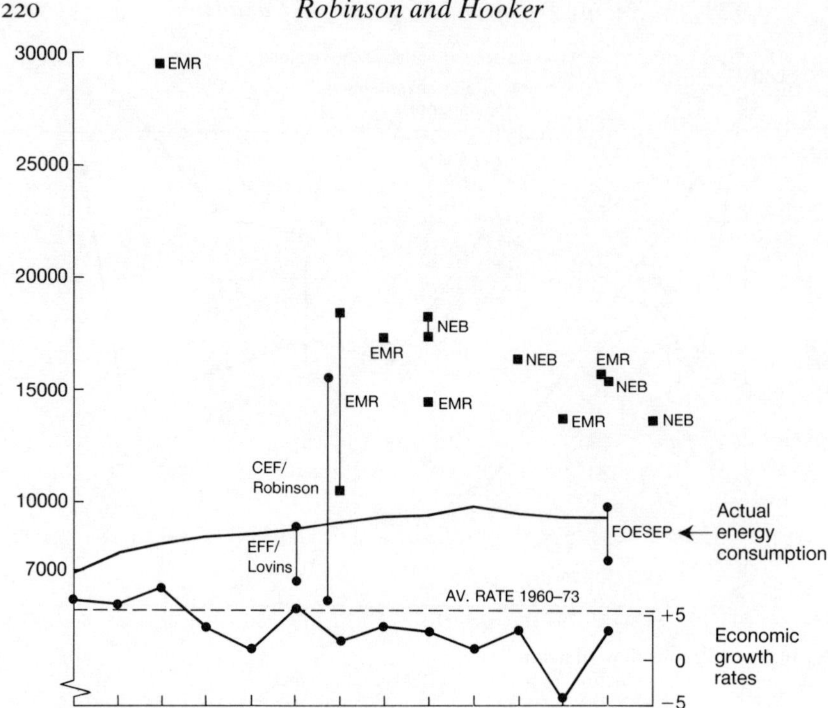

Fig. 10.3 Canada
Primary energy demand forecasts for year 2000 (in petajoules)

EMR: Energy, Mines, and Resources Canada
EEF/Lovins: Exploring Energy-efficient Futures
CEF/Robinson: Canadian Energy Futures
NEB: National Energy Board
FOESEP: Friends of the Earth

time and in subject, yet there have also been significant differences in the way each agency has undertaken and used energy forecasts, differences partly rooted in their different mandates.

The Department of Energy, Mines and Resources: evolution of policy and modelling

The Department of Energy, Mines and Resources (EMR) was formed in 1966 out of the old Department of Mines and Technical Surveys. Its formation put the responsibility for federal energy policy (as opposed to federal energy regulation) under one roof. It was not until the early 1970s however, that EMR published a comprehensive two volume review of the whole Canadian energy system titled *An Energy Policy for Canada, Phase 1.* Its foreword announced: 'The purpose of this report is to define more clearly the national framework into which provincial studies fit, to identify policy choices which must be made within the federal jurisdiction, and to provide a basis for choice by the Government and people of Canada.' (EMR, 1973, vol. I. p. iii.) This approach was in stark contrast to the much more narrow and technical oil and gas focus then characteristic of the NEB. It was made possible by the much more flexible and political mandate of EMR. As a government department headed by a federal Cabinet minister, EMR, unlike the NEB, was not tied to specific regulatory functions, nor was it a regulatory agency under pressure to maintain a semblance of independence from government policy. On the one hand, therefore, EMR had more freedom to range widely with respect to analysis and forecasting, on the other they were, in principle, more tied to government policy. The effect of these two differences was to shape the way EMR was to engage in forecasting over the next decade.

A major component of EMR's 1973 report was a detailed treatment of overall energy supply and demand in Canada (not attempted since the 1969 NEB report). While a more detailed sectoral analysis was used than that employed by the NEB, the modelling methodology was similar: future energy use levels were derived in each sector separately, either by projecting activity variables (for example, numbers of households) weighted by their energy intensities, or by judgemental extrapolation for the sector (see Figure 10.3).

Despite a similarity in methods, there was an important specific difference between the EMR and the NEB forecasts connected to EMR's wider policy perspective and political sensitivity. Orientated primarily towards oil and gas exports, the NEB focused on trying to predict the most likely future, the 'reasonably foreseeable requirements' of its mandate weighing heavily on it. EMR, on the other hand, was under no such constraints in its use of forecasts. Perhaps for this reason, even as early as 1973 EMR emphasized that:

> Canada's energy requirements are determined, in part, by the kind of society Canadians want to build in the future. It is unrealistic to forecast future energy trends based on historic trends alone. Canadians can influence the magnitude and structure of energy demand. Government policy decisions . . . will affect total energy requirements . . . The standard forecast is not presented as an objective, but rather as an indication of the possible results of a 'business as usual' approach and as a reference point against which to measure the effects of changes in policies and attitudes. (EMR, 1973, vol. I. 70.)

In later years this more flexible, and more normative, approach to energy forecasting was to bear fruit for EMR, but in the mean time, less than six months after the publication of *An Energy Policy for Canada*, the department was rocked by the events of 1973 which made the forecasts and much of the analysis in that report obsolete. Within a year of the publication of the EMR study, world oil prices had surpassed levels that it had predicted might be reached by the year 2000.

The 1973 OPEC price increases meant that, for the first time, imported oil was more expensive than domestic oil. In response to these changes, the focus of federal activity shifted to the development of oil-pricing policy, the formation of a national oil company (Petro-Canada), the planning of a new oil pipeline to carry western oil to Montreal and the promotion of a Mackenzie Valley natural gas pipeline. Correlatively, energy supply perceptions shifted from the cornucopianism of the 1960s and the unthinking optimism of the EMR 1973 report, to the more sombre question of becoming 'energy self-reliant' as soon as possible and for the reasonably foreseeable future. As indicated in the previous section, this represented a major change from the export orientation of earlier federal policy toward concern with domestic availability.

In 1976, EMR released *An Energy Strategy for Canada* (1976). This document marked the emergence of EMR as a policy-making body (the 1973 report had contained no policy) and as an energy modelling and forecasting agency. In the 1976 report, EMR unveiled what was later to become known as the Interfuel Substitution Demand Model (IFSD). This was an econometric energy demand model which represented a very significant improvement in modelling and forecasting methods. First, the model began with tertiary end-use energy demand (converted via utilization efficiencies to secondary demand), and only later allocated fuels to meet this demand on an exogenously determined market share basis. Second, the model incorporated price and economic activity variables in the form of sectorally specific econometric equations. The result was a hybrid model more flexible than traditional econometric models.

Despite the new model, the results of the EMR supply–demand projections were quantitatively similar to those of the NEB published in 1975. EMR concluded that, in both the high and moderate price scenarios, total domestic energy demand would exceed domestic availability by 1980 (see Figure 10.3).

The policy outlined in the 1976 report can best be seen as a forecast-driven policy. The gloomy perceptions of future energy prospects outlined in the report were used to indicate the urgent need for the policies outlined therein. At the same time, however, the report emphasized the degree of flexibility available and the need to make policy choices: 'The scenarios sketched . . . suggest that we have substantial potential to manage our energy future—including both the supply and demand aspects—in a way that minimizes our risks, to the benefit of all Canadians.' (EMR, 1976: 122.) During 1977, EMR published another study with significant implications for demand and policy analysis. The report, entitled *Energy Conservation in Canada: Programs and Perspectives* (EMR, 1977) was written by the Office of Energy Conservation (OEC), which had been formed within EMR at the beginning of 1974.

The results of the OEC calculations were projected levels of future energy demand considerably lower than those contained in the 1976 *Energy Strategy* report or in contemporary NEB reports (see Figure 10.3). Just as important as the actual numerical results, however, was the methodological approach taken. The *Conservation* report was based upon the recognition that future energy demand was something that could be strongly and deliberately influenced by government policy. While analytically crude—the conservation constraints were simply imposed on the end-state of an otherwise independent, forecast-driven model of the economy—the approach represented a crucial methodological divide for here the economic model was made to interact explicitly, however crudely, with policy decisions. The conservation report, however, did not represent EMR's official position on future energy demand.

During the same period that the conservation report was undertaken and published, EMR's promised post-1990 study was in preparation. In 1978 the report of the Long-Term Energy Assessment Program (LEAP) was released, containing an assessment of the Canadian energy situation over the period from 1978 to 2025 (Gander and Belaire, 1978).

The LEAP report was almost completely lacking in quantitative analysis. Instead it consisted, more than anything else, of a lengthy argument concerning future energy prospects. While a single long-term energy supply and demand projection was included, it was not discussed or presented in any detail. Nor were alternative scenarios of future supply and demand presented. Instead the LEAP report painted an almost universally gloomy picture of long-term energy conditions and called for a national energy programme in order to attain 'substainable self-reliance'.

In 1979, the election of a new federal government led to a new goal of energy policy: energy self-sufficiency by 1990 (a goal the previous government had declared in 1976 to be unattainable). As part of the process of outlining the possibilities for self-sufficiency, EMR released a report in November on various energy supply–demand scenarios to the year 2000 (EMR, 1979).

This report contained a number of different scenarios of total energy demand. The scenarios differed, however, only in terms of the assumptions made about domestic energy pricing policy. The range of the projections offered was very narrow but all of the demand projections were significantly lower than the LEAP report, the NEB's 1978 oil report, and either of the NEB's 1979 gas reports (Figure 10.2). The report gave graphic illustrations of the volatility of both NEB and EMR forecasts over the past five years, and of their policy dependence. It identified supply projections as particularly uncertain, but covered this by adopting an optimistic view of Alberta tar sands development, a more 'policy-controllable' factor.

Despite its somewhat narrow econometric focus, EMR's 1979 report marked a significant stage in the evolution of federal energy forecasting and modelling in two respects: (i) unlike the NEB, who still primarily interpreted forecasting as a method of objective determination (for instance, of how much gas could be exported or whether a pipeline should be built), EMR was beginning to use forecasting to explore the range of policies available; (ii) because EMR, unlike the NEB, had an energy demand model but no energy supply models, the focus of attention appeared to be shifting to the demand side of the supply–demand equation. These shifts were to bear fruit the next year under the newly re-elected Liberal government.

The return of a Liberal government to power in difficult energy circumstances in late 1979 soon led to increased tensions over domestic energy pricing. During early 1980 several sets of price negotiations with Alberta, the main energy producing province, fell through. In response the federal government unveiled the National Energy Program (NEP) in October 1980, by far the most ambitious and far-reaching energy policy in Canadian history (EMR, 1980). The NEP represented a set of unilateral federal decisions about energy pricing, taxation, and revenue-sharing, including a set of policies intended fundamentally to alter a number of characteristics of the energy industry, energy supply and demand patterns, and the manner in which the profits of oil and gas development were divided up among the federal government, the producing provinces, and the oil and gas industry.

One of the primary objectives of the NEP was to reach oil self-sufficiency in Canada by 1990. However, the NEP did not assume any increases in oil supply over those forecast in the 1979 EMR report. Instead the NEP focused its attention on oil demand, arguing that by means of a 'massive, unprecedented commitment to oil substitution and conservation', oil demands could be significantly reduced. This 'off-oil' policy was described as the centre-piece of the NEP.

While the EMR IFSD model was used to support the NEP's position, it was not used to justify its policies, as in 1976, or to generate policy alternatives, as in 1979. Instead the goal of 'energy security' was set in advance and defined in terms of the specific targets mentioned above; only then was the model used to generate a forecast that in fact led to that overall goal. The result was a projected growth rate of 1.9 per cent for total primary energy demand to the year 1990, an all-time low for that figure over this period (Figure 10.3).

The NEP was, then, really a normative policy. It was not based upon the results of forecasts in the sense of responding to needs and problems that were projected as likely to happen. On the contrary, certain demand and supply targets were set and then policy developed that was intended to create the conditions necessary to reach those targets. In one sense therefore the NEP represents the culmination of demand modelling at EMR. Demand modelling had moved from the prediction of likely futures in 1976, through the generation of alternative scenarios in 1977 and 1978 and the analysis of the relative implications of alternative pricing policies in 1979, to the goal-orientated policies of 1980.

It is ironic then that this culmination was reached in a policy statement that was essentially forecast-free. The NEP itself contains only one total energy demand forecast, and only to the year 1990. (Oil supply and demand are also forecast, to show the effects of the NEP on oil self-sufficiency.) In contrast to previous policy statements, such as the 1976 *Energy Strategy* report, these forecasts are contained at the end of the report. They do not provide the context within which policy is to be formed but instead show the expected results of that policy. Perhaps the clearest statement of a new perceived role of forecasts is contained in a statement made in that part of the NEP that dealt with the future role of renewable energy:

> . . . while forecasts are useful tools for analysis, they can tell us only what will happen under certain conditions. The conditions—the policies—are the keys. Many thoughtful and concerned Canadians believe that we should alter the forecast, and that we should decide soon on a preferred energy future, and establish the conditions that will take us there. (EMR, 1980: 65.)

We are here in a different world from that of the NEB, as reflected for example in their total energy report of 1981. The focus here upon 'altering the forecast' and on 'establishing the conditions' of 'a preferred energy future' leads to a totally different attitude toward policy and forecasting. The NEP report in fact concluded with the suggestion that it should be seen as part of a longer-term policy process intended to foster the transition to a more sustainable energy system.

The period following the release of the NEP was marked by an unprecedented outpouring of anger on the part of industry and the producing provinces.[7] This unhappiness was made worse by the effects of the most severe economic recession in Canada since the 1930s, which itself occurred in the context of a serious global recession.

In the context of these unsettled conditions EMR issued an 'update' to the NEP in 1982 (EMR, 1982) which can best be seen as evidence of a federal retreat on energy policy issues. Battered by the recession, falling prices and revenue projections, and a level of opposition much greater than had been anticipated, the government concluded agreements with the producing provinces that substantially blunted the edge of the NEP. Gone, in the *Update*, were the brave words about changing the forecasts, choosing preferred energy futures, and forging a transition to a sustainable energy system.

The federal government pulled in its horns in terms of forecasting as well as policy. In July 1983, EMR released what was intended to become a semi-annual series of energy demand projections. The 1983 report emphasized that it was not an official policy document and noted that 'for the

Summer 1983 forecast the off-oil market share adjustments specified and used in the NEP and the NEP *Update* forecast were not used' (EMR, 1983: 9). In other words, the new forecasts were freed from the normative targets set in the NEP, on both the demand and supply sides. The results were higher demand and lower supply forecasts than in 1982 (see Figure 10.3).

Despite the use of essentially the same IFSD model as in the NEP reports, the 1983 EMR projections represent a return to non-normative methods of energy forecasting. After almost a decade of movement toward an increasing use of forecasts as planning tools, this development reflects a reversal of that trend and a retreat from activist energy policy-making on the part of the federal government. It also reflects a general reduction in the perceived urgency of energy issues over the past several years as prices have fallen and reserves have increased. These shifts have allowed energy forecasting to be detached from specific policy goals by a government which currently prefers a more passive approach. From the point of view of the relationship between energy modelling and policy development, this report marks a return to the relationship characteristic of EMR's 1979, or even 1976, reports.

Within EMR there was also some technically based unease with the normative use of the IFSD model in the 1990 report. Analysts argued that the chosen parameter values were not selected because they were 'most likely' and hence rejected the exercise as meaningful. But there was also foundation for the caution in the model structure itself. Because behavioural econometric relationships are built into the model, it cannot easily be used to reach end-points different from those implied by the values chosen for input assumptions. And these input assumptions typically are not defined in terms of, or compatible with, detailed physical end-use characteristics or behaviour. There is, for example, no way to reflect price changes and other policy measures in the choice of the physical stocks of energy-using technologies, household habits, etc. to ensure that the conservation constraints are met, resulting employment patterns are assumed, and so on. In this sense one is 'flying blind' in trying simply to find a few aggregated parameters which will yield appropriate future levels of energy demand. Thus, although the IFSD model can be forced into a normatively orientated option analysis mode, it fits awkwardly, betraying its extrapolative–predictive econometric origins.

The rise and fall of EMR's interest in normative forecasting methods was not unconnected with events outside official forecasting circles. By the time that the NEP was written, alternative modelling and forecasting perspectives originating from extra-governmental analysts had been articulated for some time. In general these perspectives led to analyses which challenged both the methods and results of traditional analyses.

Alternative approaches: the conservation argument

By the mid-1970s, a significant alternative perspective in energy policy-making was beginning to emerge in Canada. Focusing on energy conservation and renewable energy sources for its specific policy recommendations, the movement emphasized the virtues of reducing economic growth, rejecting consumerism, waste, and conspicuous luxury and returning to a simpler and more socially orientated society. This shifted philosophical and policy emphasis led in turn to successive shifts in energy modelling and its uses, shifts that became increasingly fundamental. An interesting feature of this development is the close link it had with official forecasting activities. Briefly told, the story unfolds as follows.

In 1974 in response to increasing public debate there was created within EMR an Office of Energy Conservation (OEC) with Dr David Brooks as head. A controversial resource economist and environmentalist, Brooks believed in the technical potential and intrinsic social value of conservation, identifying zero energy growth as a socially desirable and practically achievable goal.

The OEC served as the official locus of conservation-orientated analysis within EMR. As noted above, following release of the 1976 federal energy policy the OEC published a conservation report in 1977 (EMR, 1977; see Figure 10.3). Despite using the same demand model as that used to generate the scenarios in the 1976 report, the conservation report came to much more optimistic conclusions about the potential for energy conservation.

Not all of the alternative forecasting analysis occurred within the OEC. During 1975 and 1976 Friends of the Earth energy analyst Amory Lovins was undertaking work in Canada and the United States that was to have a significant effect on the direction and approach of energy analysis in both countries. Faced at the time with a complete absence of any energy policy planning which would take his conservation and renewable energy-based perspective seriously into account, Lovins reversed the usual extrapolative procedure of energy demand forecasting and, sketching two contrasting energy-lifestyle scenarios far into the future (2025–50), a high-growth fossil-nuclear-electric and a low-growth conserver-renewable scenario, asked how he might 'get there from here' in each case. This 'looking backwards' approach shifted the focus from the *prediction* of likely futures, towards testing the *feasibility* of alternative scenarios. The argument ran that both the high and low growth futures were in fact technically and economically feasible but that each required markedly different public policies to be pursued over the intervening half century.

Lovins first applied this new approach in a study done for the Science Council of Canada, publishing the paper on the Canadian situation in *Conserver Society Notes* (Lovins, 1976). A subsequent book (Lovins, 1977) expanded upon the notion of a soft energy path as an integrated set of energy policies based on conservation and renewable technologies that were diverse, flexible, and matched in scale, distribution, and quality to tertiary end-use needs.

Lovins work directly inspired two subsequent energy studies in Canada. The first of these was a demand study prepared by David Brooks and two EMR colleagues as a background paper for EMR's Long-Term Energy Assessment Program (LEAP, discussed above). This study used both a looking backwards (or backcasting—see Robinson, 1982a) and a more conventional forecasting approach, and argued that low demand growth futures were both feasible and attractive (Brooks, Erdmann, and Winstanley, 1977).

The second study was a somewhat more detailed scenario analysis coordinated by one of us (JBR) resulting in the 1977 report 'Canadian Energy Futures—An Investigation of Alternative Energy Scenarios' (Robinson *et al.*, 1977). This study used a preliminary end-use model and a qualitative backcasting methodology to prepare contrasting fossil-nuclear-electric and conserver-renewable scenarios for a roughly fifty-year future. Like previous backcasting analyses the report suggested the relative feasibility and desirability of low demand growth scenarios.

The two national backcasting scenario analyses of 1977 were followed by a series of provincial studies. The first was a study of the province of Ontario prepared for the Royal Commission on Electric Power Planning (Crowe, *et al.*, 1978). At the same time a province-by-province conserver-renewable study of the same sort was going on, as a first initiative of the Canadian branch of Friends of the Earth, founded by David Brooks in 1978. A separate working group in each province prepared future scenarios using quasi-quantitative end-use structural models, with primary aims being to test the feasibility of provincial energy self-reliance and to identify important inter-provincial problems and trade-offs. These studies appeared in *Alternatives* during 1979 and 1980. In addition, a 1980 policy analysis co-authored by one of us (CAH) had also investigated a conserver-renewable future for the province of Ontario (Hooker, *et al.*, 1980).

In the fall of 1980 Friends of the Earth Canada was awarded a major contract by the federal government to undertake a detailed province-by-province analysis of the technical and economic potential for conservation and renewable energy in Canada to the year 2025. This was the first fully integrated and comprehensive study of these alternatives ever undertaken in Canada. Directed by Brooks and one of us (JBR) the final report was released in 1984 by EMR (Friends of the Earth, 1983–4; for a summary see Bott *et al.* or Robinson, *et al.*, 1984).

The results of the Friends of the Earth study, which suggests that a soft energy based system would not only be feasible but economically desirable, are radically different from those of conventional government studies, such as the LEAP Report of 1978 (see EMR in Figure 10.3). In large part this is due to the different approach used (backcasting as opposed to forecasting), the different model used (an input–output based materials balance model as opposed to econometric models), and the different assumptions made about the availability and cost-effectiveness of conservation and renewable energy.

It is apparent that over the past ten years official energy modelling and forecasting efforts have been paralleled by an increasingly sophisticated series of alternative analyses. Due in part to the failure of conventional forecasting techniques to predict future supply and demand patterns accurately, these alternative approaches have had some impact on both results and the methods of official analyses. EMR in particular both supported these extra-governmental efforts financially and appeared to incorporate some of the approaches and results of such analyses into their own energy modelling processes. Ironically, EMR has appeared more recently to retreat on these fronts while the NEB, after a decade of ignoring alternative views, now appears to have adopted conclusions concerning conservation potential that are much closer to those of researchers outside the government.

ENERGY FORECASTING AND POLICY-MAKING IN CANADA[8]

It is clear from the foregoing discussion that energy forecasting has played an important role in the evolution of Canadian energy policy and in specific policy decisions over the past fifteen years. Yet our analysis has also made it clear that energy forecasts, conceived as predictions of future levels of supply and demand, have been an extremely unreliable basis for policy decisions and, moreover, have served to obscure the normative bases of such decisions.

In the case of the NEB, which has narrowly interpreted its statutory mandate so as to require supply and demand forecasts for specific fuels to determine regulatory decisions, the forecasts made have been subject to large changes over very short periods. As a result, the conclusions reached by the NEB have changed drastically, sometimes over a period of less than a year. The point here is not that bad decisions have been made. It is rather that the forecasts used to justify those decisions were not a reliable indication of future events and served to obscure the policy dependence of the forecasts.

With hindsight it can be seen that the results of the forecasts were essentially reflections of highly contemporary conditions. Relatively small changes in trends (such as the finding rate of gas additions) and in important variables such as the world price of oil significantly altered the results of forecasts. And despite the development of more sophisticated techniques, the variability of NEB forecasts has not decreased. Changes in conditions during the 1980s have led to substantial changes in NEB forecasts over the last three years.

From a methodological point of view the last fifteen years have seen a substantial evolution in forecasting techniques. On the demand side relatively unsophisticated bottom-up methods of analysis gave way in the mid-1970s to the development of detailed econometric demand models. Yet this evolution has not been without its problems. Because these econometric models subsume behavioural effects, and because they are essentially price-based, they are not particularly useful tools for the detailed analysis of conservation and renewable energy. That is, these models tend to take as given precisely those behavioural relationships which are of interest in conservation and renewable energy analysis. As a result they are usually only able to treat price-induced conservation and substitution. Moreover, econometric models tend to explain energy demand as a function of relatively few, primarily economic variables such as price and income.[9]

On the supply side the last decade has witnessed the emergence of new computerized models of oil and gas deliverability developed by the NEB. However, these models are primarily useful for projecting rates of production from established reserves under specified economic conditions. Not only are the projections very sensitive to changes in these assumed conditions but these models are of no use in projecting new discoveries and connection rates. Moreover the linkage of supply and demand forecasts is problematic, as evidenced in the very complex procedures used by the NEB to assess natural gas supply in relation to projected demand. As both the NEB and EMR have emphasized, the practice of supply forecasting is relatively underdeveloped and inherently judgemental.

The last decade has also witnessed the emergence of alternative energy analyses which have differed significantly in results and approach from conventional analyses. In terms of results, official projections of energy demand and to a much lesser extent energy supply have changed significantly in the direction of these alternative analyses. The 1984 NEB report serves as a recent example: its total energy demand projections for the year 2000, for instance, are 26 per cent lower than in its 1979 gas report and only 35 per cent higher than the 'Business as Usual' scenario of Friends of the Earth (Figure 10.3).

In terms of method and approach, the record is mixed. The NEB has varied little from a basic predictive forecasting mode. EMR on the other hand undertook a significant evolution toward normative and exploratory backcasting techniques in the late 1970s, only to retreat to predictive techniques since that time. There is currently no evidence that either agency is interested in the new methodological approaches pioneered outside the government. Instead the current focus of attention appears once again to be upon the prediction of the most likely levels of future energy supply and demand.

The institutional framework in which forecasts are developed also poses problems. The institutions that administer energy policy have developed in the context of certain conventional approaches to the energy problem, which determines how these institutions formulate energy problems and their solutions. Certain assumptions and practices, long a part of the structure of these organizations, restrict the types of action open to them. As Hooker and van Hulst (1977: 47) have argued,

> Social institutions have definite structures; on the one hand those structures reveal how the culture views the making of societal decisions (i.e., reveals the social philosophy) and, on the other hand, the structures constrain the kinds of decisions the institutions are capable of making and the kinds of decision processes through which they go to make them. In particular, the institutions of our society have a characteristic structure which is revealing of the western market-democratic social philosophy and those institutional structures lead to characteristic ways of making social policy.

Some constraints on institutional behaviour are embodied in the legal and regulatory context within which an institution functions. For example, the NEB interprets its statutory mandate in such a way as to require the use of forecasting in deciding upon export and major facility applications.

Externally imposed obligations, however, are not the only factor in determining institutional behaviour. Probably of greater importance are the ingrained beliefs and assumptions of the individuals in position of power within the institution. Members of the NEB, for example, subscribe to certain assumptions and values that influence their activities. As Lucas and Bell (1977: 40) remark:

> There seems little doubt, however, the Board members' perspectives, perhaps shaped by experience and training, largely coincide with those of industry officials. All tend to view the rules of the game in very much the same way. There are no Board members with overriding radical views of the public interest of humanist approaches to energy issues. The same unstated assumptions concerning energy development and use seem to be in the minds of both groups.

In this connection the difference between the NEB and EMR is of particular interest. There was a gradual shift in the EMR approach from modest scenario analysis in 1976 to the analysis of alternative pricing options in 1979 and on to normative projections in 1980 and 1982, a shift which did not occur in the NEB approach, despite the NEB's adoption of the EMR demand model in 1977.[10] Part of the reason for this shift may well have been the in-house work at EMR of analysts from the Office of Energy Conservation, who introduced and applied the normative approach to analysis pioneered by Amory Lovins. In particular the backwards-looking analyses done by Brooks and Robinson for the Long-Term Energy Assessment Program in 1977 appear to coincide with a shift toward more normative analysis at EMR while the thrust of subsequent work of this type finds an echo in statements in the 1980 National Energy Program.

The contrast between the methods and orientation of EMR and the NEB points to an important lesson: to the extent that forecasting is perceived as a neutral or objective technique, the normative assumptions that are necessarily built into it are ignored or suppressed. Another lesson concerns the influence of personnel on the model–policy relation. The evolution of a strongly goal-orientated energy policy within EMR reflected a strong nationalist and interventionist orientation characteristic of top EMR bureaucrats during the late 1970s. The EMR shift evidently represented the culmination of an attempt to develop a nationalistically orientated industrial strategy for Canada (Laxer, 1983: 191), an attempt which was to end in partial failure for the reasons discussed above.

The tendency of policy-makers such as the NEB to use energy forecasts to justify major policy decisions results from the insistence that energy forecasts are, in principle, an accurate representation of the future supply and demand of energy. In reality, the future level of supply and demand depends substantially on the actions of governments and individuals. This circular causation makes conventional forecasts dangerous: to the extent that they form the basis of policy decisions, they confirm the belief that these decisions are the result of the future supply and demand situation rather than the cause of it. This reversal of cause and effect has two consequences. First, it obscures the nature of the real relationship between policy-making and the future supply and demand of energy. By basing present decisions on the apparent uncovering of future events, an appearance of inevitability is created that de-emphasizes the importance of present choice and further lessens the probability of developing creative policy in response to present problems. Second, it allows decision-making institutions to assume a false cloak of objectivity.

In conclusion, it has been suggested in this paper that the history of energy forecasting and policy-making in Canada over the past fifteen years is such as to call into serious question the role of energy demand and supply forecasts as reliable bases for policy decisions. Instead of providing useful predictions of future levels of energy supply and demand, such forecasts are seen to be methods of projecting certain assumptions and judgements into the future, judgements which reflect current policy goals and contemporary supply and demand conditions. Forecasts used as predictive tools are seen to be related in a particularly complex way to the policy-making process. They are both cause and consequence of policy decisions. Forecasts do not reliably reveal the future but they often justify the attempt to create (or prevent) a particular future.

The role that energy forecasts play as instruments of justification suggests that their ostensible purpose of predicting the future may be fundamentally misrepresented or even misconceived by many analysts and policy-makers. Under conditions of great uncertainty and when considerable choice as to the future direction of energy policy is available, the most useful, or at least most honest role of forecasts may not be to predict the future but to test the feasibility and implications of different futures. This implies a much different role for forecasts than that conventionally cited, a role much closer to that claimed for the backcasting approach used by extra-governmental 'soft energy path' advocates.[11]

Of the two federal agencies, EMR has come closest to using forecasts in such an overtly normative way. However as noted above, that agency has since retreated from such usage. Yet if energy forecasting is to regain any credibility as an analytical tool and to avoid great variability and reversals characteristic of the past fifteen years, then its purpose and its methods need to be fundamentally rethought.

Notes

1. For a detailed discussion of the development of the Canadian fuel policy from Confederation to the end of the 1970s, see McDougall (1982).
2. The federal government also has available certain overriding declaratory powers under the constitution whereby federal jurisdiction can be asserted over provincial resources when this is deemed necessary in the national interest. In practice, however, it is politically very difficult to use such declaratory powers, which have only been exercised in the energy field with respect to nuclear power.
3. For a discussion of the structure and history of the NEB, see Lucas and Bell (1977) and McDougall (1973). More recently see McDougall (1982).
4. If the NEB approves a project that approval must in turn be approved by Cabinet. A board decision not to approve a project, however, is final.
5. For a discussion of the NEB's northern pipeline hearings, see Robinson (1980).
6. Strictly speaking, this holds for the gas supply forecast only for the years after 1990.
7. The NEP was roundly denounced as a federal power grab in a number of academic, industrial and provincial circles and the Alberta government actually began an oil-supply boycott beginning at about 10 per cent of total domestic demand and planned to escalate to 30 per cent within a year. (Note that under the Canadian constitution the provinces, not the federal government, own the natural resources within their boundaries.) See the discussion in Foster (1982) and Laxer (1983).
8. A more extensive treatment of this topic is contained in Robinson (1982b).
9. This is less true of the EMR IFSD model than of the models typically used by industry. The IFSD model is based on tertiary energy use and incorporates a certain amount of physical stock and efficiency data. The use made of this detailed structure, however, depends on how the model is deployed; the 1983 EMR projections focus pretty exclusively on price issues.
10. While the NEB is not in a position to formulate policy goals, nothing prevents it from engaging in wide-ranging scenario analysis and sensitivity analysis for the purpose of uncovering and assessing the feasibility of alternative energy scenarios.
11. For a discussion of the backcasting alternative, see Robinson (1982a) and Robinson and Hooker (1985).

Bibliography

Bott, R., D. Brooks, and J. Robinson (1984), *Life After Oil—A Renewable Energy Policy for Canada*. Hurtig Publishers, Edmonton.

Brooks, D. B., R. Erdmann, and G. Winstanley (1977), *Some Scenarios of Energy*

Demand in Canada in the Year 2025. Report of the Demand and Conservation Task Force, Long-Term Energy Assessment Program, Energy, Mines and Resources Canada, Ottawa.

Crowe, R. *et al.* (1978), *Energy Planning in a Conserver Society: The Future's Not What It Used To Be*. a Report prepared for the Ontario Royal Commission on Electric Power Planning, Energy Probe, Toronto.

Energy, Mines and Resources Canada (1973), *An Energy Policy for Canada—Phase 1*. Vol. I—Analysis. Information Canada, Ottawa.

—— (1976), *An Energy Strategy for Canada—Policies for Self-Reliance*. Ottawa.

—— (1977), *Energy Conservation in Canada: Programs and Perspectives*. Report EP 77–7, Ottawa.

—— (1979), *Canadian Oil and Gas Supply/Demand Overview*. Ottawa, Nov.

—— (1980), *The National Energy Program*. Ottawa.

—— (1982), *The National Energy Program—Update '82*. Ottawa.

—— (1983), *Long-Term Energy Supply–Demand Outlook—Summer '83 Forecast*. Energy Strategy Branch, Ottawa.

Foster, P. (1982), *The Sorcerer's Apprentices—Canada's Super-Bureaucrats and the Energy Mess*. Totem Books, Don Mills, Ontario.

Friends of the Earth Canada (1983–4), *2025: Soft Energy Futures for Canada*. 12 vols., Energy, Mines and Resources Canada, Ottawa.

Gander, J. and F. Belaire (1978), *Energy Futures for Canadians*. Report of the Long-Term Energy Assessment Program, Department of Energy, Mines and Resources, Ottawa.

Helliwell, J. (1979), 'Canadian Energy Policy'. *Annual Review of Energy*, 4: 175–229.

Hooker, C. A. and R. van Hulst (1977), *Institutions, Counter-Institutions and the Conceptual Framework of Energy Policy-Making in Ontario*. Report prepared for the Ontario Royal Commission on Electric Power Planning, Toronto.

—— and R. van Hulst (1980), 'Institutionalizing a High Quality Conserver Society'. *Alternatives*, 9: 25–36.

—— *et al.* (1980), *Energy and the Quality of Life*. University of Toronto Press, Toronto.

Laxer, J. (1983), *Oil and Gas*. James Lorimer and Co., Toronto.

Lovins, A. B. (1976), 'Exploring Energy-Efficient Futures for Canada'. *Conserver Society Notes*, 1: 5–16.

—— (1977), *Soft Energy Paths*. FOE/Ballinger, New York.

Lucas, A. and T. Bell (1977), *The National Energy Board: Policy, Procedure and Practice*. Law Reform Commission of Canada, Ottawa.

McDougall, I. (1973), 'The Canadian National Energy Board: Economic Jurisprudence in the National Interest or Symbolic Reassurance?' *Alberta Law Review*, 11: 329–38.

McDougall, J. N. (1982), *Fuels and the National Policy*. Butterworths, Toronto.

National Energy Board (1969), *Energy Supply and Demand in Canada and the Export Demand for Canadian Energy 1966–1990*. Ottawa.

—— (1970), *Report to the Governor in Council*. Ottawa, Aug.

—— (1971), *Reasons for Decision*. Ottawa, Nov.

—— (1972), *Potential Limitations of Canadian Petroleum Supplies*. Ottawa.

—— (1974), *Report to the Honorable Minister of Energy, Mines and Resources in the Matter of the Exportation of Oil*. Ottawa, Oct.

—— (1975a), *Canadian Natural Gas Supply and Requirements*. Ottawa, April.

—— (1975b), *Canadian Oil Supply and Requirements*. Ottawa, Sept.

—— (1977a), *Canadian Oil Supply and Requirements*. Ottawa, Feb.

—— (1977b), *Reasons for Decision—Northern Pipelines.* 3 vols., Ottawa, June.
—— (1978), *Canadian Oil Supply and Requirements.* Ottawa, Sept.
—— (1979a), *Canadian Natural Gas Supply and Requirements.* Ottawa, Feb.
—— (1979b), *Reasons for Decision in the matter of Applications under Part VI of the National Energy Board Act.* Ottawa, Nov.
—— (1981), *Canadian Energy Supply and Demand 1980 to 2000.* Ottawa, June.
—— (1983), *Reasons for Decision in the Matter of Phase II—The License Phase and Phase III—The Surplus Phase of the Gas Export Omnibus Hearing.* Ottawa, Jan.
—— (1984), *Canadian Energy Supply and Demand 1984 to 2005.* Ottawa, Aug.

Robinson, J. B. (1980), 'Policy, Pipelines and Public Participation: The National Energy Board's Northern Pipeline Hearings', in O. P. Dwivedi (ed.), *Resources and the Environment: Policy Perspectives for Canada.* McClelland and Stewart Ltd., Toronto.
—— (1982a), 'Energy Backcasting: A Proposed Method of Policy Analysis'. *Energy Policy*, 10: 337–44.
—— (1982b) 'Backing Into the Future: On the Methodological and Institutional Biases Embedded in Energy Supply and Demand Forecasts'. *Technological Forecasting and Social Change*, 21: 227–40.
—— and C. A. Hooker (1985), 'Energy Policy and Energy Modelling in Canada'. Report prepared for the Structural Analysis Division, Statistics Canada, Ottawa.
—— *et al.* (1977), *Canadian Energy Futures: An Investigation of Alternative Energy Scenarios, 1974 to 2025.* Workgroup on Canadian Energy Policy, Faculty of Environmental Studies, York University, Toronto.
—— *et al.* (1984), 'Determining the Long-Run Potential for Energy Conservation and Renewable Energy in Canada'. *Energy*, 10 (1985): 689–705.

APPENDIX

Table 1 Canadian Energy Supply and Demand, 1983 (petajoules)

A. Energy production

	Primary Production	Converted to Electricity	Secondary Production	Net Exports	Energy Industry Use	Non-energy Use	Secondary Use
Coal	1081	827	254	33	—	5	216
Oil	3421	70	3351	235	217	347	2552
Natural Gas	2683	75	2608	949	105	149	1405
Electricity	1113[a]	—	1461	130	110	—	1221
Wood	483	2	481	—	15	—	466
Total	8781	974	8155	1347	447	501	5860

B. Energy use

	Residential	Commercial	Industrial	Transportation	Total
Coal	3	1	212	—	216
Oil	437	335	334	1446	2552
Natural Gas	469	387	549	—	1405
Electricity	374	290	548	9	1221
Wood	110	—	356	—	466
Total	1393	1013	1999	1455	5860

Source: Statistics Canada. Quarterly Report on Energy Supply/Demand in Canada, 1984–IV.

Notes

a. Primary (hydro and nuclear) electricity only

Table 2 A Comparison of Natural Gas Demand and Supply Forecasts prepared by the National Energy Board from 1969 to 1984 (petajoules)

A. Demand

Total Domestic Requirements[a]
(Petajoules)

	Year Forecast Was Made									
Consumption in:	1969[b]	1970[c]	1971[d]	1975[e]	1977[f]	1979[g]	1979[h]	1981[i]	1983[j]	1984[s]
1980	1950	2237	2516	2662	2187	2067	2022	—	—	—
1985	2481	2680	3089	3215	2570	2536	2478	2573	2311	2131
1990	3026	3092	3557	3764	2934	2845	2719	2874	2667	2575
1995	—	3638	4101	4399	3361	3280	3085	3143	3116	2699
2000	—	—	—	—	—	3793	3519	3585	3525	2921

Total Requirements including Exports
(Petajoules)

	Year Forecast Was Made								
Consumption in:	1969[b]	1970[c]	1971[d]	1975[e]	1977[f]	1979[g]	1981[i]	1983[j]	1984[s]
1980	3474	3291	3570	3716	3241	3213	—	—	—
1985	4112	3625	4034	4160	3515	3512	4316	3601	3127
1990	4861	3319	3784	3991	3161	3203	3118	2884	4419
1995	—	3644	4107	4406	3368	3286	3155	3122	3066
2000	—	—	—	—	—	3793	3585	3525	3088

B. Supply

Total Domestic Capability[k]
(Petajoules)

Year Forecast Was Made

Production in:	1969[l]	1975[m]	1977[n]	1979[o]	1979[p]	1981[q]	1983[r]	1984[s]
1980	3474	3210	3238	3898	4353	—	—	—
1985	4112	3088	3322	3645	4875	4537	5347	5161
1990	4861	2453	2797	3145	3983	4105	4827	4844
1995	—	1842	2058	2470	3154	3153	3698	4054
2000	—	—	—	2044	2456	2460	2905	3122

Notes

a. Includes fuel, losses, and reprocessing shrinkage; for years before 1979, units converted from bcf at 1000 Btu/cf at a conversion rate of 1055 Joules/Btu
b. National Energy Board (1969: 150); shrinkage reduced in domestic requirements forecast to reflect reduction in exports
c. National Energy Board (1970: 3–19); shrinkage added as per percentages derived from 1969 report
d. National Energy Board (1971, Appendix I, p. 1); shrinkage added as per percentages derived from 1969 report
e. National Energy Board (1975a: 21)
f. National Energy Board (1977b, Vol. 1, pp. 2–42)
g. National Energy Board (1979a: 131) 'expansion sales' in Quebec and the Maritimes included
h. National Energy Board (1979b: 4–11); Quebec and Maritimes demand included
i. National Energy Board (1981: 426); Quebec and Maritimes demand included
j. National Energy Board (1983: 32); Quebec and Maritimes demand included
k. Excluding frontier areas
l. National Energy Board (1969: 150); excluding field and plant use
m. National Energy Board (1975a: 59)
n. National Energy Board (1977b: Vol 1, App. 2–3, p. 11)
o. National Energy Board (1979a: 30)
p. National Energy Board (1979b: 3–26)
q. National Energy Board (1981: 170)
r. National Energy Board (1983: 21)
s. National Energy Board (1984: A–108)

Table 3 A Comparison of Oil Demand and Supply Forecasts prepared by the National Energy Board from 1969 to 1984 (thousand barrels per day)

A. Demand (West of the Ottawa Valley)

	Year Forecast Was Made							
Demand in:	1969[a]	1972[b]	1974[c]	1975[d]	1977[e]	1978[f]	1981[g]	1984[h]
1980	1144	1005	1185	1162	1116	1048	—	—
1985	1378	1193	1425	1353	1208	1148	1101	875
1990	1669	—	1700	1756	1308	1239	1126	828
1995	—	—	2020	1737[i]	1428	1323	1189	859
2000	—	—	—	—	—	—	1340	892

B. Supply (excluding frontier areas)

	Year Forecast Was Made							
Supply in:	1969[a]	1972[b]	1974[c]	1975[d]	1977[e]	1978[f]	1981[g]	1984[h]
1980	3017[j]	2464	1710	1575	1533	1709	—	—
1985	3275	2528	1330	1200	1057	1289	1208	1548
1990	4115	—	1385	1250	959	1331	1053	1299
1995	—	—	—	1450	1032	1392	914	1471
2000	—	—	—	—	—	—	758	1436

Notes:

a. National Energy Board (1969: 55–7); figures estimated from total Canadian demand numbers on the basis of the ratio of oil products demand in the Maritimes and Quebec relative to the rest of Canada
b. National Energy Board (1972, Table 1)
c. National Energy Board (1974, App. 3–II and 2–III)
d. National Energy Board (1975b: 30, 87)
e. National Energy Board (1977a, App. G & L)
f. National Energy Board (1978, App. G & L)
g. National Energy Board (1981, App. I and p. 146); oil converted to barrels at 0.159 cubic metres per barrel
h. National Energy Board (1984, pp. A–160–1); oil converted to barrels at 0.159 cubic metres per barrel
i. Production in 1994
j. Unconstrained supply

Table 4. Forecasts of Total Primary Energy Demand in Canada for Year 2000 (petajoules)

year	ident	status	high	medium	low	average/ forecast
1973	a	est	0	29450	0	29540.0
1976	b	n.est	8000	0	6000	7000.0
1977	c	n.est	15800	0	5600	10700
1977	a	est	18254	0	11098	14671
1978	a	est	0	17100	0	17100.0
1979	a	est	0	14440	0	14440.0
1979	d	est	18013	0	17395	17704.0
1981	d	est	0	16176	0	16176.0
1982	a	est	0	13749	0	13749.0
1983	d	est	0	15115	0	15115.0
1983	a	est	0	15488	0	15488.0
1983	e	n.est	9957	0	7460	8708.5
1984	d	est	0	13397	0	13397.0

Key
a EMR: Energy, Mines and Resources Forecast
b EEF/Lovins: *Exploring Energy Efficient Futures*
c CEF/Robinson: *Canadian Energy Futures: An Investigation of Alternative*
d NEB: National Energy Board Forecasts
e FOESEP: Friends of the Earth, Canada BU: Business as Usual; CS: Consumer saturation

11

Energy-Policy Modelling in the US: Competing Societal Alternatives

Martin Greenberger and
William W. Hogan

INTRODUCTION

POLICY analysis recently reached its fiftieth birthday in the United States, having got started in earnest during the depression years of the 1930s (Greenberger *et al.*, 1976: 106). As an expanding art form, policy analysis has grown steadily in scope and volume. Modulating this growth have been intervals of rampant expansion, followed by periods of disillusionment and retrenchment, producing an effect not unlike the peaks and valleys of the business cycle. The pattern may be more pronounced in the United States than in other countries for several reasons.

Economic modelling has been institutionalized as a business in the United States. Several commercial firms offering economic modelling and forecasting services set up shop during the late 1960s (Greenberger *et al.*, 1976: 216–24). These firms are now in their mature phase, subject to the same ups and downs as other service and consulting firms. The slump they are currently experiencing may be due largely to the strong trend away from central time-sharing toward personal computing (*Business Week*, 1984: 123–4). It could also be the outcome of the dismal record of forecasts compiled by business economists in recent years (Clark and Malabre, 1984: 1 and 8). Or it may simply be the result of cost-cutting by customers responding to general economic conditions.

Also contributing to fluctuations in modelling is the fact that periods of national crisis and uncertainty tend to spark modelling, whereas periods of composure tend to dampen it. A complex, unfamiliar situation like rapidly rising energy prices and abruptly constrained energy supplies alarms officials and provokes the public. There is a search for explanations, whether rational or not, and a sense of urgency. The upshot is more studies, more modelling, and more scapegoats (Greenberger *et al.*, 1983: 1–7).

Highly charged situations carry large stakes in a plural democracy. The truth is that modelling elicited in such circumstances can be counter-productive. The need is for patient reasoning and diplomatic compromise, not modelling results that accentuate disagreement (Greenberger *et al.*, 1976: 333). It is a Catch-22 situation. When most in demand, models are least likely to be accepted or believed. Their success, on one measure, can be their downfall on another.

Aggravating the problem is the over-emphasis given to forecasts and point estimates. The real value of modelling is in the understanding and deepening of insights it provides, not in the numbers it produces (Hogan, 1978; Greenberger and Richels, 1979: 467–500). Yet the numbers are what attract attention. They are what journalists and policy-makers want. They become the material for headlines and the reasons for rejecting a study whose chain of reasoning might actually be useful. The effect is a weakening of credibility for analysis.

THE POLITICIZATION OF MODELLING

Modelling has become highly politicized in the United States. Politics impinges on policy analysis in two important ways. First, contending parties can appropriate even the most objective of studies for their own political use. In this way, a study may take on an unintended partisan cast, thus complicating the responsibilities of analysts and tending to undermine their neutrality.

It has been said there is a model or modeller for every point of view. To the extent this cynicism is justified, it may not actually be so bad in a plural democracy. But it does impose a special requirement on the vitality of the market-place of ideas and on the analytical processes of counter-modelling and critique (Greenberger *et al.*, 1983: 288–99).

The second way that politics can affect analysis is in widening the forms of investigation. Implicit assumptions, preliminary results, indeed the entire framework of analysis may undergo basic revision.

The seventies was an extraordinary period in the application of analysis to US policy problems. The uncertainty of world energy supplies during those years, the two unexpected jumps in crude-oil prices, and the changing prospects for nuclear energy gave rise to an unprecedented number of studies. Many of these studies became tools or victims of the severe political infighting that raged throughout much of the decade. Analysts took their share of the heat and professional reputations suffered. Yet there was a positive side. Policy analysis, as a profession, became politically wiser and considerably content-richer.

There arose understanding among disciplines where it had not previously existed. Charles J. Hitch expresses it well:

> What was accomplished by all the frenzied activity in energy studies in the 1970s? Substantively, I would say, a good deal. The participants . . . learned from each other and ended the decade in far greater agreement than they were in when the oil embargo was imposed in 1973. The chorus of dissidence . . . had largely subsided by the latter years of the decade . . . A special and most important case of the participants learning from each other involved the technologists and the economists . . . The energy studies of the 1970s, by demonstrating that both kinds of talent were needed, succeeded where earlier attempts had achieved only modest and mainly temporary results. Economists learned that their models and their advice were pretty empty without a lot of technological content. Engineers and scientists learned about the behavior, uses, and failures of markets; about elasticities of demand and supply and how potentially massive their effects can be; and about cost/benefit and cost/effectiveness concepts which can be useful operational tools in helping to choose among policies. . . . Economists and engineers have not only talked and worked together; each has started to think and talk a little like the other. Economists who have acquired a respectable knowledge of energy technologies are not likely to return to games played with empty boxes (Greenberger *et al.*, 1983: xvii–xix).

The third wave of analysis

The energy studies of the seventies were preceded by two prior waves of policy analysis in the United States. We do not count the uses of analysis during Franklin D. Roosevelt's terms of office (1933–45). They were not sustained or systematic. It was rather with the Kennedy Administration's vigorous pursuit of problem solving (1961–3), in our opinion, that policy analysis first came into its own.

The two main thrusts of this first wave were the use of economic analysis by the Council of Economic Advisers and the use of systems analysis by the Department of Defense. For economic analysis, it was the beginning of a steady growth in volume and visibility that continues today. For systems analysis, based on economics and operations research, it was the start of a programming function that came to span the gap between planning and budgeting in defence, a three-phase operation known as the 'planning-programming-budgeting system' or PPBS.

Use of PPBS blossomed in defence during the 1960s. Impressed, President Lyndon Johnson promoted a massive effort to introduce the methodology throughout the federal civilian agencies. But wholesale assaults made on the traditional ways of conducting government business were hasty and ill conceived. Meeting great resistance and suffering from its association with the unpopular war in Vietnam, PPBS waned in use most everywhere except in defence by the early 1970s.

The second wave of policy analysis arose in the sphere of social and urban problems, reflecting Johnson's emphasis on the Great Society. The second wave carried over from the first a systematic attention to policy alternatives (Rivlin, 1971). It added environmental impact statements and ambitious urban models. Extending from the late 1960s to the middle 1970s, its decline was marked by growing disillusionment with large social programmes and by the demise of the New York City-RAND Institute, an intrepid organizational experiment to apply policy analysis to the problems of a very complicated and politicized metropolis (Greenberger *et al.*, 1976: 231–317).

Once again, as one wave was coming to a close, another was already taking form. The third wave, the wave of energy studies, also incorporated traces of its predecessor. The earliest of the energy studies, for example, the Ford Energy Policy Project, was initiated largely as an environmental analysis (Greenberger *et al.*, 1983: 89–92). By 1980, the third wave was itself subsiding. As energy use fell back and the budget-minded Reagan Administration took office, there was more oil on the world market than interested buyers. Activity in the Department of Energy began to be phased down, and government spending was sharply curtailed on solar, conservation, and other energy programmes. It was the start of a dry season for energy research and policy.

The energy studies of the seventies

This most recent wave of analyses in the United States produced a profusion of energy studies. The studies, triggered by growing concern about the energy future, ran through the decade of the seventies. We examined fourteen of the most significant of these studies in a five-year investigation published in 1983 (Greenberger *et al.*, 1983: 89–92). We explored the meaning of the studies both to the understanding of energy problems and to the practice of policy analysis in general.

The fourteen studies, covering the principal policy issues debated during the 1970s, probably had a decided influence—directly or indirectly—on the form and content of the debate. The analyses might not always have been as incisive, helpful, or timely as one might have wished, but the fault was often more with how they were used than with what they did or did not say. We found the differences between content and use, and between objectives and effects, often striking as we considered the studies individually.

Table 11.1 lists the fourteen studies. Half of them were undertaken specifically at the behest of government; another four were conducted at the initiative or with the primary sponsorship of private foundations. Staffing of the studies was varied in nature, ranging from the use of analysts internal to the responsible organization, to the assembling of teams of researchers and expert panels from outside, to the employment of external

Table 11.1 Classification of the Fourteen Studies

	Publ. Year	Origin[a]	Staff[b]	Subject[c]	Level[d]
Ford Energy Policy Project	1974	F	C,T	D,E	1
Project Independence Report	1974	G	S	P	1
Rasmussen Reactor Safety Study	1975	G	T	N	1
ERDA–48 AND–76–1	1975/76	G	C,S	A	1
Synfuels Commercialization	1975	G	C	A	3
MOPPS	1977	G	C,S,T	A	1
Ford-MITRE Study	1977	F	X	E,N	1
Lovins's Analysis	1976	I	S	A,D	1
WAES Study	1977	I	T,X	W	2
CIA Assessment	1977	G	S	W	1
CONAES	1979	G	T,X	A,D,E,N	1
Stobaugh and Yergin	1979	U	T	A,D,W	2
RFF-Mellon Study	1979	F	S	A,D,E,P	3
Ford-RFF Study	1979	F	X	A,D,E,P	3

Notes

a. Origin. F: Foundation sponsored; G: Government commissioned; I: Individual initiative; U: University based.

b. Staffing. C: Consultants or contracts; S: Staff members or self; T: Team assembled temporarily; X: expert panel.

c. Subjects: A: Alternate technologies; D: Demand and conservation; E: Energy/Economy relationship; P: Price and markets; N: Nuclear power; W: World oil situation.

d. Level of politicization and controversy: 1. highest; 2. moderate; 3. lowest.

consultants and research contracts. Issues covered by the studies included conservation and demand, nuclear power, the prospects for alternative sources of energy, the relationship between energy and the economy, the workings of energy prices and markets, and the future of world oil.

As Table 11.1 indicates, no fewer than nine of the fourteen studies were highly controversial and politicized in their execution, reception, or use. The tone was set by the first, the clairvoyant Energy Policy Project of the Ford Foundation. Begun in 1972 in response to environmental concerns then at a peak (energy was not yet front-page news), the project released its final report in the autumn of 1974 as the public was reeling from the initial pangs of the energy crisis (Energy Policy Project of the Ford Foundation, 1974). The document was received with praise by environmentalists and conservationists and with condemnation by irate supporters of business-as-usual, who viewed the report writers as misguided advocates of restraint on economic growth. Subsequent events have been kind to the project. Its promotion of the need for moderation in energy use was prophetic. Its heresy became the new orthodoxy within three years (Freeman, 1983: 9–14).

In 1974, the United States was outraged at what it regarded as blackmail by Arab oil-exporting countries. President Nixon vowed to make the nation energy self-sufficient by 1980! To help plan the drive to independence, the White House approved a major in-house analysis by the newly formed Federal Energy Administration (FEA). The FEA effort, the second study we reviewed, developed and used the Project Independence Evaluation System (PIES), a large collection of data and computer models. The 1974 Project Independence Report, prepared with the help of PIES, pointedly disavowed Nixon's original self-sufficiency goals (Federal Energy Administration, 1974). The report focused instead on alternative strategies for dealing with the problem of energy security and drew attention to the effects of energy prices. The Republican administration employed PIES repeatedly over the next two years to justify and buttress White House positions on price deregulation and other issues, greatly irritating many Democratic members of Congress and consumer advocates in the process.

Nuclear power, the bright hope of the sixties, might have been expected to lead the march in the attack on energy problems. But public concerns about nuclear safety were already mounting by 1972. The Atomic Energy Commission (AEC), eliminated in a bureaucratic reorganization a few years later, was at the time responsible for overseeing the civilian nuclear energy programme. With the hope of easing safety concerns, the AEC created an ambitious research project to demystify the complex and emotional subject of risk in nuclear reactors. The project was placed under the direction of MIT nuclear engineer Norman Rasmussen. Released in final form in the fall of 1975, the project's highly controversial report—the third study we reviewed—was used extensively thereafter in campaigns to reassure the public (US Nuclear Regulatory Commission, 1975). Some believe it should have been used instead to reduce chances of an accident like that at Three Mile Island (Kemeny, 1979).

In its bureaucratic response to the energy crisis, the government created the Energy Research and Development Administration (ERDA) as a successor to the AEC. By statutory requirement, the new agency was under pressure to produce an R&D plan in the second quarter of 1975. Rushing to meet the deadline, drafters of the plan (known as ERDA-48) slighted conservation options that appeared promising in modelling runs financed by ERDA. The report's supply focus upset environmentalists and advocates of energy moderation (Energy Research and Development Administration, 1975). Stunned by the criticism, ERDA produced a second, more conservation-orientated and politically wise report (ERDA 76–1) the following year (US Congress, Office of Technology Assessment, 1976). These two reports constitute the fourth study we reviewed.

One supply alternative favoured by ERDA enjoyed strong political support within Congress and industry at the time. Although its cost was uncertain, synthetic fuel from domestic coal and oil shale could potentially provide a needed supplement to domestic resources in oil and gas. President Ford set out an exuberant Project-Independence type goal for synfuels in his 1975 State of the Union address. A White House task force was set up to examine the goal and recommend a realistic programme for Congressional approval. The cost–benefit analysis done for the task force by an outside consulting firm was our fifth study (Synfuels Interagency Task Force, 1975). Its modelling effort concluded that 'no programme' was the best option, but it was a more optimistic compromise recommendation that was forwarded to Congress.

ERDA went out of business in 1977, the year the incoming Carter Administration created the Energy Department that absorbed it. Just before making its exit, the agency organized its most comprehensive study yet, the Market Oriented Program Planning Study (MOPPS). MOPPS was a comprehensive modelling study with human actors. It was intended to winnow ERDA's list of increasingly expensive technological programmes and mediate among the agency's competing programme managers. It was not supposed to become the centre of a highly publicized conflict with the new administration over estimates of future gas supply; but this is what happened. MOPPS, our sixth study, was never officially released (Energy Research and Development Administration, 1977).

Nuclear anxiety soared to new heights in 1974 with the startling announcement that India had exploded an atomic bomb with plutonium from the spent fuel of a research reactor supplied by the Canadians. The hazards of plutonium production and the breeder programme quickly became urgent issues for those concerned about nuclear proliferation, including presidential candidate Jimmy Carter. Carter's views conflicted with those of certain American allies abroad and many members of industry at home. He was glad to have his positions reinforced by the Ford-MITRE nuclear energy policy study, a second effort sponsored by the Ford Foundation (Nuclear Energy Policy Study Group, 1977). Several contributors to the Ford-MITRE study joined the Carter team in key nuclear policy positions. This was our seventh study.

For understandable reasons, those who opposed nuclear development typically favoured conservation and renewable sources of energy. This school of thought was called the 'soft path' by the provocative Amory Lovins, *enfant terrible* to the nuclear and electric utility industries. Lovins was an articulate analyst/polemicist who preached widely and (as viewed by the establishment) radically to audiences in many countries. His ultra-low projections of energy consumption were considered extreme at the time, but not today. Lovins (quoting Niels Bohr who said, 'It is difficult to make predictions, especially about the future') had little use for forecasting models he considered elaborate extrapolations:

> Such models have trouble adapting to a world in which, for example, real electricity prices are rapidly rising rather than slowly falling as they used to . . . Extrapolations have fixed structure and no limits, whereas real societies and their objectives evolve structurally over decades and react to limits. Extrapolations have constants, but reality only has slow variables . . . Extrapolations assume essentially a surprise-free future even when written by and for people who spend their working lives coping

with surprises such as those of late 1973. Formal energy models can function only if stripped of surprises, but then they can say nothing useful about a world in which discontinuities and singularities matter more than the fragments of secular trend in between. Worst, extrapolations are remote from real policy questions. (Lovins, 1979: 64.)

Lovins's way of dealing with the future was an extension of the scenario method used in the Ford Energy Policy Project (EPP). He used the term 'scenario' in the specific sense adopted by the film industry to mean a description sufficiently vivid for the audience to be able to imagine itself participating in the events described. Lovins credited the EPP with having incorporated fundamental attitudes toward energy into its scenarios. To these he added attitudes about centrism, autarchy, vulnerability, and technocracy—the root issues, he charged, that establishment studies almost always neglect. By working backward from a desired future rather than forward into a gap—thus 'turning divergences into convergences'—and by being explicitly normative, Lovins felt he was able to obtain insights unavailable in traditional modelling exercises (Lovins, 1979: 65–6). The writings and arguments of Lovins formed our eighth study.

The 1970s brought the United States to the realization that domestic oil production would never again be able to meet domestic consumption. It was alarming to see much of the world's oil located in some of the most politically volatile and unfriendly regions of the globe. With many of the advanced economies considerably less energy self-sufficient than the United States, people worried about the problems ahead and wondered what could be done about them. How long would the free world's available oil hold out? Carroll Wilson of MIT brought together a group of people from the industrialized countries to address these concerns and pool national statistics. Wilson's international Workshop on Alternative Energy Strategies (WAES) was the ninth of our studies (WAES, 1977).

WAES assumed that the Soviet bloc, on net, would be neither importers nor exporters of oil. An assessment by the Central Intelligence Agency disagreed (CIA, 1977). CIA economists estimated that Communist countries in the aggregate could be transformed within eight years from a body exporting one million barrels of oil a day to one importing several times that amount. These dire predictions, originally presented to Carter in a classified briefing, were publicized by the president to underscore the need for his National Energy Plan. The political ploy backfired. The CIA study was our tenth.

The energy studies repeatedly illustrated the difference between outcomes and objectives—between the sponsor's initial intentions and the study's subsequent course. Nowhere was the contrast sharper than in the work of the Committee on Nuclear and Alternative Energy Systems (CONAES), set up by the National Academy of Sciences at the request of ERDA. Asked to focus on the breeder programme, CONAES ranged across almost the entire scope of the nation's energy problems, bringing together hundreds of experts of varied disciplinary backgrounds and ideological views. Over budget and ridden with strife, CONAES had a terrible time obtaining closure. Yet it provided a valuable training ground for very talented people and helped significantly in the rethinking of attitudes on energy supply, energy use, and their relationship to the economy (National Academy of Sciences, 1979).

CONAES was study eleven. One of its working committees, the hard-working Modelling Resources Group (MRG), devoted itself to analysing the results of several different models of the energy economy as they bore on central policy issues (National Academy of Sciences, 1978). The comparative modelling efforts of the MRG were parallel and complementary to those of the Energy Modelling Forum (EMF), established by the Electric Power Research Institute at Stanford University in 1976 (Sweeney and Weyant, 1981: 295–318).

Study twelve appeared just as gasoline lines were beginning to form once more in the aftermath of the Iranian revolution. Well written, skilfully communicated and marketed, and favoured by fortuity, the study report became a national bestseller to the great surprise of its co-editors, Robert Stobaugh and Daniel Yergin (Stobaugh and Yergin, 1979). In their advocacy of conservation, Stobaugh and Yergin echoed points made by the Ford Policy Project and Amory Lovins. In their espousal of solar energy, they again lined up with Lovins. But Stobaugh, leader of the project, was an expert on the oil industry and a professor at the Harvard Business School, an institution honoured by a community not known to be enthusiastic about either conservation, solar, or Mr Lovins. Although acceptance of the report's position came from a broader cross-section of the political spectrum than for any previous energy study, many remained unpersuaded. The call for conservation and solar—music to the ears of the soft-path constituency—did not avert a subsequent cut-back in solar support by the Reagan Administration. But it may have helped prevent complete emasculation of the conservation programmes.

Resources for the Future (RFF) is a research organization with a carefully cultivated reputation for scholarly and objective analysis. It has generally avoided taking sides in partisan political disputes. So it was no surprise to find an RFF energy study (funded by the Mellon Foundation) declining to propose any one policy plan that might be interpreted as a prescription for solving the energy problem. Instead, the study presented information and analysis for assessing a wide range of policy options. RFF-Mellon was our thirteenth study (Schurr *et al.*, 1979).

The last of our studies was the third and final energy project of the Ford Foundation. Headquartered at RFF, it was conducted as a panel of author-experts in the pattern of the earlier Ford-MITRE effort. It brought together much of the best thinking at the end of the decade (Landsberg, *et al.*, 1979).

In pursuing our investigation of these studies, we interviewed hundreds of energy policy analysts and policy-makers. We gave an attitude questionnaire to 150 of them to explore their perspectives on key energy issues and to probe the underlying nature of the energy debate. Two core viewpoints emerged: 'traditionalist' and 'reformist'. Traditionalists were growth-orientated. They favoured nuclear power and were sceptical about the near-term promise of solar energy. They were convinced that the marketplace could help solve the energy problem and they thought deregulation of oil and natural gas was the key to efficient allocation. Reformists, on the other hand, were very sensitive to environmental concerns and believed that environmental protection laws should be enforced vigorously. Reformists favoured a resource-conserving ethic. They were troubled by the profound implications of today's energy decisions for future generations and did not regard the American standard of living as dependent upon substantially more energy in the future. Reformists were against primary reliance on nuclear power and were for greater emphasis on renewable sources of energy such as solar and biomass. Not surprisingly, traditionalists and reformists had rather different opinions on the fourteen studies.

Report card on the studies

To obtain opinions on the studies in a systematic way, we designed a 'study' questionnaire that listed the studies by name in the order their final reports were released. For each, we asked about the *quality* of the analysis, the *attention* it attracted in the public media and government circles, and the *influence* it had (or was expected to have) on public policy considerations. Of the 150 people who took the attitude questionnaire, 135 also took the study questionnaire. This was sample I. To form a control group, we gave the study questionnaire to 54 additional policy analysts and policy-makers in the energy field. This was sample II. Both samples were taken early in 1980.

The ratings were averaged within each sample separately and converted to letter grades. The two samples were significantly different in organizational profile. For example, whereas 30 per cent of the respondents in sample I were government workers and 7 per cent were from industry, in sample II 17 per cent were from government and 17 per cent from industry. We expected the results from the two samples to have noticeable disparities. They did not. Although there were wide variations from person to person, the two sets of averages were very similar. We took the near identity of results as an indication that the ratings were reasonably representative of the impressions of energy experts at the time.

Table 11.2 Grades for the Fourteen Studies

	Quality			Attention			Influence		
	Tot	Tr	Re	Tot	Tr	Re	Tot	Tr	Re
Ford Energy Policy Project	C	D	A−	A	A	B	A	A	A−
Project Independence	D	C	E	B	B	B	C	C	D
Rasmussen Study	B	A	D	A	A	A	A	A	A
ERDA-48 and 76-1	D	D	E	D	D	C	D	D	D
Synfuels Commercialization	C	C	D	D	D	D	D	D	B
MOPPS	C	C	D	D	D	D	E	E	E
Ford-MITRE Study	B	B	B	C	C	D	A	A	A−
Lovins's Analysis	D	E	A−	A	A	A	A	A	A−
WAES Study	C	C	B	C	C	C	C	C	B
CIA Assessment	C	C	B	B	B	B	B	B	A
CONAES	B	B	C	C	C	C	C	D	D
Stobaugh and Yergin	C	D	A	A	A	A	A	A	A
RFF-Mellon Study	A	A	B−	D	D	D	D	D	E
Ford-RFF Study	A	A	A	D	D	D	C	D	C

The grades in the first column under each category are based on the results of samples I and II combined. Only in a small number of instances did these two sets of results show any significant differences. The second grade under each category is based on the average scores given by Traditionalists (Tr), the third grade by Reformists (Re). Grades run from a high of A to a low of E.

Reassured, we combined the results from the two samples to obtain a single set of letter grades for the fourteen studies. The grades, as set out in Table 11.2, suggest that attention and influence tend to be linked in people's minds. The only noticeable exception was the Ford-MITRE study, which scored low for attention but high for influence. The reason may be found in the fact that Ford-MITRE meshed well with Carter's position on nuclear energy. Several of the study's participants assumed important roles in the administration and were able to put into effect some of the study's main recommendations. On the other hand, the study did not receive much coverage in the press, and those seeking copies complained that the report was hard to come by.

Table 11.2 bears out a suspicion we had before sending out the questionnaire. Impressions of influence and quality were inversely related. In no case did a study scoring an A for quality also score an A for influence. The only two studies receiving an A for quality scored C and D for influence. Conversely, the five studies scoring A for influence obtained two Bs, two Cs, and one D for quality.

Ford-RFF was the top choice of both samples I and II in the quality dimension. It was the only study to receive an A by all occupational groups and was rated first by both traditionalists and reformists, ahead even of the studies most naturally akin to these two viewpoints. In second place on the quality dimension was RFF-Mellon. Again, its position was the same in the replies of both samples I and II. The ranking of the two RFF studies was not affected by removal of the replies from respondents associated with RFF.

With respect to attention received, Stobaugh–Yergin and Rasmussen shared top honours. Also high in this category were Lovins and Ford EPP. As for influence, it was a close race among five of the studies: Lovins, Ford EPP, Stobaugh–Yergin, Rasmussen, and Ford-MITRE. *Energy Future*, the Stobaugh–Yergin book, having just been out a few months at the time of the questionnaire, had already attracted a great deal of attention. People seemed to feel it was going to have a strong influence on government policy. (Perhaps it would have, had Carter been re-elected.)

The low quality ratings received by a few of the studies were unexpected and perplexing, but seemed to be influenced by the perspective of the respondent. Lowest in both samples I and II was Lovins; next lowest were ERDA and Project Independence. But the 45 traditionalists and 15 reformists among the 135 respondents in sample I rated Lovins and Project Independence very differently. Traditionalists gave Project Independence a middle ranking and Lovins a definite last. Reformists, on the other hand, placed Lovins fairly high and put Project Independence at the bottom. Both played down ERDA. Table 11.2 summarizes the grades also by viewpoint.

In last place for attention received was the Synfuels Commercialization study. It was the least known of the analyses. Next lowest were ERDA, MOPPS, and RFF-Mellon. These same four studies were also considered to have had the smallest influence—MOPPS, never officially released, the least of all.

We partitioned responses according to whether respondents came from government, universities, or industry. The largest occupational groupings in sample I were government (with forty respondents) and universities (with forty-one respondents). Government responses conformed quite closely to the overall results and reflected no noticeable biases. Somewhat more surprising, the same was true of the university group—an indication, perhaps, of the increasing secularization of academe.

Of the 189 people in the two samples, about half had been associated with at least one of the fourteen studies, over ten per cent with two or more of the studies, and about four per cent with as many as three of the studies. We were curious to see how study-group members felt about the studies in which they participated.

Participants in every study without exception clearly demonstrated pride in their work and, on average, scored their study higher in quality than did others. The same tendency (except for Ford-RFF) showed up in ratings of influence, but to a lesser degree. Self-evaluations of attention displayed a different pattern, with participants giving MOPPS and the two RFF studies no higher attention score than non-participants.

In general, respondents exhibited less divergence of opinion in rating attention than quality or influence. Attention is a more concrete concept. It can be measured by column inches in the *New York Times*, coverage on the evening news, references by colleagues, and requests for interviews. When the report is commercially published, the attention it receives can be measured by sales of the book. The attention scores correlated almost perfectly with sales (Greenberger *et al.*, 1983: 80). No such tangible indices exist for quality and influence.

PROJECTING ENERGY

The most significant change that took place in energy modelling in the 1970s was a pronounced shift from supply-side to demand-side analysis. This was accompanied by increased attention to the energy price variable, and by an attempt to integrate the supply and demand relationships. Gradually, the importance of price was conceded, and it was recognized, grudgingly by some, that alleviation of the energy problem would come more on the demand than supply side of the energy equation. A significant revision also occurred in the economic impact attributed to energy scarcity, and the focus changed from single fuels to energy balances.

The Ford Energy Policy Project started the debate in 1973–4 with its contention that strong measures to curtail spiralling energy use were necessary and would not injure the economy. Subsequent analyses such as the CONAES study continued to argue that healthy economic growth was possible without energy consumption continuing to grow at historic rates. Finally, several years after the first major increase in crude oil prices, just after decontrol of domestic oil prices in 1979, demand began to moderate. As it did, forecasts, which were already declining, started to turn down decisively, as seen in Figure 11.1. To one commentator, there arose a 'new trend in energy prophecy'—a 'scramble for low demand estimates' (Marshall, 1980).

The energy/GNP ratio

In the early 1970s, the energy/GNP ratio seemed relatively constant. People in the electric business and elsewhere in industry spoke of a 'lock-step relation' or 'linkage' between energy and GNP. Those who subscribed to this linkage notion firmly believed that continued economic growth implied and required corresponding growth in the consumption of energy. According to this line of reasoning, even a modestly rising GNP led to very high forecasts (by today's standards) of future US energy demand: 200 quadrillion (quads) Btus a year and more by the year 2000. For comparison, US energy use in 1981 was about 75 quads.

The growth rate in real GNP of 4.3 per cent between 1960 and 1969 was still in people's minds at the beginning of the seventies. With the energy/GNP ratio fixed, a 4.3 per cent growth rate would transform a 1971 energy demand of 69 quads into 234 quads at the turn of the century. Some analyses went further. The much publicized *Limits To Growth* study in 1972, for example, was thought to have projected an energy consumption of 300 quads in 2000. This led to a warning that society was in danger of exhausting its resources and suffering economic collapse, if it did not asphyxiate itself first with its own pollution (Meadows *et al.*, 1972).

Not only is the simple direct proportion between energy consumption and national income not an immutable scientific law, it is not even a good approximation. In its 1972 forecast of 192 quads in 2000, the Bureau of Mines assumed a GNP growth rate of over 4 per cent.[1] But it also projected a long-term decline in the energy/GNP ratio, corresponding to a downward trend observed historically (Dupree and West, 1972: 12–13). Like the *Limits to Growth* study, the bureau's analysis did not seem to give much weight to the possibility of rising prices (although it did have the cost of fuel growing faster than that of other commodities).

The Bureau of Mines' original projection base can be lowered from 192 to 85 quads through successive stages of 151, 125, and 107 quads by adding income and price effects and altering the GNP and price assumptions (Greenberger, 1981: 1–24). This parallels the lines along which thinking and perceptions actually evolved. Indeed, people today believe the United States may be using well under 85 quads of energy by the end of the century.

The price elasticity

Most of the energy studies took it for granted that the GNP growth rate was a primary determinant of energy demand. If it was the *only* determinant, demand could basically be extrapolated. This approach worked reasonably well during the 1950s and 1960s when there were no abrupt discontinuities in energy prices, but not after 1973. Increasingly, it became apparent that energy models must allow for price-induced conservation through specific design changes, such as insulation in homes, heat exchangers and cogneration in industry, and diesels in cars instead of internal combustion engines. These changes became cost-effective as energy became sufficiently expensive.

Economists employed the price elasticity of demand to summarize conservation possibilities. To estimate price elasticity, they assumed that GNP and other factors influencing energy demand were unaffected by energy price increases. There are wide variations in the numerical value used. Within the CONAES Modelling Resource Group (MRG) alone, price elasticities ranged from virtually zero to 0.50. Some price elasticities were estimated from time series data collected on the United States as a whole, others from regional and international cross-sections of data, and others from pooled combinations of both.[2]

We should not be surprised, therefore, to find a wide range of demand estimates emerging from these studies. Some of the variation is due to differences in assumptions about GNP growth, mandatory conservation, or energy prices, but most of it seems attributable to differences in the price elasticity of demand for energy. In the Ford-MITRE and MRG studies, the price elasticities were evaluated explicitly. For the rest, they were implicit, and one can only surmise qualitatively whether they were high or low.

In Figure 11.1, note that the three alternative scenarios of the Ford Energy Policy Project (historical growth, technical fix, and zero energy growth) cover a sufficiently wide range so that they bracket the estimates of demand for primary energy in the year 2000 by almost all of the subsequent studies. It is interesting to compare the date of release of a study with the size of the year 2000 projections it produced. Several studies not counted among the fourteen we reviewed are included in Figure 11.1. There is a clear downward drift. This is the result of not only mounting energy prices, lowered expectations for GNP growth, and observed moderations in energy consumption, but also of the influence of one study on another.

Non-government studies like those of the Ford Energy Policy Project, Lovins, the Institute of Energy Analysis, and the CONAES Demand Panel led the way down. Government projections followed quickly. The result was a fall in ten years from a plateau of almost 200 quads to one less than half as high.

The Ford Energy Policy Project was the vanguard. It employed the Hudson–Jorgenson model in an attempt to capture consumer responsiveness to higher energy prices and effectively buttress conclusions already reached based on technical feasibility, mandatory conservation, and public regulation:

> Our calculations assume that in the future, users in the industrial sector will be more aware of energy costs (and therefore more responsive to using energy in an economically efficient way), and that the market imperfections that inhibit investments to save energy in the residential, commercial, and transportation sectors can be removed by specific government actions . . . We believe that if the nation adopts energy conservation as a goal . . . we can . . . alleviate concerns about supply, environment, and foreign policy without appreciable energy price increases. (Energy Policy Project of the Ford Foundation, 1974, ch. 6, App. F)

While acknowledging the potential usefulness of market mechanisms, the Ford project argued that market imperfections created a need for new regulatory institutions. Political conservatives saw this as a proposal for expanding the role of government and rejected it out-of-hand. By way of contrast, the ideologically middle-of-the-road Ford-MITRE and MRG studies showed little inclination to advocate non-price routes to energy

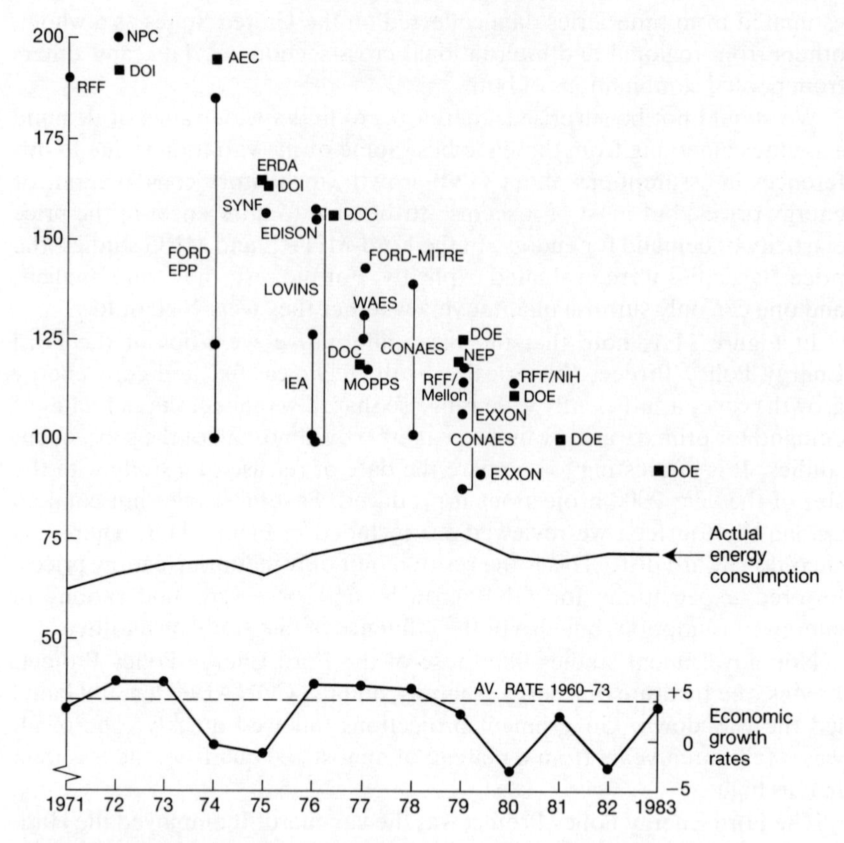

Fig. 11.1 United States
Energy forecasts for year 2000 (in quad. Btu)

RFF: Resources for the Future
DOI: Department of the Interior
NPC: National Petroleum Council
FORD EPP: Ford Foundation, Energy Policy Project
AEC: Atomic Energy Commission
ERDA: Energy Research and Development Administration
SYNF: Synfuels Commission
LOVINS: Amory Lovins
DOC: Department of Commerce
EDISON: Edison Electric Institute
IEA: Institute for Energy Analysis
WAES: Workshop on Alternative Energy Strategies
MOPPS: Market Oriented Program Planning Study
CONAES: Committee on Nuclear and Alternative Energy Systems
RFF/Mellon: Resources for the Future, founded by the Mellon Foundation
DOE/EIA: Department of Energy/Energy Information Administration
EXXON: Exxon Oil Company
NEP: National Energy Plan
FORD-MITRE: Nuclear Energy Policy Study sponsored by Ford Foundation
RFF/NIH: Third Energy Project of the Ford Foundation, headquartered at the Resources for the Future

National Development of Energy Forecasts for Year 2000

conservation, except to suggest shifting consumption away from manufactured goods toward services. They did not postulate any widespread adoption of an energy-conserving ethic as did the FORD EPP.

Energy-economy interactions

A framework in which the price of energy and the rate of GNP growth enter as independent variables is often described as a partial equilibrium model. In such a description, the energy sector is assumed to have only minor impact upon the long-run rate of GNP growth. A general equilibrium model, by allowing for indirect impacts of the energy sector on GNP, requires more data input.

None of the fourteen studies reviewed was designed to analyse the short-term effects of energy shortages on employment, inflation, or balance-of-payments. Except for the CIA study, all were intended to provide insights into the more distant future under the assumption of gradual and foreseeable energy price increases. Only the Hudson–Jorgenson model allowed for feedbacks from the energy sector to the balance of the US economy. Hudson and Jorgenson concluded that energy conservation was possible within the existing structure of the economy, and the results of reduced energy use in higher inflation and reduced real incomes and output were significant but not catastrophic (Energy Policy Project of the Ford Foundation, 1974: 49).

The other studies assumed that general equilibrium effects were negligible and that GNP growth would be virtually unaffected by energy scarcities. They did not undertake a detailed analysis of the assumption that energy demands and GNP growth could be decoupled. Even the MRG only made rough side calculations and approximations. Yet, many CONAES participants were under the impression that the MRG performed an explicit analysis of energy–economy decoupling. In fact, the MRG could not have performed such an analysis. The Hudson–Jorgenson model was virtually the only tool available in the United States for this purpose at the time (1976), and it was not among the models represented in the MRG.

After 1976, the Energy Modelling Forum conducted intermodel comparisons of energy–economy feedback effects using six models, including both the Hudson–Jorgenson model by itself and the equilibrium framework created by linking it to the Brookhaven model (Hogan *et al.*, 1978). This was said to be the first time that a macro-economic growth model had been linked to an interindustry sectoral model and subsequently linked to an energy technology-orientated resource-allocation model.

By mid-1977, Charles J. Hitch, then president of Resources for the Future, was able to organize a symposium to compare several alternative approaches to modelling energy–economy interactions (Hitch, 1977). By that point, a considerable degree of professional consensus had emerged.

The general equilibrium studies reached conclusions consistent with the position of the MRG that the feedback effects would be small unless the energy curtailment was large and the price elasticity substantially below 0.5. Thus did the old heresy of 1974 become the new orthodoxy.

CONCLUSION

Underlying the energy modelling of the seventies in the USA were implicit values associated with broad social goals. Several energy studies made these goals explicit in their efforts to establish criteria for ranking options. For the Ford EPP, it was saving the environment and reducing energy use. For Project Independence, in the wake of the oil embargo, it was energy independence through the reduction of oil imports. For the Synfuels Commercialization study, it was a classical cost–benefit framework.

To understand the outcome of the studies and the differences among them, we must first observe their predisposition and focus. The EPP was led by a lawyer with a background in regulation. It emphasized the role of government intervention. ERDA analysts were working in an agency populated with engineers. They viewed energy policy in technological terms and searched for the most efficient means of increasing supply. Project Independence economists were surprised by the sudden rise in energy prices. They fixed on economic issues: pricing and consumer response.

Meanwhile, CIA analysts schooled in the intelligence tradition saw energy problems in terms of institutions and international power. WAES measured world supply–demand balances. The Synfuels Commercialization study formulated event trees laced with compound probabilities. Rasmussen's risk analysis of reactor safety took a similar approach. Ford-MITRE had a disarmament orientation. Stobaugh–Yergin stressed opportunities for conserving energy. CONAES and MOPPS brought interest groups with obviously divergent views into their analyses. Ford-RFF and RFF-Mellon had strong economic bents and placed much weight on issues of techno-economic efficiency and environmental impact. Amory Lovins, separating himself from the other studies, questioned the basic social structure he charged they implicitly assumed.

Most important were not the forms of the analyses as much as the questions they asked and the unspoken restrictions they imposed on possible answers.

In some of the studies, the evalution of policy alternatives was the main activity and consumed much of the resources. The FEA organized its entire analytical effort around models and data-collection procedures for evaluating alternative equilibrium market solutions. Analysts compared large numbers of alternative paths for expanding supply or reducing demand—expressed within model equations solved by speedy computer algorithms—and examined thousands of possible energy flows among hundreds of regions and fuel types. In these simulations, modification in the construction schedule for the Alaskan oil pipeline would cause adjustments to estimated electricity prices in New England, and changes in environmental standards in New York would alter the projected demand for Colorado coal.

The results of some of the early Project Independence runs seemed counter-intuitive. They suggested that more stringent 'new source performance standards' for burning coal would increase the total level of sulphur pollution. Re-examining the model solutions, analysts found that the tighter standards on new sources made older plants more attractive economically, thereby delaying their retirement. Since the older plants were greater polluters, the logic was indeed correct.

Policy evaluation did not always lead to policy recommendations. The Rasmussen study estimated the likelihood of a variety of failure events occurring in a nuclear reactor and characterized the nature of these failures, but it did not propose means for their avoidance. Project Independence analysts held back from suggesting action after describing the implications of a range of government regulations and initiatives. Neither did the CIA study make any explicit recommendations, although its gloomy picture of the future if the world failed to reduce the demand for oil had obvious policy implications, and was so used. A similar message came from WAES. The RFF-Mellon team pointedly abstained from proposing policy solutions.

Yet many of the studies did make policy recommendations, and members of the studies sought to publicize these recommendations through press conferences, talks, and testimony to Congress. Examples of public advocacy are found in the EPP, Ford-MITRE, Lovins, and Stobaugh–Yergin studies, among others. Members of the CONAES, Ford-MITRE, and EPP studies would serve afterwards in senior government positions where they were able to execute policy initiatives and help implement the results of their analyses.

Professors of policy analysis would like to tell their students that an analysis done professionally attracts attention and gets used. We found, instead, that the highest quality analyses received least attention and exerted the lowest immediate influence on policy.

If quality does not explain acceptance, what does? Timing, for one thing. The timing of the Ford EPP could not have been better. With its concept of 'Zero Energy Growth', the project was prescient. (Even its estimate of 1980 energy consumption turned out to exceed actual use by more than 20 per cent (Hogan, 1983: 5).) The project's emphasis on conservation and its use of scenarios exerted an influence on many subsequent studies, despite the highly controversial nature of its conclusions at the time.

Policy studies can be very useful to a young administration with few public positions, vested interests, or political investments. In 1977, the incoming Carter Administration embraced the Ford-MITRE report, the CIA analysis, and the ideas of Amory Lovins. Findings coming out of CONAES, particularly on conservation, also got a hearing.

Analysts and politicians march to different drummers. By the time an analysis is complete, political attention may have turned to matters other than those that seemed pressing when the analysis began. An analysis not ready in time to serve the purpose for which it was intended may lose its audience.

In 1975, Norman Rasmussen was buried under an avalanche of responses to the draft version of his Reactor Safety Study which he had circulated for review the previous year. Nuclear safety was a central issue in congressional consideration of the proposed renewal of the Price–Anderson amendment which was about to get underway. Rasmussen succeeded in getting the final version of his report released just in time to be included. He was later criticized for responding inadequately to peer criticism; but his study got used.

Also important in determining acceptance is whether the results of an analysis accord or clash with the beliefs of those in power. Ford-MITRE was in harmony with the Carter Administration's position. MOPPS collided with it. The president warmly received the Ford-MITRE report and presented it proudly to the Japanese prime minister. The MOPPS report was recalled and studiously ignored.

Research is more likely to gain approval and be put to use when presented in a concise, comprehensible format than in a voluminous, highly technical tome. The complex and massive Rasmussen study, which its critics called 'inscrutable', was never applied operationally. Its popularized executive summary, on the other hand, was widely cited.

It has been said that policy analysts do best to stay away from wholesale changes in social values if they are to have an impact. Though Amory Lovins would disagree that his ideas required a revamping of social mores, he could be considered the exception that proves the rule. Analysts generally do assume the status quo and tend to concentrate on matters where consensus seems feasible.

There was no clear division of labour during the seventies between the analyses sponsored by foundations—done through universities and private research groups—and that undertaken by government agencies, except for the fact that the latter efforts tended to be mission-orientated and were often used to support the views of the current Administration. Examples were Rasmussen and CIA (but not Synfuels or MOPPS).

This does not mean that the foundation-sponsored analyses did not have an agenda. As we have seen, the first two Ford studies clearly did, but it was their own agenda, not necessarily that of the government. As it turned out, EPP was counter to conventional thinking, whereas Ford-MITRE became a resource for the new administration.

There never was a period of true national agreement on energy in the USA during the seventies. None of the energy studies, singly or in combination, whatever their virtues, inspired a real consensus, despite a felt urgency within the country to reach one. Energy problems, 'marked by a high degree of uncertainty and large political or economic stakes' were an example of the 'kinds of political circumstances in which policy modelling has risen to prominence: the big, unprecedented issues, the crisis in public policy that tend to create a market for modelling'. A paradox of the democratic society is that these are just the conditions under which analysis is most likely to be disputed or ignored. Our investigation into the energy studies of the seventies reconfirmed this lesson from earlier research (Greenberger *et al.*, 1976: 331, 333).

The differences between policy analysis in Europe and the United States are found in the dissimilar roles of government and the traditions of private research and individual initiative. There is less government dominance of policy analysis in the United States and a looser coupling with the many independently established research organizations. There is the beginnings of a policy analysis industry and an unusually active market-place for models and ideas.

The upshot is what is probably a much untidier methodological development in the United States than in other countries. Trend extrapolation, economic modelling, scenario analysis, and judgement calls all proceeded simultaneously during the energy modelling days of the 1970s, even within the same study. Nevertheless, methodological patterns may be discerned. There was a definite movement through successive studies away from simple extrapolation towards more complicated models, and away from an exclusive supply-side preoccupation with single fuels towards attention to supply–demand interactions and energy balances.

These shifts parallel what occurred in Europe. The path may have been different, but the results were similar.

One of the major functions of the energy studies—a function we believe they served both well and not well enough—was to bring questioning and logic to the discussion of energy problems. The studies exposed mismatches of policy and fact and helped to correct misunderstanding of the relationships governing energy use and energy markets. Attitudes on the issues gradually changed over the course of the studies, but not without dissension and loud protest.

The subject matter of the fourteen studies we examined span the range of all the principal energy questions debated in the USA during the 1970s. Directly or indirectly, the studies influenced how these questions were framed, argued, and at times resolved. The evidence is that despite the controversy and misinterpretation they sometimes engendered, the studies as a group had an overall salutary effect in clarifying the energy problem. Analysis served as a healthy antidote to the nigh irresistible urge to look for whipping boys and chase witches during a time of national crisis.

Notes

1. The Bureau of Mines' two reports entitled, 'United States Energy through the Year 2000', were influential in early discussions of the energy crisis. The first, by Walter G. Dupree, jun. and James A. West in December 1972, a year before the Arab oil embargo, predicted gross energy inputs increasing from 69 quads (quadrillion Btus) in 1971 to 96 quads in 1980 and 192 quads in 2000, with 80 quads in the year 2000 attributed to electrical generation. Although assuming rising real fuel prices and a declining energy/GNP ratio, the consumption figure was already 20 per cent too high in 1980 and seems headed for much wider discrepancy by 2000. In 1975, the Bureau of Mines, in a revised report by Dupree *et al.*, scaled its estimates downward to 163 quads of total demand in 2000, still well above what now seems probable.
2. For one attempt to compare these alternative approaches, see EMF4 Working Group (1981).

Bibliography

Business Week (1984), 'How Personal Computers Are Changing the Forecaster's Job', *Business Week*, 1 Oct: 123–4.

CIA (1977), 'The International Energy Situation: Outlook to 1985', ER 77–10240, Washington, DC, April.

Clark, L. H., jun. and A. L. Malabre, jun. (1984), 'Business Forecasters Find Demand Is Weak in Their Own Business', *Wall Street Journal*, 7 Sept.

Dupree, G. W., jun. *et al.* (1975), 'United States Energy through the Year 2000', Revised report. Bureau of Mines.

—— and J. A. West, (1972), 'United States Energy through the Year 2000', Bureau of Mines, Dec.

EMF 4 Working Group, Energy Modelling Forum (1981), 'Aggregate Elasticity of Energy Demand'. *The Energy Journal*, 2(2), Nov.

Energy Policy Project of the Ford Foundation (1974), *A Time to Choose: America's Energy Future*. Ballinger Publishing Company, Cambridge, Massachusetts.

Energy Research and Developmment Administration (1975), *A National Plan for Energy Research, Development, and Demonstration*, Government Printing Office, Washington, DC, two vols.

—— (1977), 'Market-Oriented Program Planning Study' (MOPPS), Sept. (unpublished).

Federal Energy Administration (1974), *Project Independence: A Summary*. Government Printing Office, Washington, DC, Nov.

Freeman, D. S. (1983), 'Still a Time to Choose . . . Ten Years Later'. *The Energy Journal*, 4(2).

Greenberger, M. (1981), Hearings Before the Subcommittee on Investigations and Oversight of the Committee on Science and Technology, US House of Representatives, Ninety Seventh Congress, First Session, 1–2 June, No. 14, US Government Printing Office, Washington, DC.

—— and R. Richels (1979), 'Assessing Energy Policy Models: Current State and Future Directions'. *Annual Review of Energy*, 4.

——, M. A. Crenson, and B. L. Crissey (1976), *Models in the Policy Process*. Russell Sage Foundation, New York.

——, G. D. Brewer, W. W. Hogan, and M. Russell (1983), *Caught Unawares: The Energy Decade in Retrospect*. Ballinger Publishing Company, Cambridge, Mass.

Hitch, C. J. (ed.) (1977), *Modelling Energy-Economy Interactions: Five Approaches*. Resources for the Future, Washington, DC, Sept.

Hogan, W. W. (1978), 'Energy Models: Building Understanding for Better Use'. Paper presented at the Second Lawrence Symposium on Systems and Decision Sciences, Berkeley, Calif. Oct.

—— (1983), 'Patterns of Energy Use'. Brookings Conference on Institutional Barriers to Energy Conservation, Washington, DC, 6–7 Oct.

——, D. Jorgenson, A. S. Manne *et al.* (1978), *Energy and the Economy*. EMF1 Final Report, Energy Modelling Forum, Stanford University.

Kemeny, J. G. (chairman) (1979), 'Report of the President's Commission on the Accident at Three Mile Island'. Government Printing Office, Washington, DC.

Landsberg, H. H. *et al.* (1979), *Energy: The Next Twenty Years*. Ballinger Publishing Company, Cambridge, Massachusetts.

Lovins, A. B. (1979), *Soft Energy Paths: Toward a Durable Peace*. Harper Colophon Books, New York.

Marshall, E. (1980), 'Energy Forecasts: Sinking to New Lows'. *Science*, 208, 20 June: 1356.

Meadows, D. H., D. L. Meadows, J. Randers, and W. W. Behrens (1972), *Limits to Growth*. Potomac Associates, Washington, DC.

National Academy of Sciences CONAES MRG (1978), *Energy Modelling for an Uncertain Future: The Report of the Modelling Resource Group*. National Academy of Sciences, Washington, DC.

National Academy of Sciences (NAS) (1979), *Energy in Transition: 1985–2010*, Final Report of the Committee on Nuclear and Alternative Energy Systems (CONAES), National Research Council, NAS, Washington, DC.

Nuclear Energy Policy Study (NEPS) Group (1977), *Nuclear Power: Issues and Choices*. Ballinger Publishing Company, Cambridge, Massachusetts.

Rivlin, A. (1971), *Systematic Thinking for Social Action*. Brookings Institution, Washington, DC.

Schurr, S. H. *et al.* (1979), *Energy in America's Future: The Choices Before Us*. Johns Hopkins Press, Baltimore, Maryland.

Stobaugh R. and D. Yergin (eds.) (1979), *Energy Future: Report of the Energy Project at the Harvard Business School*. Random House, New York.

Sweeney, J. L. and J. P. Weyant (1981), 'The Energy Modelling Forum: Past, Present, and Future', in I. Kavrakoglu (ed.), *Mathematical Modelling of Energy Systems*. Sijthoff and Noordhoff, Rockville (Md.), pp. 427–55.

Synfuels Interagency Task Force (1975), *Recommendations for a Synthetic Fuels Commercialization Program*, vol. ii: 'Cost/Benefit Analysis of Alternative Production Levels'. Government Printing Office, Washington, DC, Nov.

US Congress, Office of Technology Assessment (1976), *Comparative Analysis of the 1976 ERDA Plan and Program*. Government Printing Office, Washington, DC, May.

US Nuclear Regulatory Commission (1975), 'Reacter Safety Study: An Assessment of Accident Risks in U.S. Commercial Nuclear Power Plants'. WASH–1400, NUREG 75/014, Oct.

WAES (1977), *Energy: Global Prospects 1985–2000*. Report of the Workshop on Global Energy Strategies. McGraw Hill, New York.

PART VI

Comparative Notes on Politics and Methodology

12

Energy Forecasting and Political Structure: Some Comparative Notes*

Thomas Baumgartner and Atle Midttun

INTRODUCTION

THE development of energy forecasting in the 1970s and early 1980s showed a politicization of the forecasting activity. Political conflict and institutional confrontation came about as actors who defended their vested energy-interests were opposed by groups that were more concerned about the ecological impact of energy futures, and the logistical feasibility of mega-projects and exponential growth. The confrontation was not only between conflicting interests, but in many cases also between different values, paradigms, or cultures. These different underlying perspectives in many cases led to distinctions in forecasting approaches, methods, and models. In this process modelling became a part of the political process and politics became part of modelling, breaking down the traditional boundary between science and politics.

Our case studies illustrate how confrontation over energy forecasting took place on many levels of societal planning and decision-making. The first section of this chapter gives a summary of these developments. The purpose of this section is to give an overview of the variety of channels and ways of politicizing energy forecasting and of using energy forecasts for political and institutional purposes. The second section of this chapter explores the differences between our national cases in terms of political and institutional characteristics of national energy forecasting and the consequences for forecasting results. In this section we move beyond the case studies and also include a more general discussion of national decision-making and energy forecasting.

POLITICAL AND INSTITUTIONAL ASPECTS OF ENERGY FORECASTING

In most of our cases, forecasting before the 1974 energy crisis provided an apparently scientific tool to support 'technocratic' decision-making in democratic societies. But the scientific legitimacy of modelling and forecasting was undermined as counter-experts and political oppositions challenged the established forecasting monopolies of government and fuel industries and introduced competing forecasting paradigms and methodologies throughout the 1970s and early 1980s. Our cases illustrate this transformation on several levels of the planning and decision-making process.

The West German, British, and Dutch cases all show how challenges from outside oppositions affected political and administrative decision-making. Although in different ways and to a different extent, these cases illustrate how ecologically orientated actors were able to manœuvre into bargaining positions with the political and administrative energy establishment. This, in turn set the stage for both methodological and institutional innovation in energy-forecasting.

The Danish and Norwegian cases indicate the importance of organizational factors in energy forecasting. In both cases the organizational restructuring of the energy field also led to methodological changes in energy forecasting, and to substantial revisions of forecasting results. Because such factors as organizational goals, professional structure, and inter-organizational networks vary, organizations are likely to emphasize different methodologies and data in their predictions of the future.

Our case of French electricity forecasting gives a good illustration of technocratic/élitist and self-fulfilling aspects of energy-forecasting. The technocratic/élitist character is produced by the efficient administrative centralism of the French state planning system as well as by the integration achieved by the common background of state planners as graduates from the élite engineering schools. The self-fulfilling character of French electricity forecasts came about by committing Electricité de France both to energy forecasting and to expansive nuclear production together with a commercial strategy to get rid of the energy surplus.

Both the US and Canadian cases give examples of energy forecasting in political contexts of weaker central state influence than most of our European cases. The impact of independent research groups, private foundations, and universities on energy forecasting and the loose coupling between forecasting and political decision-making in the US produced a variety of forecasts and forecasting methodologies at an early stage. Energy studies and modelling in the US were a laboratory for academic virtuosity rather than a playground for explicit political values and interests. Even though a number of studies influenced government thinking, the ties to the political system were too unstable for decisive political control over energy forecasting.

Our Canadian case, on the other hand, exemplifies the problems of forecasting in a federal system with strong regional interests. Federal energy modelling remained limited essentially to oil and gas issues involving international trade. Electricity/hydropower is considered to be the responsibility of regional bodies. The interplay between federal forecasting, policy, and private market reactions to such policies in Canada gives a good example of how forecasting, policy-making, and implementation of energy policy can produce self-fulfilling and self-denying prophecies.

The IIASA case illustrates how complex modelling may represent a subtle politicization of energy forecasting at a fundamental, methodological level. The complexity of IIASA's modelling, with approximately 1,600 constraint variables and 2,600 activity variables became so overwhelming and impenetrable that it diverted attention from the highly questionable and politically loaded input variables—variables that are demonstrated by Keepin and Wynne to have been crucial in determining the modelling output. Only minor changes in these variables have been shown to have major effects on the forecasts.

The challenge from outside 'opposition'

West Germany, Britain, and the Netherlands all faced considerable challenge from outside the regular political system. From the unchallenged high-growth energy forecasts of the early 1970s, these countries all moved into a situation with a number of competing low-growth forecasts, promoted by ecological groups outside the regular political scene and the state apparatus.

The Enquête Commission in West Germany, the Windscale Inquiry in Britain, and the Maatschappelijke Discussie in the Netherlands all marked turning points in national energy forecasting. The politicization of energy decisions, especially nuclear decisions, and the emergence of ecologically orientated movements had created a general distrust of official energy policies and the forecasts behind them, threatening to a greater or lesser extent the legitimacy of the national decision-making systems.

In West Germany, the repeated attacks on nuclear projects in the administrative courts effectively blocked nuclear expansion and eventually led to the formation of a parliamentary committee, the Enquête Commission, to solve the conflict. Faced with highly divergent interests and cognitive frameworks, the commission had to take up some of the deeper justifications for the nuclear programme and thus ended up with a review of forecasting, engaging in an energy scenario analysis of its own.

In Britain, the public inquiry over the construction of a nuclear reprocessing plant at Windscale (Sellafield) gave rise to a broader debate on alternative energy futures. Although the latter was strictly speaking outside the mandate of this type of inquiry, the strength of the ecological movement and the support of a friendly energy minister helped to expand the scope of the inquiry and to initiate subsequently a number of alternative energy forecasts.

The Dutch Maatschappelijke Discussie (Public Debate) was inspired by the British Windscale inquiry, and was also triggered by nuclear projects. However, in line with Dutch political tradition, this debate was better incorporated into the regular political system than in Britain, and involved consensus-making negotiations between established political élites and the alternative movement.

These alternative forecasting exercises resulted in a set of low energy forecasts, initially far below those produced by government and industry. However, the latter were gradually adjusted downwards as the economic crisis continued and the political negotiations between growth-orientated and zero-growth political milieux proceeded.

But there also exist differences between the positions of the alternative scenarios in the three countries. In West Germany, the ecological movement maintained an independent modelling capacity, and was therefore able to produce scenarios even below the low scenario of the Enquête Commission.

The debate in Britain led to a forecast convergence. But this was more apparent than real. Leach's original low energy future and the alternative scenarios were based on policy intervention. The latest CEGB scenarios with their wide range mostly reflect uncertainty over future economic growth rates, not policy alternatives. They imply that a resumption of economic growth would quickly reveal that energy consumption growth would again follow. The Friends of the Earth scenarios have now been eliminated from public discussion through the judgement of official experts, who see them as technically and socially irresponsible.

In the Netherlands, the energy forecasting establishment and the ecological opposition had much closer contact and reached a compromise through political negotiation. This compromise led to the ecological forecast being included as a low scenario in the official energy prognosis; but the ecologists had to pay the price of using official economic data in order to gain this privilege.

The 'alternative' energy forecasts had interesting consequences both in methodological and political terms. Methodologically, many of them challenged the traditional objective–scientific paradigm and introduced an explicit value-based type of forecasting. Some of them also introduced new frameworks such as the bottom-up and backcasting approaches, which allowed for a wider range of possible alternatives than traditional forecasting.

Politically, this development in forecasting helped to expand the energy debate from the question of siting and the development of single projects to the general social and economic context for such decisions. By invading the field of forecasting, the alternative movements gained access to a strategic political tool in a scientifically orientated society. However, by emphasizing the inherently normative and value-laden nature of forecasting, this invasion also many times delegitimized forecasting as a 'scientific' activity and thus undermined its use in political bargaining. Sometimes however it merely widened the earlier agreed framework of analysis— for example to give greater weight to structural and technological change.

Forecasting as intra-governmental negotiation

The political negotiations over forecasts of energy futures in Denmark and Norway were never as open as in Britain, West Germany, and the Netherlands. Long social-democratic traditions in state administration and lack of knowledge and resources outside the state apparatus to develop alternative forecasting models may be at least part of the explanation as to why conflicts over forecasting were to a large extent internalized in the state apparatus.

In Norway, the competition between the Watercourse and Electricity Board (NVE) and the Ministry of Environment (MD) allied with the Central Bureau of Statistics (SSB) was a fight over methodology, output, and professional paradigms. The supply-orientated NVE had traditionally projected very expansionist energy futures based on simple trend extrapolations and wishful thinking from regional utilities and the electro-metallurgical industry. The generation and flow of information was therefore under the exclusive control of actors within the electricity supply system. This was used strategically in controversial energy decisions as a part of intra-governmental struggles, notably between NVE and MD. It was also used strategically *vis-à-vis* political decision-making bodies and the public at large.

The engagement of the SSB in energy forecasting, with its traditional macro-economic modelling, resulted from the new competition over resource management created by the establishment of MD in 1972. For the MD, lower energy consumption fitted in with central goals in other parts of its field of responsibility. In a situation with heavily subsidized electricity prices, an application of economic investment criteria would lead to reduced expansion. The legitimacy of economic science, and the possibility of an alliance with the Ministry of Finance, made it easier for the MD to achieve its goals through economic arguments than through, for example, ecological ones. MD was here to a large extent the spokesman for the environmental movement. However, the movement would surely wish to go considerably further in establishing an energy policy along ecological guidelines.

The dispute over Danish energy forecasting is closely linked to the increased assertion of state autonomy *vis-à-vis* the Danish energy system. Whereas the forecasting for the Danish energy plan of 1976 was heavily influenced by the utilities and industry, forecasts for the 1981 plan were negotiated within central government. This also led to a substantial downward revision.

Nearly every single parameter in the 1981 energy forecast was the object of dispute: the energy intensities were challenged, the economic scenario was questioned, and the methodological approach itself was attacked.

The struggle over methodology was to a large extent a disguised contest over growth rates. The utilities preferred aggregate macro-economic models that produced electricity consumption growth rates of 4 per cent. The Energy Ministry, on the basis of sector models, arrived at only 2 per cent. The reactions from the utilities to this latter model were harsh, but the critique was subtly pulverized through bureaucratic procedures in the ministry.

The socio-economic inputs into the 1981 energy forecasts were clearly dictated by the Ministry of Finance, although formally arrived at through discussions between several ministries. The economic forecast was probably undertaken before these discussions took place.

The success of the Ministry of Energy in gaining autonomy *vis-à-vis* the utilities also rested on its use of new expertise in estimating specific energy consumption. By involving research councils, private consultants, and industrial groups, the ministry became independent of the utilities and their estimates.

Although energy forecasting in Norway and Denmark was never as politicized as in Britain, West Germany, and the Netherlands, intra-governmental tension due to conflicting mandates between government agencies led to substantial revisions of energy forecasts. This lack of politicization, however, led to considerably less variety in forecasting methodology and less public discussion of the value assumptions behind forecasting and forecasting methodology than in countries where establishment forecasts were more heavily challenged by outsiders.

Technocratic élitism and self-fulfilling forecasts

French electricity forecasting resembles the Norwegian and Danish in so far as the main part of the forecasting exercise remained within government. In addition, French electricity forecasting also escaped the considerable intra-governmental conflicts that characterized the Scandinavian development. Electricity forecasting in France, in other words, appears as the prime example of technocratic élitism among our cases, and is also an eminent example of self-fulfilling forecasting.

Electricity forecasting in France is undertaken by EDF, the public-sector electricity monopoly. It is addressed to the Ministry of Finance, which has to approve EDF's budget and investment plans. The Planning Bureau mediates between EDF, the ministry, and other energy suppliers, all of them large and almost exclusively within the public sector. But the quarrel remains entirely within the family of graduates from the élite engineering schools.

Forecasting up to 1968 followed the rule of thumb established in the immediate post-war years of a doubling of electricity consumption every ten years. In the 1960s, economic growth, industrial production, and household consumption were used to modify the trend. This was thought to represent progress and a switch from a technical to an economic approach. Nobody questioned the absence of price variables, even as late as the summer of 1973: the belief in the indefinite fall of the relative price of electricity was universal. Cheap nuclear power was going to make it possible. (And in fact, later studies confirmed the absence of a price effect during this period.)

The balance of payments impact of increasingly cheap oil had in the meantime changed the logic of EDF's operation. The forecasting methodology remained unchanged, but the forecasts now acquired a normative content. The French state saw in the substitution of oil with nuclear electricity the solution to the balance of trade and exchange rate crisis and a chance to return to industrial grandeur. In 1970, EDF was allowed to adopt a more aggressive commercial strategy and promote the consumption of electricity.

The oil crisis led to a multiplication of forecasting methodologies and increasingly complex models. But the commitment to nuclear electricity expansion remained dominant. This meant that energy conservation was never seriously modelled. This would have involved the need for policies to influence demand, and not satisfy it through capacity expansion in the electricity sector, the chosen political goal. Everybody played the scenario game, though. All possible futures were covered, thus assuring the public that all hypotheses had been studied. Low energy growth scenarios were, however, presented in sufficiently negative tones to ensure general rejection. The scenario with the most optimistic view of future electricity consumption was usually adopted instead.

More recently in face of the stagnating economy and low growth in energy demand, France, with its heavy commitment to nuclear electricity production has faced an energy surplus. In this context, electricity forecasting was virtually turned into an electricity sales programme, and EDF has received the order to strengthen its domestic sales 'pitch' and to extend its commercial strategy to the international arena. It is to sell as much electricity in Europe as it can, thus trying to ensure that demand matches its continuing nuclear capacity expansion.

Energy forecasting in the context of weaker state influence

Both the US and Canadian cases give examples of energy forecasting in a political context of weaker state influence. In the US, this is probably due to structural characteristics of its political system, such as the individualistic and *ad hoc* character of presidential elections, and the lack of a strong party machinery; but it is also the result of a strong market ideology. In Canada, the autonomy of the provinces, which is also substantial in energy questions, prevents the emergence of a strong and consistent national energy policy.

The US case exhibits a greater variety of modelling efforts than most of our other national case studies. This is probably due to the tradition of policy modelling and the impressive resources and know-how available for methodological development in a large industrial nation, as well as very good availability of—and access to—energy data.

A second feature of the US energy forecasting scene when compared to the European development is the loose coupling between forecasting and political decision-making. Presidential shifts and reorganization of energy agencies seem to take place faster than the finalizing of major energy forecasting projects. This means that the links between policies and modelling tend to become rather accidental and far less planned than, for example, under more centralized, bureaucratized, and stable political conditions like in Norway and France. A large proportion of the studies in the US are initiated by universities and research foundations outside direct government control.

This institutional independence of government policies allows important departures from the conventional line. From its concern with environmental issues in 1974, the Ford Energy Policy Project was amazingly farsighted both in promoting energy conservation and in the analytical techniques it employed. The extreme scenarios of this study bracketed almost all subsequent modelling results.

After the 1974 energy crises, a series of projects were set in motion in the US to investigate the consequences of a range of future national energy strategies. One of these was undertaken by the Energy Research and Development Administration (ERDA), itself a bureaucratic response to the energy crisis. ERDA's first report in 1975, based on a very expansive energy future, was heavily criticized by environmentalists and advocates of energy moderation. Stunned by the criticism, ERDA produced a second, more conservation-orientated, politically wiser report.

A second study financed by the Ford Foundation illustrates how the connection between forecasting and policy can be almost accidental. The hazards of plutonium and the breeder programme had become urgent issues for those concerned about nuclear proliferation, including presidential candidate Jimmy Carter. Carter's view conflicted with those of certain American allies abroad and many members of the nuclear industry at home. He was glad to have his positions reinforced by the Ford-MITRE nuclear energy policy study. Several contributors to the Ford-MITRE study later joined the Carter team in key nuclear policy positions.

Amory Lovins presented a methodological critique of forecasting and conceptualized different perspectives and approaches that were highly controversial yet influential in North America and western Europe. Based on the Ford Energy Policy Project, he expanded the societal dimensions to be taken into account, and launched backcasting as a strategy for policy modelling. This involved defining desirable future utopias, and working backwards by means of consequence analysis, a fundamental break with the then current forecasting paradigm.

The energy studies in the US repeatedly illustrated a break between outcomes and objectives—between the sponsor's initial intentions and the study's subsequent course. Nowhere was the contrast sharper than in the work of the Committee on Nuclear and Alternative Energy Systems (CONAES), set up by the National Academy of Sciences at the request of ERDA. Asked to focus on the breeder programme, CONAES ranged across almost the entire scope of the nation's energy problems, bringing together hundreds of experts with varied disciplinary backgrounds and ideological views. Over budget and ridden with strife, CONAES had a terrible time obtaining closure. Yet it provided a valuable training ground for talented people, and helped significantly in the rethinking of attitudes on energy supply, energy use, and their relationship to the economy.

The most significant change that took place in energy modelling in the 1970s in the US was a pronounced shift from supply-side to demand-side analysis. This was accompanied by increased attention to the energy price variable and by an attempt to integrate the supply and demand relationships. Gradually, the importance of price was conceded, and it was recognized, grudgingly by some, that alleviation of the energy problem would come more on the demand than supply side of the energy equation.

The methodological developments and diverse pluralism of the US energy forecasting scene gave influential impulses to forecasting in most European countries as well as Canada. A great deal of methodological innovation originated in the US. The loose connection between politics and forecasting there gave more room for unconstrained exploration than in most European countries.

Energy forecasting in Canada has been influenced by the distribution of energy policy-making competence between governmental levels and institutions with different mandates on the federal level. Energy falls solely within federal jurisdiction—apart from taxation—if energy resources are traded across provincial and national borders. Federal energy policy has, therefore, been primarily concerned with gas and oil policy, given the location of these resources in the west and the north and of their markets in the east and the south. Except with respect to export, electricity has essentially remained a provincial matter, as grids are not strongly interconnected across boundaries. The focus of energy policy-making on oil and gas, and there mostly on questions of the sharing of rent, has restrained the emergence of federal efforts at comprehensive energy modelling. In particular, the explicit modelling of the effects of alternative energy policies has been hampered by the fact that such policies did not fall solely within the jurisdiction of the federal government. While the most comprehensive energy modelling was done at the federal level up to the late 1970s, the results were none the less applied essentially to oil and gas issues. Over the same period, provincial governments and individual specialists have also become active in energy modelling.

On the federal level, it was the National Energy Board (NEB) which initially came to dominate energy forecasting and methodology development from its creation in 1959 till around 1974. NEB is a regulatory institution, essentially restricted to approving projects for pipeline construction and oil, gas, and electricity exports. Its mandate required it to link this approval to the availablity of gas and oil surplus and to Canadian demands. This made the development of supply and demand projections for oil and gas an important part of the NEB's activity. But as a regulatory body removed from policy-making, the NEB felt comfortable using extrapolation and econometric techniques. This methodological stance was maintained despite an awareness that the quality of the projections depended on policy decisions. The projections and their effect on pipeline approval affected both future demand, the search for new oil and gas, and the policy response of the federal government.

This unresolved dilemma of the regulatory mandate explains the low quality of the board's projections and the frequent reversals in projected supply shortfalls and surpluses. In general the results of forecasts seemed to be closely linked to the energy conditions and policy context prevailing at the time the forecasts were prepared.

The Department of Energy, Mines and Resources (EMR) is the other federal actor in energy forecasting. Free from regulatory tasks, it has a certain analytical freedom, but is, on the other hand, more tied to government policy and the oil and gas focus of federal policy. Consequently, it developed its analytical approach working, as early as 1976, backwards from tertiary end-use to demand for the different primary energy resources. EMR was also aware that future demand and supply were dependent on policy choices, but generally remained content to elaborate forecasts on the basis of policy-as-is, then using the forecasts predictively to justify policy decisions.

The only EMR forecasting exercise that proceeded to show the effects of strong conservation policies on future demand—elaborated in 1976 by the Office of Energy Conservation—was disavowed as contrary to official, government policy. The next innovative step came with the National Energy Program (NEP) in 1980, which was also quickly reversed in the face of political opposition and changing economic and energy contexts. The NEP reveals the links between the environmental movement and federal government agencies. Amory Lovins's work was supported by the Science Council of Canada while Brooks and Robinson's early work was in the first case done within and later funded by the federal energy department. The large Friends of the Earth scenario study was also funded by that department. The NEP was an exercise in normative forecasting, setting the goal of energy independence and then working out an energy demand path and appropriate policies to realize this goal. But the required choices went against the interests of both the oil and gas industry and the provincial governments. The revised forecast of 1983 therefore retreated from a normative approach, consequently resulting in higher demand and supply forecasts.

The development of alternative energy modelling techniques and of alternative forecasts had to take place in the private sector, but contrary to the experience in many countries, it benefited from moral and financial government support.

The Canadian case therefore incorporates elements from our other cases. With respect to the development of alternative modelling approaches, it has similarities with the market-driven process of the US. Moreover, as in West Germany and the Netherlands, the alternative analysis in Canada appears to have influenced the approach and results of some official forecasts.

Nevertheless, governmental actors played an important role similar to our Norwegian and Danish cases. But while in Norway and Denmark there was a struggle between national decision-makers, in Canada the federal structure and the strong position of the provincial governments led to an inter-governmental negotiation and reconciliation process. However, the fact that the various governmental actors have different energy interests and different mandates and jurisdictions over energy matters led many of them to take narrow perspectives on the energy problem and consequently on energy forecasting methodology. The hostile reaction to the National Energy Program strengthened this tendency by leading to a methodological retreat on the part of the few methodologically innovative official forecasts.

Modelling and the politicization of methodology: the IIASA world model

The analysis by Keepin and Wynne of the IIASA World Energy Model, one of the major world modelling and forecasting projects of the 1970s, illustrates how cognitive prejudices subtly infiltrate a supposedly sophisticated and objective modelling exercise.

As pointed out in chapter 3 the IIASA model consisted of a number of submodels, and was extremely complex. One of the submodels, for example, required the input of approximately 1,600 specified constraint and 2,600 activity variables. The modelling effort went on for almost eight years, involving 225 person-years of work and cost about $10 million. The presentation of the results (Häfele, 1981) implied that this apparatus had been fully used to generate what necessarily had to be 'objective' results.

The critical analysis by Keepin and Wynne points at a quite different reality. This mathematically based critique claims that the models actually used were analytically empty, that no machine-processed iteration between the submodels was performed, and that no serious sensitivity analysis of the model was made (despite claims to the contrary). Keepin's own sensitivity test revealed results that were extremely sensitive to minor changes in assumptions, contradicting the claims for 'robust conclusions'. By diverting attention from highly questionable input assumptions, the vast modelling exercise thus served an important legitimation function.

The main conclusion of the IIASA study was that the transition to fast breeder reactors and large-scale solar and coal synfuels must be made, and could be achieved by 2030 if the world began now to act decisively and to install the necessary plants (Häfele, 1981).

It is useful to know in this context that the director of the study is one of the 'fathers' of German fast breeder technology, who returned to West Germany to head the Jülich nuclear research facility, a facility which is to help to bring about the future so confidently forecast by the IIASA team.

The IIASA modelling exercise reveals a subtle politicization of forecasting at a fundamental, methodological level. This type of politicization does not even have to be conscious, but can simply result from neglecting to expose one's basic assumptions to critical discussion. And it is hard to discover because the mathematical system is highly complex and accessible to only a few insiders.

The complex mathematics, the use of computers, and the large staff of scientists helped to legitimate the results scientifically. Yet the model output neatly fits in with the expansionist and technological vision of the research and industrial complex from which the team leader came and to which he again returned. This type of transfer of political negotiation over energy futures to mathematical modelling raises a series of interesting questions about the role of modelling in politics, as well as about some methodological and philosophical problems that we will touch upon in the next chapter.

COMPARISON OF NATIONAL FORECASTING DEVELOPMENTS

Our case material was collected as a set of studies illustrating examples of the interplay between politics and energy forecasting. Although they represent national developments, each case cannot be said to be fully representative of the politics of energy forecasting in the country it is taken from. Some studies only focus on a selected part of the national development and leave other aspects uncovered. The French, Danish, and Norwegian cases, for example, cover only electricity forecasting.

Nevertheless it is tempting to draw some comparative lessons. In order to allow this, we have, in this section, included a broader discussion on national energy forecasting and national economic development that goes beyond our case material.[1] We here draw on additional information about national energy forecasts as well as general knowledge of political and administrative traditions in the countries involved.

The overall picture when comparing our eight national cases, is one of a dramatic decline in the forecasts of energy demand throughout the period we are studying. However, the magnitude as well as the timing of this downward adjustment vary from country to country. One possible explanation would be to see this development as a mere reflection of changes in energy consumption throughout the 1970s and early 1980s.

Table 12.1 Average Energy Consumption Growth Rates

	Canada	Denmk	France	Germy	G. Brit.	Netherl.	Norway	USA	All
for 71 to 73(a)	6.13	−1.83	3.74	2.84	−0.20	6.46	3.17	3.05	2.92
for 74 to 82(b)	1.58	−0.64	1.49	−0.32	−1.02	−0.40	2.93	−0.28	0.42
a–b	4.55	−1.19	2.25	3.16	0.82	6.86	0.24	3.33	2.50

Source: OECD/IEA Energy Balances of OECD countries 1982 and 1983 figures are taken from Energy Policies and programmes of IEA countries, 1984 Review, OECD/IEA, Paris (Canadian, British and US figures are weighed on the basis of comparison of 1973 figures in the two documents).

From an economic point of view, changes in economic growth would be a natural starting point in trying to explain this development. On the other hand, the discussion in the previous section of this chapter emphasizes that energy forecasting in most cases was heavily politicized. We would therefore also expect to find political and institutional factors contributing to the explanation. We shall start with the energy consumption and economic growth hypotheses and then proceed to the broader political and institutional discussion.[2]

Energy forecasting and changes in energy consumption

One possible way of explaining the decline in the rates of growth in forecast energy demand is to see it as a mere reflection of declining rates of growth in energy consumption throughout the 1970s and early 1980s.

As Jørgen Randers (1984) points out, forecasts—at least for commodity market developments—often reflect marginal changes in actual market behaviour. In the case of energy forecasts for year 2000, however, we are dealing with long-term developments and would expect the general and persistent decline of forecasts to reflect long-term changes in energy consumption.[3]

Table 12.1 presents the development of energy consumption for the eight countries included in our study. We can observe very different developments among the countries. The decline in average energy consumption growth rates from the early 1970s—before the energy crisis—to the late 1970s and early 1980s ranges widely from −1.2 to 6.9 per cent per annum. The corresponding variety of decline in establishment forecasts for year 2000 during the same period is from 34 to 68 per cent (see Table 12.2).

The energy consumption hypothesis seems to predict well in six of the eight cases, but with Norway and West Germany as important exceptions. The very low Norwegian score on decline in establishment forecasts is not matched by corresponding scores on change in energy demand. The same goes for West Germany, where very high scores on decline of

Table 12.2 Span of Energy Forecasts for Year 2000, from Early 1970 to 1980
Differences between high initial and low forecasts in % of high

	Canada	Denmk	France	Ger.	GB	NL	Norway	USA	Average
ESTAB.	54.5	51.9	50.4	68.0	46.5	59.3	33.9	51.4	51.99
NON-EST.	74.7	66.2		73.5	67.6	76.6		52.1	68.45
ALL	74.7	66.2	50.4	73.5	67.6	76.6	33.9	52.1	61.88

Source: See national chapters.

establishment forecasts is not matched by corresponding changes in energy consumption.

Energy forecasting and economic growth

Economic growth in almost all the countries included in our study has declined in the aftermath of the oil crisis of 1973–4, and has since then on average stayed below the high growth averages of the 1960s. The average GDP growth rate of 4.8 per cent per annum for all of our countries for the period 1960 to 1973 has fallen to an average of 2.1 for the ten years since then. This low economic growth has of course induced a lower growth in energy consumption. It has also affected the expectations about the future energy growth potential of our countries. It is therefore tempting to explain the fall in energy forecasts during the period covered by our case studies by this fall in the actual growth experience and reduced growth expectations. Economists, especially, are tempted to argue that lower energy forecasts therefore are a mere reflection of the crisis in the industrialized economies, and that it is unnecessary to introduce political processes as an explanatory variable.

Closer inspection of the economic growth experience since the 1974 energy crisis in the eight countries included in our analysis reveals that there is considerable correlation between the decline in growth rates from the 1960s and the downward readjustment of energy forecasts. But there are also important exceptions (see Table 12.3). The economic growth hypothesis also seems to predict well in six of the eight cases. The exceptions here are West Germany and France. The relatively low score on decline in establishment energy forecasts does not correspond to France's relatively high ranking on change in economic growth. For West Germany, the case is the other way around.

As shown in Figures 12.1 to 12.8, the economic growth rates and energy consumption patterns reveal short-term fluctuations that are not reflected in the energy forecasts.

Economic stagnation and the decline in energy consumption undoubtedly explain some of the general downward readjustment of energy fore-

Table 12.3 Average Economic Growth Rates

	France	Ger.	Norway	Denmk	NL	GB	USA	Canada	Average
from 60 to 73(a)	5.7	5.3	4.7	4.7	5.1	3.2	4.0	5.4	4.8
from 74 to 83(b)	2.2	1.6	3.9	1.7	2.0	1.1	2.1	2.2	2.1
a–b	3.5	3.7	0.8	3.0	3.1	2.1	1.9	3.2	2.7

Basis 1980
For France, Norway, Denmark, Netherlands, UK we use GDP figures.
For Germany, USA and Canada we use GNP figures.
Source: IFS Yearbook 1984.

casts. However, other economic and socio-cultural factors that have resulted from, or have been induced by, the energy crisis, like higher energy prices and greater energy conservation, shifts in the structure of production, and the emergence of new lifestyles, have also to be taken into account.

Moreover, our case studies definitely indicate that the translation of the experience of lower economic growth and declining energy consumption into corresponding adjustments to growth rates and growth expectations in energy models and forecasts has been neither an automatic process, nor a purely technical affair. It has instead, in most cases, been the result of a political struggle and of a larger confrontation of conflicting paradigms and social movements and institutions.

This means that understanding the development of national energy forecasts also requires looking into the particulars of the national energy debates and the interplay between the different political, administrative interest groups and social movement representatives involved in energy forecasting. It also means that political and administrative traditions in decision-making in general, and in energy matters in particular, may become important explanatory variables.

Forecasting and its political and institutional context

In most cases, forecasting before the 1974 energy crisis provided an uncontroversial and apparently scientifically based 'extra-political' tool to justify potentially controversial decision-making in democratic societies. But the scientific legitimacy of modelling and forecasting was undermined as counter experts and political oppositions challenged the established forecasting monopolies and introduced competing forecasting paradigms and methodologies throughout the 1970s and early 1980s.

Behind this recurring theme, however, the political and institutional context of energy forecasting varied considerably between the countries

National Development of Energy Forecasts for Year 2000

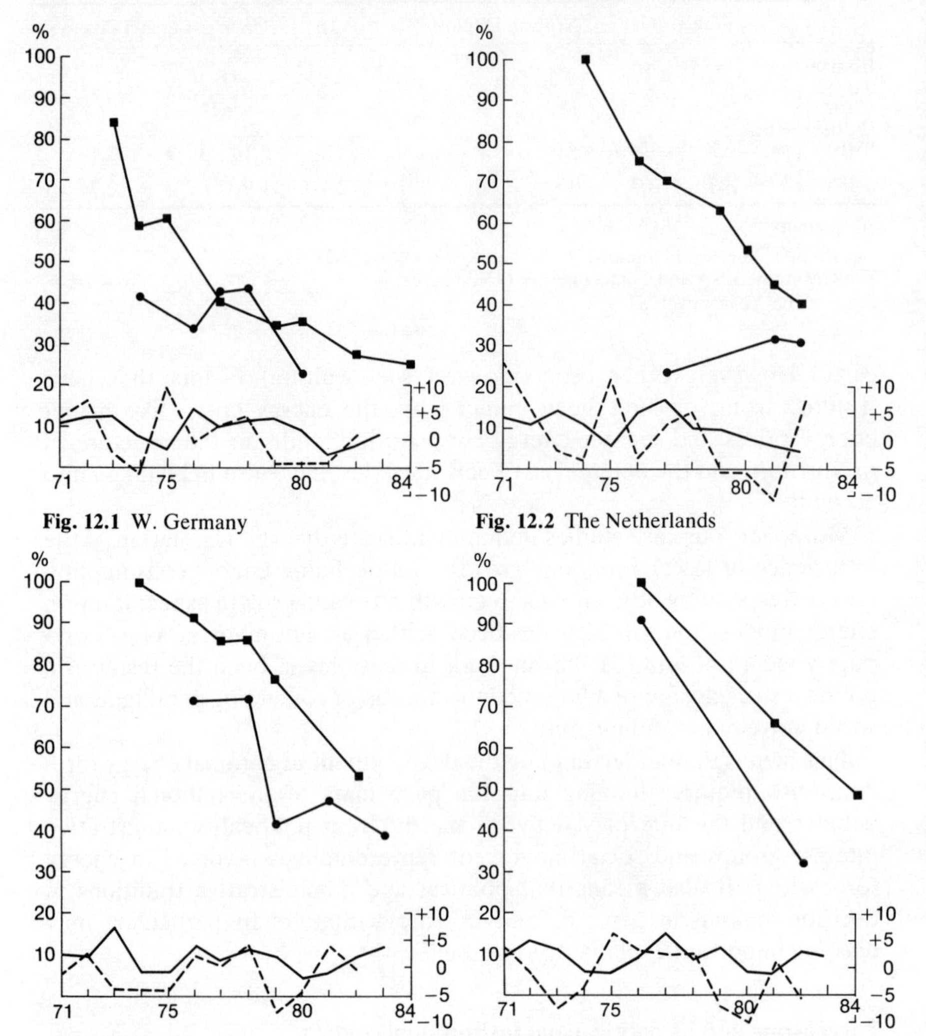

Fig. 12.1 W. Germany

Fig. 12.2 The Netherlands

Fig. 12.3 Great Britain

Fig. 12.4 Denmark

■ Establishment/official[a]
● Non Est./Non off.
—— Economic Growth Rates
- - - Energy Consumption Rates

Source: see chapters 4 to 11 and Appendix to chapter 12

a. Establishment forecasts refer to forecasts made by the energy establishment, including government, energy producers, and research institutes closely linked to them. Non-establishment forecasts refer to forecasts made by groups outside the energy establishment. For the US we find the distinction official/non-official more appropriate. Official forecasts refer to government, or directly government-sponsored forecasts. Non-official forecasts refer to other forecasts.

National Development of Energy Forecasts for Year 2000

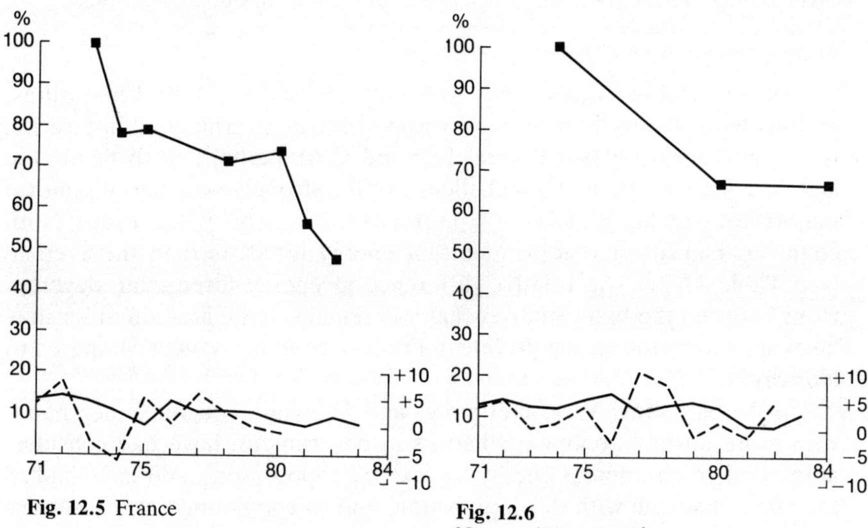

Fig. 12.5 France

Fig. 12.6
Norway (Electricity)[b]

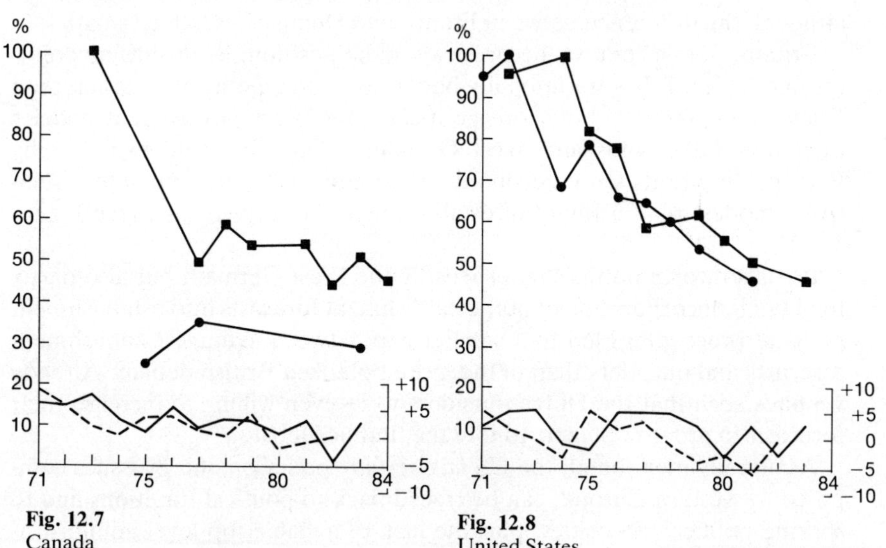

Fig. 12.7
Canada

Fig. 12.8
United States

■ Establishment/official[a]
● Non Est./Non off.
——— Economic Growth Rates
– – – Energy Consumption Rates
Source: see chapters 4 to 11 and Appendix to chapter 12

b. As forecasts of general energy consumption did not exist in Norway before the 1980s, we base our conclusions on comparisons of electricity forecasts for this country.

included in our study. This variety may to some extent help to explain the divergent patterns of energy forecasts that we find in our case studies.

Political structure and energy forecasting

On the political level, countries with long and stable political majorities, such as Norway and France, have escaped extra-governmental opposition in energy forecasting (see Figures 12.5 and 12.6), perhaps partly due to the lack of political support for a challenge to the strongly established political majorities, whether socialist or conservative, on energy policy issues. Both countries had lower readjustments of energy forecasts than the average (see Table 12.2). The relative difference in energy forecasting development between the two countries that still remains to be accounted for may be explained by the strong decline in French economic growth compared to Norway.

The Netherlands, West Germany, and Denmark, on the other hand, with more unstable political alliances in government, have had considerable extra-governmental energy forecasting 'opposition', and have had to take up a dialogue with this 'opposition' and its competing lower forecasts (see Figures 12.1, 12.2, 12.4), a fact that may account for the stronger decline in establishment forecasts than in the case of Norway and France, although the difference between France and Denmark is indeed small.

Britain, in this context, occupies a middle position. With shifting political majorities, it has traditionally had a less stable political situation than France and Norway, but more political support for government policies than the Netherlands and West Germany. This may help explain why Britain, despite its strong economic recession, still comes up with a relatively modest readjustment of establishment energy forecasts (see Figure 12.3).

It is also worth noting that especially the West German, but also partly the Dutch, incorporation of non-establishment forecasts into establishment political proceedings led to a smaller gap between average establishment forecasts and outsiders than in the more polarized British debate. Already we have seen that the Dutch outsiders were even willing to increase their forecasts in order to be able to join the 'Public Debate'.

As already mentioned, the US *laissez-faire* policy on energy issues compared to western Europe, can be traced back to political traditions and to shifting political majorities, plus the lack of a stable top-level administration. This made for a less government-dominated energy debate than in most European countries in our material. The US also differs from the French and Norwegian situation, where energy policy largely was left to permanent administrative élites. The US energy debate did not produce the sharp division between establishment and non-establishment forecasts so typical of most European cases. In line with US tradition, forecasting took place in a market-like arena with academic institutions and research foundations playing a major role. This may help to explain why the US has the smallest differences between official and non-official forecasts and why it does not occupy an extreme position among our cases in changes in energy forecasts (see Figure 12.8).

We have previously described the Canadian political scene as characterized by strong regional governments with political majorities that vary from those at the federal level. Together with ethnic controversies, this makes for a difficult political climate for development of long-term policies on controversial issues. In addition Canada is in net surplus as an energy producing country with large current surpluses of all conventional sources of energy. As a consequence, Canada is the only case where establishment forecasts, made by the same agency, have been adjusted upwards (see Figure 12.7), reflecting both the emergence of these surpluses and the negative reactions of provinces to the normative policies and projections which the federal government tried to impose in 1980. Both the energy surplus, and the strength of provincial interests with wishes for an expansive energy future may furthermore help to account for the relative lack of harmonization between establishment and non-establishment in the Canadian case.

Administrative structure and energy forecasting

On the administrative level, the degree of centralized planning traditions seems to have a bearing on the institutionalization of forecasting and on the energy forecasting results. Countries with strong centralized planning traditions like Norway and France seem to have escaped extra-governmental opposition. The combination of robust political majorities and a well integrated central planning apparatus seems to have a preventative effect on 'alternative' energy forecasting.

The shift in control over forecasts from engineers to the planning profession *par excellence*, the national economists, caused a certain amount of intra-governmental turbulence in Norway; as opposed to the more stable development in France, where the different professions which graduate from the same *grandes écoles* are probably better integrated. However, the conflict over energy forecasting in Norway still remained within the governmental apparatus, in line with the Norwegian tradition of institutionalized interest mediation.

Countries with less developed central planning and a less uniform corps of planning professionals, like Britain and West Germany, had considerable extra-governmental opposition. The West German system of court hearings, however, compared to the British public inquiry system, proved to be more open to mobilizing political opposition within government. The lack of a strong planning apparatus would probably be more critical for West Germany with its weaker party coalitions and federal political set-up—a fact that may help to explain the considerably stronger German revision of energy forecasts, despite rather similar decline in economic growth in the two countries.

Contrary to its strong tradition in centralized economic planning, the Netherlands, as already noted, had considerable extra-governmental opposition in energy forecasting. It seems that the brittle corporative political structure here overrules the administrative planning capacity and provides a basis for unauthorized opposition and a relatively dramatic revision of energy forecasts.

Denmark, on the other hand, with its far weaker tradition in central planning, would be expected to exhibit more outside opposition in the energy forecasting field. The shifting and complicated nature of political constellations behind Danish governments over the last years also points in this direction. The lack of alternative forecasting resources in a small country might help to account for this and for the relatively modest downward revision of Danish energy forecasts.

Both Canada and the US lack strong central planning ambitions. In the US, this results from an ideological commitment to a market-orientated society and the continuous shifting of the top political leaders.

In Canada, the problematic balance between central and regional political bodies and ethnic groups has already been mentioned as a factor that inhibits strong central 'steering' both at the political and administrative levels. In both cases the administrative structure corresponds to the political, and only serves to emphasize the consequences for energy forecasting spelt out above.

CONCLUDING REMARKS

The preceding comparative analysis has highlighted the need to approach the complexity of development in energy forecasting from different angles. Economic factors are insufficient to explain the differences between the various national developments. On the other hand no simple institutional or political theory fits the data either.

The diversity of national forecasting experiences can only be fully understood with reference also to specific institutional characteristics of the countries involved. We see such factors as mediators that transform broader economic and political forces into actuality. The relatively low decline in French energy forecasts despite dramatic decline in economic growth has, for instance, to be explained in terms of the strength and autonomy of the French planning apparatus. On the other hand, the market orientation and the continuous shifting of the top political leaders of the US has allowed economic growth and energy consumption development to penetrate energy forecasting more directly.

Besides insight into political and economic developments of a number of western nations, our case material also illustrates a variety of methodological approaches to forecasting. With the dramatic changes in political and economic conditions, and with the great attention devoted to it, the energy sector has become a pioneering field for long-term forecasting. In the next chapter we shall explore some aspects of this development and some methodological and political lessons to be learned from it.

Notes

*For bibliographical references in this chapter, see the bibliography following ch. 13.

1. It must be emphasized that it is very difficult to be comprehensive about energy forecasts in this period. The proliferation of energy modelling and modelling organizations was indeed overwhelming. We do claim, however, to cover most of the important official energy forecasts and other forecasts that were central to the energy debate.
2. The analysis is based on correlation between the span of energy-forecasts, and energy consumption—and economic growth rates using linear regression.
3. Short-term forecasts like those of Canadian oil and gas supply will of course have to be explained by short-term developments. See ch. 13 for further discussion.

Appendix

Primary Energy Consumption Growth-Rates

	Canada	Denmk	France	Germy	G. Brit.	Netherl.	Norway	USA	Average
1971	9.35	4.12	4.75	3.83	−0.15	14.56	4.20	5.88	5.82
1972	6.48	−0.86	8.62	6.95	3.36	6.19	5.05	2.75	4.82
1973	2.56	−8.76	−2.16	−2.25	−3.82	−1.37	0.26	0.53	−1.88
1974	0.81	−0.11	−5.73	−6.12	−4.40	−3.31	1.23	−2.77	−2.55
1975	3.88	6.91	5.73	9.37	−3.29	11.65	4.34	5.82	5.55
1976	4.06	3.84	1.04	−0.98	2.31	−2.94	−2.66	2.66	0.92
1977	1.97	2.53	6.84	2.70	0.04	3.03	10.72	3.60	3.93
1978	0.34	2.67	3.27	4.43	4.11	5.40	8.48	−0.13	3.57
1979	4.75	−7.50	0.03	−4.39	−8.02	−5.54	−1.24	−3.89	−3.22
1980	−1.37	−9.93	−0.72	−4.20	−4.29	−5.79	1.47	−2.13	−3.37
1981	−0.48	1.56		−4.38	4.37	−10.59	−1.40	−5.83	−2.39
1982	0.25	−5.68		0.68	0.00	4.52	5.44	0.14	0.76

Source: OECD/IEA Energy Balances of OECD countries.
1982 and 1983 figures are taken from Energy Policies and Programmes of IEA countries, 1984 Review, OECD/IEA, Paris.
(Canadian, British, and US figures are weighed on the basis of comparison of 1973 figures in the two documents.)

Economic Growth Rates for the Countries Involved in our Study

	France	Germy	Norway	Denmk	Netherl.	UK	USA	Canada	Av. All
1960	7.2	16.2	5.2	6.4	7.6	4.6	2.1	2.9	6.5
1961	5.5	4.9	6.6	6.1	3.1	3.3	2.6	2.8	4.4
1962	6.7	4.4	7.5	5.4	3.8	0.9	5.8	6.8	5.2
1963	5.3	3.1	3.9	0.7	3.6	4.3	4.0	5.2	3.8
1964	6.5	6.6	5.0	8.9	8.4	5.2	5.3	6.7	6.6
1965	5.3	5.5	5.3	4.8	5.2	2.3	6.0	6.6	5.1
1966	5.2	2.7	3.8	5.3	2.7	2.0	6.0	7.0	4.3
1967	4.7	−0.1	6.3	3.7	5.3	2.7	2.7	3.3	3.6
1968	4.3	6.1	2.3	3.8	6.4	4.2	4.6	5.7	4.7
1969	7.0	7.5	4.5	6.5	6.2	1.3	2.8	5.1	5.1
1970	5.7	5.0	2.1	2.3	6.5	2.3	−0.2	2.9	3.3
1971	5.4	3.2	4.5	2.5	4.2	2.6	3.4	6.8	4.1
1972	5.9	4.1	5.2	5.4	3.3	2.1	5.7	6.1	4.7
1973	5.3	4.5	4.1	3.8	5.6	7.6	5.8	7.5	5.5
1974	3.2	0.5	4.0	−0.7	3.4	−0.9	−0.6	3.6	1.6
1975	0.2	−1.6	5.4	−1.0	−1.1	−0.8	−1.2	1.2	0.1
1976	5.2	5.5	6.8	6.5	5.3	3.7	5.4	5.8	5.5
1977	3.0	2.8	3.6	2.3	7.8	1.2	5.5	2.0	3.5
1978	3.8	3.5	4.6	1.8	2.5	3.5	5.0	3.6	3.5
1979	3.3	4.0	5.0	3.7	2.4	2.0	2.8	3.2	3.3
1980	1.0	1.9	4.3	−0.5	0.8	−2.3	−0.3	1.0	0.7
1981	0.2	−0.3	0.8	−0.9	−0.8	−1.0	2.5	3.3	0.5
1982	2.0	−1.1	0.9	3.4	−1.5	2.2	−2.1	−4.4	−0.1
1983	0.3	1.2	3.2	2.5	1.0	3.6	3.7	3.0	2.3

Basis 1980
For France, Norway, Denmark, Netherlands, UK. We use GDP figures
For Germany, USA and Canada we use GNP figures
Source: IFS Yearbook 1984

Danish Energy Forecasts

		Actual Cons.	Forecast 1985	Forecast 1995	Forecast 2000	Unit	Definition
1976	Energy Plan	(1975)			*	Mill	Net Energy
	MOT	111.5	132.8	175.4	203.47	G. Cal	
1976	Alt. Plan	(1975)			*	Mill.	
	OOA	111.5	128.9	162.5	183.63	G. Cal.	
1981	Energy Plan	(1981)		863	922	PJ	Gross Energy
	MOE	722					
1982	Alt. Plan	(1982)					
	OOA	705			440	PJ	Energy Supply
1984	Risø Report				650	PJ	Gross Energy Cons.

* Have been estimated on the basis of 1985 and 1995 figures, assuming stable growth rates.

Danish Energy Forecasts for 2000, Weighted and Unified*

	Actual Cons. mtoe**	Act. Cons. Forec.	Act. Cons Weight	Act. Cons. w. Weight	Forec. 2000	Forec. 2000 in % of highest forec.	Status
1976 Energy Plan	(1975) 17.8	(1975) 111.5	1.0	111.5	1.825	100.0	Est.
1976 Alt. Plan	(1975) 17.8	(1975) 111.5	1.0	111.5	1.647	90.2	N.est.
1981 Energy Plan	(1981) 17.33	(1981) 722.0	0.974	741.2	1.244	68.2	Est
1982 Alt. Plan	(1982) 17.6	(1982) 705.0	0.989	712.8	0.617	33.8	N.Est.
1984 Risø Report					0.877***	48.1	Est.

* We are here assuming that the relations between the different types of energy consumption in these reports remain constant throughout the period forecast, in relative terms.

** Figures taken from OECD/IEA: Balances of OECD countries, and Energy Programmes for IEA countries 1984 Review.

*** We are assuming that the Risø report uses the same definition of energy rules as the 1981 Plan.

13

Modelling and Forecasting in Self-Reactive Policy Contexts: Some Meta-Methodological Comments

Thomas Baumgartner and Atle Midttun

INTRODUCTION

Classic sociology has always been aware of the close link between applied science and the field to which it is applied.[1] Our material reconfirms this message: energy forecasting does influence the system that is modelled and forecast—thereby to some extent also undermining its own accuracy and ultimately even its bases of legitimacy. Mainstream social science methodology, especially the influential discipline of economics, has however tended to ignore this knowledge. This methodology is still generally based on the traditional natural science paradigm that sees the scientist as a neutral observer who does not affect the object he is observing.[2]

Few social scientists are taught to consider the probable consequences of social prediction pronounced as the findings of 'science', even though science and technology have been shown to play an increasing role as legitimating ideologies in modern society (Habermas, 1968a). The fact that energy modelling and forecasting traditionally have been the domain of engineers with their natural science background, and of economists with their aspiration to natural science respectability, has probably strengthened the tendency to overlook the issue of self-fulfilling and self-falsifying consequences in this area of activity.

The problem of modelling self-reactive systems is of course a difficult one, as it encompasses a number of interaction patterns and raises some serious methodological problems. The answer to this situation cannot simply be to include in the model and its forecasts an element that represents the likely reaction of the forecast population to the forecast itself. This solution is only a partial one, as the new advanced forecast will itself be subject to social reaction. It therefore ends up in an infinite regress, where modelling and forecasting become more and more inclusive, but never inclusive enough. Furthermore, as the modelling of reactions to modelling and forecasting would involve mapping activities of an increasingly more unpredictable kind, such an approach would thereby also soon get caught up in its own predictive weakness.

This self-destruction of the image of the objective and neutral modeller calls for a more fundamental paradigmatic debate. The social sciences do, after all, include alternative perspectives, such as those advocated by Marcuse, Habermas, and the psychoanalytic tradition in pyschology originating with Freud, which have shifted the focus from prediction—and hence control—to choice and liberation. In this perspective the self-reflective nature of social actors, which creates the problem of predictability in forecasting, is transformed into an asset and a focal point. The role of the social sciences, including modelling and forecasting, is to make people aware of options and alternative paths of development (Marcuse 1964; Habermas 1968b; Freud 1917).

We have examples of highly sophisticated modelling and forecasting being used in this liberating way even in the early 1970s, although without explicitly revealing the break with traditional objective methodology.[3] But these were exceptions that confirmed the rules.

The subsequent emergence of normative futures analysis, in its initial form of scenario analysis and in its further development in the form of backcasting as well, made the normative and didactic functions of forecasting more explicit, and can be seen as a contribution to the exploitation of the liberating potential of modelling and policy analysis. This is amply demonstrated in our case material.

This insight, of course, leads on to a number of methodological questions. Once the role of the forecaster as an objective observer is abandoned, we are faced with problems of verification that involve evaluating the forecaster's own societal commitments along with traditional criteria of scientific evaluation. As we will argue later such commitments may, in fact, be embedded in the forecaster's scientific training and professional paradigms.

Once we move forecasting out of its detached scientific ivory tower, we approach the broader question of forecasting and democracy. Societal complexity, longer lead times, and the economies of scale lead to increased needs for long-term planning and decision-making; this requires information on possible future development. Forecasting, therefore, has a large potential role to play. If we accept the fact that long-term societal futures cannot be objectively and accurately predicted, but actually must be highly normative and politically loaded, the question of how such planning and prediction can be democratically controlled becomes pertinent.

In this chapter, we take up these questions and give some preliminary answers on the basis of our material on energy forecasting. We shall start with the question of how reactive structures and processes link modellers and their models and forecasts to their social environment. We shall then investigate some of the methodogical questions that this leads up to before finally concluding with the broader normative issues of forecasting, modelling, and democracy.

MODELLING AND FORECASTING A REACTIVE SOCIETY

There are a number of reasons for discussing in some detail the reactive structures and processes that link modellers—and their models and forecasts—with their social environment. First, such an analysis helps to clarify the often turbulent developments in energy modelling and forecasting described in our cases. Second, it also helps to clarify questions around the issue of truth in modelling and forecasting. And, thirdly, it provides policy analysts with some insights into the power and pitfalls of their work.

We shall begin with the conventional case of the modeller as an observing outsider, not because it is typical for energy forecasting, but because it is an ideal-typical situation to which other forecasting situations can be compared. Futhermore, this is the situation assumed in the natural science-orientated forecasting tradition that, until recently, has been the dominant paradigm in energy forecasting.

We shall then proceed to the case where forecasting is linked to the social system analysed through its impact with regard to information. A step further in societal involvement comes when forecasting is linked up with policy formulation and implementation. Finally, we shall analyse the situation where forecasting becomes part of societal conflict through competition between models and forecasts for cognitive monopoly and influence upon policy formulation and implementation.

The forecaster as an observing outsider (type 0)

The case of the modeller as an observing outsider represents an ideal-type situation. It reflects the forecasting ideal inspired by the natural science paradigm and represents the view still held by many modellers in the early 1970s. Or it represents at least the view that many were only too happy to let others have of their craft even if they themselves knew otherwise.

This situation, sketched in Figure 13.1 is characterized by:

(1) The modeller as a neutral, observing outsider. He is not part of the system he is observing.
(2) The forecasts prepared by the modeller are not known to the actors in the system analysed.
(3) The forecasts are made for purely scientific purposes with no policy consequences.
(4) The modeller refines his model by comparing past predictions with

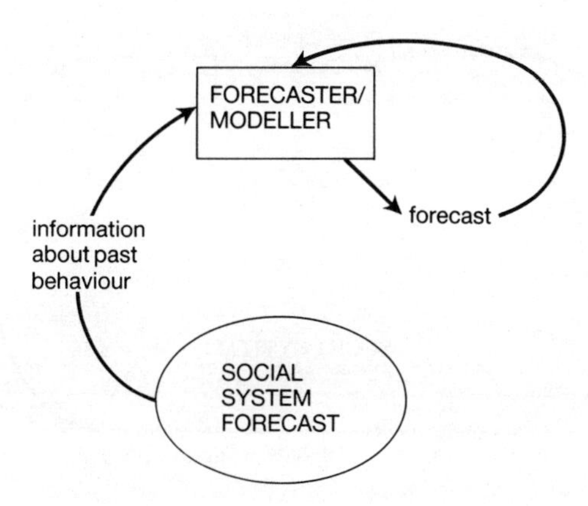

Fig. 13.1 The forecaster as an observing outsider

actual development. Nobody in the predicted system is even aware of this modification.

Needless to say, this situation does not reflect reality. It is in fact even contradictory to the *raison d'être* of modelling and forecasting in everyday life: to analyse and understand the system for policy purposes.

The forecaster as informant (type 1)

By including the information effects of forecasting upon the field analysed we take one step into the complexity of interaction between the forecast and the societal sector subject to forecasting (see Figure 13.2). The main characteristics of this situation are:

(1) The forecaster is still an observing outsider.
(2) Actors in the system studied have access to the forecast.
(3) They may use this knowledge to adjust their behaviour even though this may not have been intended by the forecaster.

This constellation may lead the forecast to have self-falsifying or self-fulfilling consequences dependent on the reaction of social actors to the forecast. Such reactions will depend on the social organization of 'competence power' and the types of exchange networks and alliances that exist. As Henshel (1982) and Schneider (1975), among others, point out, one determining element is the forecaster's prestige itself. If the forecaster is widely accepted as an authority, actors may expect his predictions to come true and adjust their behaviour accordingly. If he is held in low esteem, they may neglect his prediction or they might consciously contradict it.

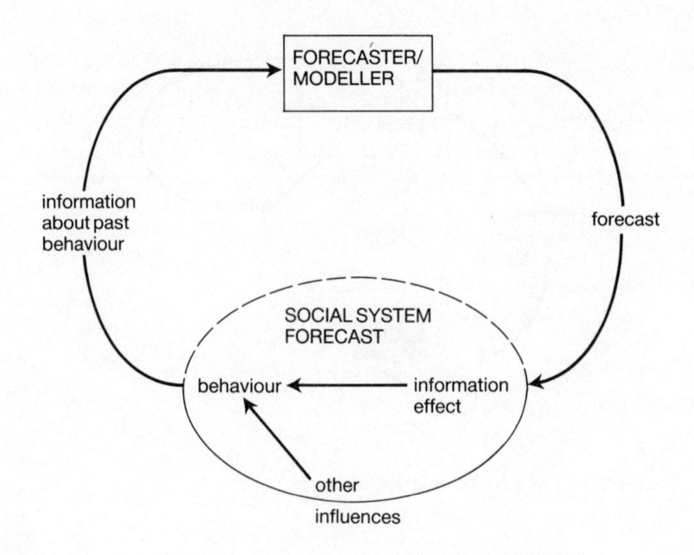

Fig. 13.2 The forecaster as informant

Greeenberger and Hogan show that in the case of the US, the political prestige of a forecaster does not necessarily coincide with his professional recognition. They might diverge because a new administration has a completely different ideological orientation. The Norwegian and Danish cases suggest that forecasting failure and inability to adopt more sophisticated forecasting methods can relatively quickly undermine the prestige of well-established forecasting monopolists, relegating them to a minor role in the new forecasting structure. The British, Dutch, and West German studies suggest that many non-establishment forecasters had to overcome a handicap of low legitimacy, which could only be accomplished by gaining political support.

Although hardly any modellers have included the information effect of their forecasts in their models, they have been quite willing, individually and as a profession, to increase their prestige by referring to past forecasting success and by stressing the use of complex, computer-based models, thus attempting to increase their influence over the development of the social system that is analysed.[4]

The Canadian case illustrates self-falsifying forecasting in the terms discussed here. The National Energy Board (NEB) repeatedly analysed the future supply shortages or surpluses of gas in the course of deciding on new pipeline projects and applications for permission to export gas. The industry, however, also took these forecasts as an indication of future price movements and market opportunities. The forecast of future supply surpluses led to the downward revision of price expectations, reductions in drilling programmes and, therefore, to reductions in future supply. In the next forecasting round this led to a forecast of supply shortages. This in turn made the industry expect high prices. They increased drilling thereby again falsifying the forecast supply situation (see Figure 13.4).

Forecasting for policy purposes (type 2)

Forecasts become integrated one step further into the social system when they are used deliberately for policy-making (see Figure 13.3). This means that:

(1) The forecast is used by policy-makers as a tool for policy formulation and therefore becomes part of policy implementation.
(2) The social behaviour modelled is thus influenced directly and deliberately by actors who have 'structuring power' on the basis of forecasts.
(3) The forecaster himself may become explicitly involved in translating his forecast into policy, but this is not a necessary characteristic of this type.
(4) This influence is generally combined with the information impact of forecasts described above, unless the forecast is deliberately kept secret. Policy implementation may of course still lead to reactions unforeseen by the forecast.

The effect of the self-reactive aspect of the forecast now depends basically on the extent to which the implementation system makes the forecast the basis of its action and the degree to which the chosen policy can be implemented. Our material provides plenty of illustrations of the complex interaction effects that can occur in such cases.

Returning to the Canadian case, we find that the information effect of the forecast was compounded by the implementation effect in response to the forecast. The NEB used the forecast of domestic supply surpluses to approve new pipeline projects and additional export contracts. This tended to increase future demand, thus decreasing projected surpluses in the next forecasting round. Projected supply shortages led to the opposite decision: pipeline plans and export applications were rejected, thus increasing future domestic supply potential. This of course again brought the NEB back to the initial forecast situation. (See Figure 13.4).

The tight coupling of the forecasting and the implementation system in France, in contrast, explains why forecasts have been rather successful in the past. The extrapolation methods of the monopoly electricity supplier Electricité de France (EDF) were a success because they were used in a period of electricity shortages. Demand was therefore growing as fast as supply could increase. The moment potential demand caught up with

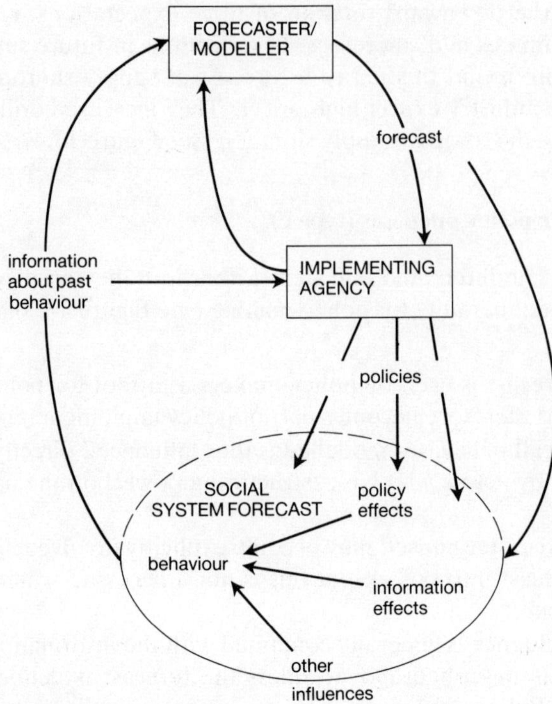

Fig. 13.3 Forecasting for policy purposes

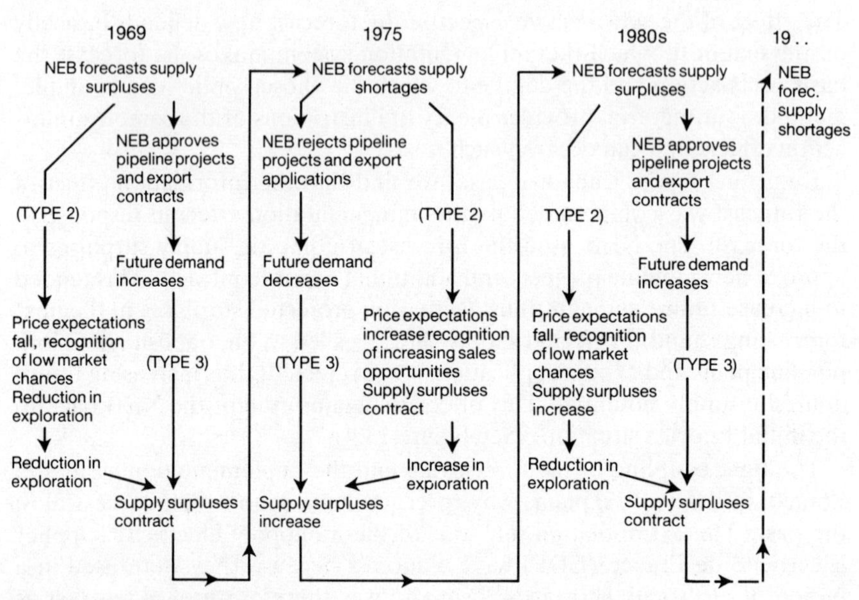

Fig. 13.4 Self-falsifying forecasts:
the case of NEB oil forecasts in Canada

actual supply, the forecasts were in danger of being proved wrong. But then EDF learned that it could use pricing and sales policies to produce the demand that its method had predicted and which, with unchanged policy stances, would not have materialized.

This system only broke down when other state interests intervened and undermined EDF's policy autonomy: pricing became controlled as part of anti-inflation policy. An ambitious nuclear power programme was launched for reasons of national grandeur, to assure national energy independence, and is being continued despite looming capacity surpluses to save the nuclear industry and to protect employment.

The links between the forecaster and the implementation system that are characteristic of the type 2 forecasting situation can be institutionalized in many ways. The forecaster may be working directly for the policy-makers, as was the case in France. But the link may also depend on loosely coupled coalitions, as was the case in the US. Here, modellers and forecasters sometimes moved over into policy-making positions if the administration and the president had a predilection for the main thrust of their modelling. There they could work directly toward the realization of the future they had predicted, or rather designed. At times, of course, the time lag between beginning and ending a modelling exercise was longer than the life of an administration. In these cases even forecasters employed directly by administrative energy agencies did not have the opportunity to try to implement their studies. The new administration was even unlikely to take notice of their modelling and forecasting work. Implementation would have to wait for better times.

The forecaster may, of course, also misjudge his power of implementation. Forecasters in the Ministry of Energy, Mines and Resources (EMR) in Canada, for example, produced low energy forecasts that required the use of prices and of policy intervention to bring about substantial conservation gains. However, the federal government does not have the constitutional power to legislate on energy consumption matters. Publication of the forecast was, therefore, taken by many provinces to mean that the federal government would try to land a political coup. The resultant storm of protest forced the forecasters to go back to their drawing boards and build a correspondingly higher forecast that would not imply such measures.

In other cases, the policy-maker and implementation agencies are in a position to dictate forecasts even prior to their publication, a power that is used particularly over forecasts of politically sensitive sectors. We have seen, for example, that forecasts in the UK went substantially wrong because the government could not be seen to admit that its preceding electoral promises for an expanding steel industry were just electioneering. But this already leads us near the third stage of forecast and society interaction.

Forecasting with modelling and forecasting competition (type 3)

A third step of integration between forecasters and society is reached when forecasters, their models and forecasts, and even their modelling methodologies become part of a struggle for model and forecast dominance and for modelling influence over policy formulation and implementation (see Figure 13.5). This situation is characterized by features such as the following:

(1) There is more than one modeller, either competing for the ear of an implementation agency or in the possession of privileged access to one or other of a range of implementing agencies.
(2) There is one (or possibly several) implementing agencies that use model-based analysis for policy formulation and implementation. Some agencies have their favoured modeller and models, and simply insist that they provide the right perspective on the social system that they try to influence. Other agencies may use competing models and in turn make up their own mind about the correct view of the system.

In this situation implementation systems faced with competing modellers have to select from among models and forecasts or possibly reconcile them.

Where there is a plurality of implementation agencies, each agency may, of course, choose its preferred model. This means that forecasters now face a social system that includes reaction to competing models and implementation systems. They will, therefore, have to model the actions of the other implementation agencies and/or the information effects of competing models in order to gain predictive accuracy. But if they try to do so, they will simply run into the problems of infinite regress that were described at the beginning of this chapter.

Thus the implementation system may well work at cross-purposes unless some superior agency, or an agency with co-ordinating responsibility for the overall performance of the system, intervenes and brings about a model selection or reconciliation. Since the agency then may also be in a position to impose the policy measures that underline the forecasts, this situation may easily end up with forecasts as self-fulfilling prophecies.

Our cases show first that implementation agencies have felt an increasing need to have privileged links with a model or have their own in-house model in order to be able to argue with strength in the larger policy-making

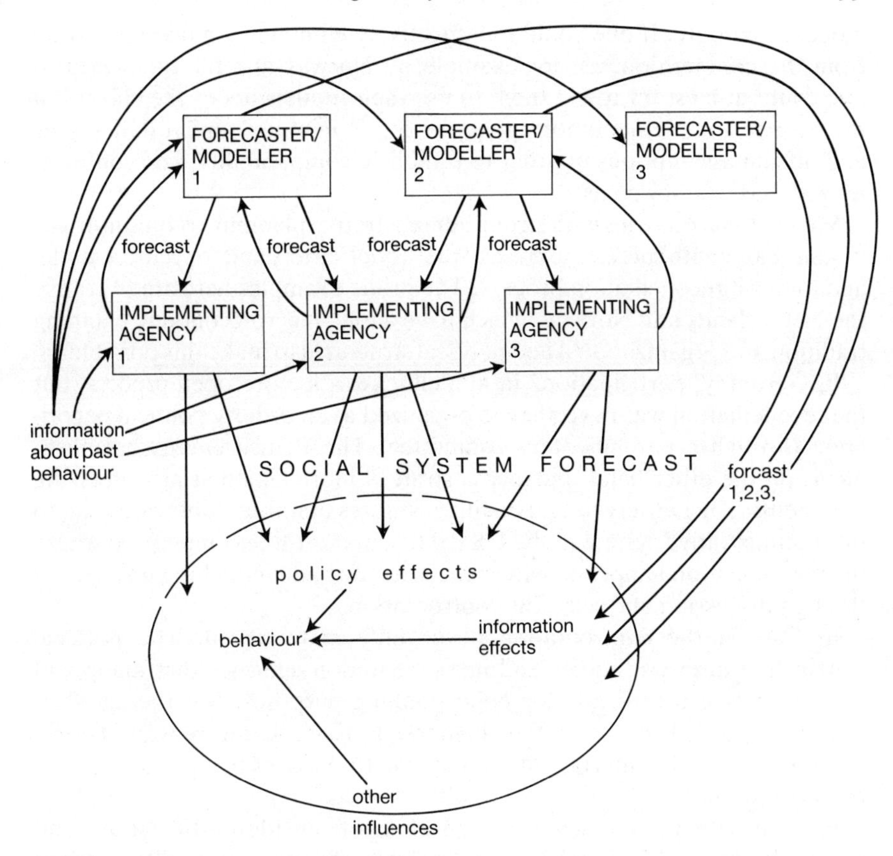

Fig. 13.5 Forecasting with modelling and forecasting competition

system. We find, secondly, that the policy system in fact tried to use model selection and reconciliation to maintain control over the planning process.

The institutional structure and the specific national characteristics of policy administration to some extent determine the span of model competition and therefore of model and forecast diversity. In addition, administrative and political styles influence the selection and reconciliation procedures between competing models. Our Norwegian case, for instance, clearly shows that new implementation agencies felt a need to have their own model to oppose policies pursued by agencies that had traditionally held a model monopoly. The impact of modelling on policy control is illustrated by the Danish utilities, which failed to improve their model-building capacity, and hence lost much of their influence over energy planning. At the same time, the new Danish Ministry of Energy secured its supremacy in energy planning by alliances with new models and modelling institutes.

Dominant modelling and policy-making agencies sometimes understood only too clearly what it would mean for their position if they lost their model monopoly. If one could not effectively eliminate a modelling group from the competition, as, for example, in Norway and the Netherlands, one could at least try to get them to use their models under the control of the strongest implementation agency or get them to use similar data inputs and official assumptions in order to establish 'comparability' and conformity with basic policy goals.

We see that countries with strong administrative planning traditions tried, in some cases with success, to control the model variety and to secure a model and forecast reconciliation process. The prime examples here are of course the Netherlands and Norway, which have a long macro-economic planning tradition and organizational and political structures to make this possible.

In Germany, participation2 in modelling was a more open process, but the reconciliation was nevertheless organized as an orderly political negotiation through a parliamentary committee. The British energy establishment, on the other hand, pursued a strategy more aimed at marginalizing opponents. The variety in forecasting institutes and their loose coupling to the administrative system in the US led to a market-like competition where the reconciliation happened within the academic and modelling fraternities through discussion and scientific confrontation.

In Canada the confrontation never fully materialized. The political system had different bodies and implementation agencies that made and used models and forecasts for policy-making purposes. But the agencies were not coupled, nor were they competing. Each therefore went its own way without really entering into a dialogue or without being forced into a reconciliation.

In France there was never any question of outside participation, and reconciliation could take place within the administrative élite without much concern for the political public. Here the third step in our typology of forecasting never occurred, at least not in public.

METHODOLOGICAL DILEMMAS AND SHIFTS IN FORECASTING PARADIGMS

As pointed out above, the 1970s and early 1980s saw a move of forecasters and modellers away from a position of unchallenged expertise to a new role that more resembled that of the politician. In most countries in our study, modellers gradually found themselves in a more or less open competition with each other, and more or less openly in alliance with interest groups or sectors of society. The recruitment of forecasters outside traditional forecasting milieux and professions emphasized this development even more. This activated a discussion of the value basis and political assumptions behind the competing forecasts.

As we have already pointed out in the introduction, many forecasters in the early 1970s projected short-term exponential growth rates into the long-term future without critical examination. By leaving important goals and political values implicit, and by neglecting the wide ramifications of the energy sector development in other societal sectors, revolutionary change was projected into the future as business as usual.

The emergence of new groups of ecologically orientated forecasters during the 1970s provoked a debate that brought these implicit assumptions out into the open. By exposing their own value assumptions, the alternative forecasters succeeded to some extent in bringing forecasting down from its pedestal of scientific objectivity. Establishment forecasters were now often forced to reveal their own value assumptions, or it at least became clear to the public that their forecasts were not neutral.

It is not necessary to revert to political ideologies to find examples of contradictory value-structures. Even a comparison of professional paradigms within the scientific community such as economics, engineering, and ecology is quite sufficient. Economic theory basically treats energy as a matter of supply and demand, where price plays a major role in determining production and consumption patterns. Future interests are decided by capital discounting, which for any realistic discount rate means that a perspective beyond seven to ten years becomes insignificant. The policy concerns of the economist are with the market balance and the efficient allocation of (some) resources through the price mechanism. In an early phase, economists tended to simplify the whole energy question by extrapolating future developments from past history.

The foremost concern of the technologist is with technical feasibility. Technological theory does not in itself contain normative standards as to which criteria to use in order to make choices between several feasible technologies. But the technology profession tends to favour complex, high-technology solutions, probably because this is a challenge to their skills. The coupling of a technological orientation to a 'green' institutional and value system and the resulting 'bottom-up' approach to energy forecasting, however, shows that the expansiveness of technological forecasts is highly dependent on the institutional context.

From an ecological perspective, the main concern of energy policy should be to provide energy solutions that fit in harmoniously with natural processes and support the variety and complex interplay of elements in natural environments. A number of the energy resources that are tapped by mankind are built up over millions of years and are, therefore, in practice non-renewable. With his ethic of maximising the stability and balance of natural systems, the ecologist typically focuses on the long-term developments and consequences.

Professional paradigms influence energy forecasting by acting as cognitive filters that, among other things, determine what variables should be included in energy modelling. As already pointed out earlier in this book, the selection of data and variables may exert a decisive influence on modelling and the types of forecasts that are produced. The lack of environmental variables and data will, for example, prevent a representation of ecological boundaries to energy development, just as the lack of socio-political variables may prevent the representation of distributional and socio-structural consequences.

The acrimony of these conflicts between professional paradigms derives from the fact that there are only meta-level arguments to resolve the issue. The formal rationality (or methodology) of forecasting—to use the terminology of Weber (1964)—cannot be justified by appeals to itself, but has to refer to a specific context of substantive rationality, to fundamental values and beliefs. At best, one could resolve the issue much later when one could compare the actual development with the fundamental assumptions behind each approach. But even this is highly problematic given the extensive involvement by forecasters in the societal development that they try to predict and the self-fulfilling and self-falsifying effects of forecasting in such situations.

From a methodological perspective, the energy crisis and the conflicts and complexity of energy forecasting in the late 1970s and early 1980s also had positive effects. It brought about a number of methodological innovations, both among establishment and alternative forecasters. Some innovations even transcended the concept of forecasting altogether and introduced new modelling paradigms, paradigms that explicitly redefined the relations between the forecaster and the social sector analysed. We shall shortly discuss three types of forecasting innovation: (1) mega-modelling; (2) scenario analysis; and (3) backcasting.

Mega-Models

One solution to the new energy forecasting situation used by some traditional forecasters with large resources was to build more complex models to represent new qualities of the social system. By this they hoped to gain greater predictive accuracy and to be able to foresee some of the socio-structural developments that had marked the energy developments of the 1970s.

This strategy, however, has at least two weaknesses. (1) Making the model more inclusive often involves the introduction of behavioural variables that are hard to measure and difficult to forecast. In so far as this also involves modelling the effects of modelling and of implementation, we have already argued that the modelling of this interaction would lead to an infinite regress of modelling complexity, and shall not repeat our argument here.

(2) In addition, this strategy is faced with methodological problems related to the working with and control of complex models and the verification of model results. With the increasing complexity of models and the technical skills required to understand their operation and to replicate the conclusions arrived at by using them, model results can hardly be checked by the scientific community (Wynne, 1984). This is an intriguing problem, because the proponents of mega-modelling usually see themselves as scientists, and present the models and their results as scientific. Yet they themselves have chosen a methodological path that ultimately undermines the possibility of scientific control.

The lack of scientific control implies that the subjective element in modelling becomes less constrained. Prejudices, values, and assumptions can be introduced unconsciously or even consciously through the estimation of growth rates, energy elasticities, technology innovation rates and so on, where these subjective factors are never explicitly spelled out. The weakening of peer and public review could lessen the pressure to document these decisions.

To the complaint of lack of scientific control, forecasters and modellers point to the extensive sensitivity analysis that is undertaken to check the robustness of their results. However, this analysis is normally undertaken within the paradigm under which the models were created. This means that some of the most important alternative assumptions are usually left out. This seems to have been the case in most of our national studies, and is well documented in the IIASA study, where, for example, highly questionable price assumptions were never questioned (see Keepin and Wynne, chapter 3).

Despite its claims to scientificity, mega-modelling therefore has several features that resemble politics. As politics, the modelling process, for example, typically has an on-stage and an off-stage version. The official version, presented to policy-makers and the public at large, generally implies that the model has been constructed on the basis of historical data and available technical and social-science knowledge, and that given the input $x, y, \ldots$ it gives the result R.

The off-stage version of actual modelling often shows that the result R has been unconsciously or consciously aimed at, and the modelling is actually an exercise in how to arrive there in a consistent way. It must be said, however, that certain modelling traditions, like system dynamics, are more open about their process than others.

Not only the IIASA study, but most of the big national econometric energy models have reached a size and level of complexity where they are vulnerable to this type of critique. In fact, as we point out in the introduction, it now requires computer programs to extract the structure of these complex models which are hardly fully understood even by the most senior insiders.

Scenario analysis

Scenario analysis and backcasting are steps in an altogether different direction. To a varying degree they both explicitly introduce normative elements into the modelling process.

By presenting a span of possible future developments, scenario analysis explicitly acknowledges the uncertainty of future developments and/or the possibility for political choice of futures. If the difference between scenarios relies on developments exogenous to the control of political decision-makers reacting to it, such as international oil prices, the step in normative direction is a small one. It was this step that most establishment forecasters took during the 1970s, often in combination with the development of mega-models. This way of modelling implies that the political decision-maker, and not the forecaster, has to choose among uncertain developments of important exogenous variables.

On the other hand, the variety of scenarios may build upon alternative policy courses that are available for political realization, such as the introduction of energy savings, the choice of decentralized versus centralized technologies and social structures, and so on. The scenario analysis in this case acquires a more normative character and calls for a political debate where one scenario is selected because of its social acceptability. In this way scenarios illustrate the consequences of different political options, which society can then choose to pursue or reject.

Backcasting

Backcasting takes one more step down the normative path, and it also breaks with the forecasting terminology. This modelling procedure was popularised by Amory Lovins, who linked it to the concept of scenario. As indicated by Greenberger and Hogan in chapter 11, Lovins used the term 'scenario' in the specific sense adopted by the film industry to mean a description sufficiently vivid for an audience to be able to imagine itself participating in the events described. The scenarios consisted of fundamental attitudes toward energy, such as centrism, autarchy, vulnerability, and technocracy—root issues, he charged, that establishment studies tended to neglect. The term backcasting arises from the method of working backward from a desired future described in a scenario to the present, thereby attempting to design an internally consistent path for societal development between the two points.

Traditional scenario analysis starts out with alternative policy sets and then calculates their implications for future energy developments. Backcasting, instead, first determines desirable, a priori, feasible future states, and then attempts to identify sets of policy measures that will get us from here to there.

By focusing primarily on possible future states, so far away in time that major societal transformation is possible, backcasting emphasizes the normative aspects and societal choices open to political decision. In this way it shifts the task of social science modelling away from prediction and over towards the choice paradigm advocated by Marcuse and Habermas among others. This shift is already contained in the policy-based version of scenario analysis.

This shift also contains a reinterpretation of the relation between forecaster and society. Energy modelling is now redefined as an element of a decision support system. This avoids some of the problems embedded in traditional forecasting and mega-modelling, such as the problem of verification and of infinite regress mentioned in the first part of this chapter. But by doing this the new approaches face a new set of problems. These problems have to do with the institutionalization of democratic procedures to define the scope and results of modelling.

FORECASTING, DEMOCRACY, AND THE LEGITIMATION OF INSTITUTIONAL POWER

The institutionalization and democratic control of forecasting becomes an important issue once we accept that modelling and forecasting are not only a prediction of future developments, but a central element in the shaping of institutional structures that support specific energy developments.

The model monopoly held by traditional experts in energy forecasting was broken during the 1970s. The clash between experts and counter-experts revealed the weak scientific nature of forecasts that presented themselves as unavoidable facts. On the one hand, this strengthened the role of counter-experts, but at the same time weakened the expert role in general. Policy-based analysis provided opportunities for non-experts to move into energy forecasting and to contribute to energy policy formulation even at the forecasting stage.

This development was made possible only because the relation between modelling and the implementation system and between the modelling/implementation system and society as a whole underwent a change. Since forecasts could no longer be legitimated by reference to purely scientific norms, they had to be legitimated politically. This, however, focused the spotlight of public attention on the interest aspects in forecasting. And it also tended to highlight the role of the implementation system to which the forecaster was linked.

The questions were no longer, 'Which energy development will take place, and what estimation techniques should be used to capture it?', but rather, 'Whose interests does this forecast serve, and who is behind it?' Energy forecasting as an activity has in other words developed from an expert to a negotiation model. Models are no longer used to detect what the future will be, but to negotiate about the future.

One of the institutional implications of this negotiation of the future is that it is now difficult to link expert-controlled forecasting models to hierarchically organized implementation systems. Instead, energy forecasts tend to be produced by several modelling groups, and implementation takes place through loosely coupled systems.

The criterion for successful modelling is no longer so much to hit the correct future, but to reach reasonable compromises between affected interests. Our cases show that a good deal of forecast harmonization took place in parliamentary and/or public debates and hearings, or came about through the less dramatic competition between bureaucracies.

In this new role, modelling and forecasting can obviously be evaluated quite legitimately by some of the same standards that we customarily apply to politics. This means asking questions such as:

(1) Are all relevant societal interests well represented?
(2) Are the basic assumptions made clear and open to public debate?
(3) How fair and open are the procedures for model and forecast selection?
(4) How impartial is the implementation system when it comes to implementing alternative energy futures?

Taking these types of questions about modelling and forecasting seriously may in fact be crucial to political democracy, because these tools have come to play an important role in legitimating if not guiding political decision-making.

The expanding role of long-term planning will increase the number of premisses for future development that are laid down before decisions about concrete projects are taken. Actual power and decision-making authority are thereby to a certain extent displaced from the level of concrete projects to planning for long-term development. Without questioning the role of expertise in long-term planning, the political system, which still seems to be more geared to dealing with short term events and single cases, risks losing out to technocrats.

In this sense, the energy debate with its politicization of modelling and forecasting has contributed to a much broader political theme, which also concerns other societal sectors. The long-term effects of the politicization of energy forecasting, however, depend on the permanent institutionalization of democratic modelling procedures. Unfortunately, there are strong forces against this. First, from a technocratic point of view, a democratization of long-term planning means that the planners would have to deal with greater complexity and unpredictability in the planning process. Planning would increasingly involve explicit negotiation, rhetoric, and public legitimation. Such processes would tend to undermine the autonomy of professionals, the scope for implicit negotiations, and the exercise of control by the decision-making élites. Second, most technocratic forecasting techniques tend to project existing social structures and technologies into the future. This outcome tends to be more palatable from an establishment point of view than the unpredictable normative alternatives. Third, traditional forecasts also fit in better with bureaucratic routines and planning procedures. From an administrative point of view, they are therefore preferable to normative public debates as inputs to the policy process. Forecasts are more foreseeable and more under control than public debates, particularly if the issues cut across traditional, political conflict dimensions.

This means that the institutionalization of democratic modelling and forecasting is brittle. It probably requires a political definition of at least a sectoral crisis important enough to create public attention. Nevertheless, the energy debate and the politics of energy forecasting have provided an important example of how to synchronize technocratic needs for long-term planning and democratic needs for public participation. This synchronization may be painful both in democratic and technocratic terms. We nevertheless believe it to be essential for achieving a balanced societal development.

Furthermore, the methodological innovations in energy forecasting have, in our opinion, brought social science modelling to a more realistic acknowledgment of its limitations.

Notes

1. Standard references are Marx (1970), Weber (1964), Merton (1948), Maruyama (1963), and, more recently, Henshel (1982).
2. Particle physics is also undermining this perspective in the natural sciences.
3. Meadows *et al.* (1972) were early users of this effect, unfortunately without being clear on this purpose of their study. Failure to make clear that they were not forecasting in the usual sense, and therefore could legitimately use another methodology, may well have been responsible for a lot of misunderstanding and acrimony in the ensuing debate about the limits-to-growth and the use of system dynamics.
4. The IIASA world modelling exercise is a case in point.

Bibliography for Chapters 12 and 13

Freud, S. (1917), *Vorlesungen zur Einführung in die Psychoanalyse.* Vienna.

Habermas, J. (1968a), *Erkentnis und Interesse*. Frankfurt.

—— (1968b), *Antworten auf Herbert Marcuse.* Frankfurt.

Häfele, W. (1981), *Energy in a Finite World: A Global Systems Analysis*, vols. i and ii. Cambridge, Mass.: Ballinger.

Henshel, R. L. (1982), 'The Boundary of Self-Fulfilling Prophecy and the Dilemma of Prediction'. *British Journal of Sociology*, 33(4).

Marcuse, H. (1964), *Negations, Essays in Critical Theory.* Boston: Beacon Press.

Marx, K. (1970), *The Capital.* Moscow: Progress Publishers.

Maruyama, M. (1963), 'The Second Cybernetics: Deviation Amplifying Mutual Causal Processes'. *The American Scientist*, 51.

Meadows, D. H., D. L. Meadows, J. Randers, & W. W. Behrens, (1972), *Limits to Growth*. London.

Merton, R. (1948), 'The Self-Fulfilling Prophecy'. *Antioch Review*, 8.

Randers, J. (1984), 'Praktisk prognosearbeid'. *Working Paper No. 84/7*. Norwegian School of Management.

Schneider, J. B. (1975), 'The Self-Fulfilment of Long-Range Transportation Forecasts'. *Traffic Quarterly*, 29.

Weber, M. (1964), *The Theory of Social and Economic Organization*. New York: Macmillan.

Wynne, B. (1984), 'The Institutional Context of Science, Models and Policy: The IIASA Energy Study'. *Policy Sciences*, 17: 277–320.

Name Index

Abma, E., 93
Albrechtsen, E. H., 20
Altner, G., 75, 78, 79, 80, 82
Andersen, F. M., 176
Andersen, H., 176
Araj, K. J., 55
Armstrong, J. S., 16, 26
Atz, H., 81
Anken, M., 178
Aushubel, J., 52

Bacharach, P., 13
Ball, J., 3–4
Baratz, P., 13
Basile, P., 36, 55
Baumgartner, T., 9, 151
Bazelon J. D., 54
Bech, O., 164, 176
Becht, H. Y., 106
Belaire, F., 223
Bell, T., 231, 233
Benn, T., 117–18, 125, 127, 129
Berger, P. L., 13
Bergougnoux, J., 191, 197
Bjerkholt, O., 138
Bohr, N., 247
Boiteux, M., 185, 200, 202, 203
Bons, C. P., 105
Bott, R., 228
Boutillier, M., 9
Brookes, L. G., 122
Brooks, D. B., 228, 276
Brown, G. E. Jr., 54
Bråten, S., 13
Buckley, W., 115, 129
Bunyard, P., 120
Burn, T., 113, 114
Burns, T. R., 9, 13, 15–16

Carter, J., 248, 251, 252, 274
Cassette-Carry, M., 204
Chadwick, M. J., 54
Chapman, P. F., 91, 116, 118–19, 129
Chateau, B., 187, 195–6, 202
Chesshire, J., 115, 121, 129
Chomsky, N., 204
Christakis, A. N., 25
Clark, L. H., 241
Conrad, J., 77, 82
Cook, P. L., 114, 115, 129
Coyne, P., 111
Criqui, P., 196
Crowe, R., 228

de Bruyne, D., 91
Decelle, A., 203
de Gaulle, Ch., 181, 183
Dejou, A., 202
de Man, R., 103, 104
Destival, C., 202
Dethomas, B., 193
DeVillé, Ph., 9
Dickler, R. A., 61, 62, 63
Dietrich, O., 165, 167, 170–2, 177, 178
Driehuis, W., 106
Dupperin, C. M., 203
Dupree, W. G., 262
Dupuy, G., 195
Dybkjaer, L., 178

Ek, A., 143
Erdmann, R., 228

Feoktistov, L. P., 54
Feyerabend, P., 199
Finon, P., 202, 203
Ford, A., 67
Ford, G., 246
Foster, P., 233
Freeman, D. S., 245
Freud, S., 291
Frisch, J. R., 203
Fulkerson, W., 55
Furet, F., 181

Gander, J., 223
Garnåsjordet, P. A., 143
Gass, S. I., 54
Gershuny, J., 26
Gibrat, R., 204
Gilli, M., 3
Goldman, A. J., 53
Gorz, A., 198
Gouni, L., 202
Greenberger, M., 26, 54, 241–4, 253, 254, 261, 294, 304
Greenhalgh, G., 54

Habermas, J., 201, 290, 291, 305
Häfele, W., 5, 34–5, 37, 39, 44, 47–9, 51–5, 277–8
Haley, K. B., 55
Helliwell, J., 172
Henshel, R. L., 293, 307
Hervik, A., 143
Hitch, Ch. J., 243, 257
Hogan, W. W., 242, 257, 259, 294, 304
Hooker, C. A., 228, 231, 233
Hucting, R., 92, 95, 96, 98
Huisman, H., 91
Hutber, F., 122

Inglehart, R., 145

Jaeger, J., 54
Johnson, L., 243–4
Jonas, P., 122

Kalecki, M., 201, 204
Keck, O., 82
Keepin, B., 38, 43, 47–9, 52–3, 55, 277, 303
Kemeny, J. C., 246
Knutsen, O., 145
Kononov, Y., 37
Krause, F., 67, 71, 72, 78
Kristensen, O. P., 176
Kuzmin, I. I., 54

Landsberg, H. H., 250
Lawson, N., 125
Laxer, J., 232, 233
Leach, G., 91, 111, 116, 118, 120–5, 128–9
Lecomber, R., 123
Legasov, V. A., 54
Lehmbruch, G., 13
Leyral, R., 202
Lijphart, A., 107
Lindman, H., 54
Long, T., 54
Longva, S., 143, 145
Lovins, A., 55, 91, 124, 227–8, 247–9, 260, 274, 276, 304
Lucas, A., 231, 233
Lucas, J. D. N., 202
Lunde, T., 151

McDonald, A., 47
McDougall, J. N., 233
Malabre, A. L., 241
Marcuse, H., 291, 305
Marshall, E., 253
Marshall, W., 116
Maruyama, M., 307
Marx, K., 307
Massé, P., 182
Meadows, D., 5, 36, 55, 254, 307
Mendershausen, H., 202
Merton, R., 307
Mesarovicz, M., 71
Messmer, P., 185, 196, 202
Meyer-Abich, K.-M., 61–3
Miall, H., 126
Miller, M. M., 55
Monnier, E., 202
Morthorst, P. E., 176

Nadel, M. V., 13
Nelkin, D., 202
Nilsson, P., 164, 169
Nixon, R., 245
Nordhaus, W. D., 52, 55
Nordstrand, R., 177

Odell, P. R., 5, 88, 107
Offe, C., 105
Olivier, D., 126
Olson, J. P., 13

Packer, G. S., 54
Parry, A. M., 55
Patterson, W. C., 106, 115, 118, 122
Péan, P., 203
Pearce, D., 117–19, 124
Pestel, E., 71
Peters, H. R., 81
Pollack, M., 202
Por, A., 37
Posner, M. V., 114
Potma, Th. G., 91, 106
Price, S., 123
Prigogine, I., 198
Puiseux, L., 187, 199, 201–3

Radkau, J., 63, 82
Randers, J., 279
Rasmussen, N., 246, 260
Reichel, P., 81
Richels, R., 242
Ritschard, G., 3
Rivlin, A., 244
Roberts, F. S., 54
Robic, A., 202
Robinson, J. B., 228, 233, 276
Rogner, H. H., 48, 53
Rolighed, A., 176
Roosevelt, F. D., 243
Rose, D. J., 55
Rosing, K. E., 5
Rotty, R. M., 54
Rourke, F. E., 13
Rudzinski, K., 62
Ruske, B., 63
Ryan, C. J., 25

Sachs, I., 204
Sartre, J.-P., 199
Sassin, W., 47, 54, 55
Saumon, D., 187, 202, 203
Schiffer, H. W., 81
Schmitt, D., 81
Schmitter, P. C., 13
Schneider, J. B., 293
Schrattenholzer, L., 37, 55
Schulze, D., 19
Schurr, S. H., 25, 249
Serini, J.-P., 203
Simmonot, P., 203
Slesser, M., 129
Spencer, K., 126
Starr, Ch., 54
Stengers, I., 198
Stobaugh, R., 16, 141, 249
Suleiman, E. N., 181
Surrey, A. J., 114, 115, 121, 129
Sweeney, J. L., 249

Tempest, P., 125
Teufel, D., 63
Thompson, M., 3, 11, 81, 203
Togsverd, T., 177
Touraine, A., 202

van Dellen, H., 106
van den Beld, C. A., 86
van den Hoeven, E., 98
van der Sluijs, H., 91
van Dieren, W., 106
van Hulst, R., 231

Weber, M., 302, 307
Weiss, B., 178
Wenk, E. jun., 26
Wessels, T., 81
West, J. A., 262
Weyant, J. P., 249
Williams, R., 113, 117, 129
Wilson, C., 248
Winstanley, G., 228
Wynne, B., 49, 53, 277, 303

Youngblood, A., 67
Yergin, D., 249

Subject Index

actors 11, 13, 19, 232
  British 114ff
  Danish 164–5
  Dutch 87ff
  French 187
  German 63ff
  Norwegian 141ff
  US 244ff
alternative forecasts 7, 26, 71ff, 111ff, 118f, 156, 173–4, 188, 226, 227ff, 230, 247–8, 252, 269–270
analogue models 188–9
art of modelling and forecasting 3–4, 53
assumptions 5, 11, 16–17, 38–9, 42–4, 47–8, 122, 146, 157, 161, 167, 168ff, 200, 216–17, 231, 242, 255, 277–8, 301, 303 *see also* economic growth, growth, elasticities, prices and costs
  acceptability of 94
  about environment 34
  estimates of 48
  plausibility of 100
  specification of 36, 95–97, 100, 105, 144, 155, 168ff, 272
  about technical constraints 54, 169–170, 216

backcasting 228, 230–2, 270, 274, 304–5 *see also* end-use models

churches 93
coal 41–4, 45–6, 49–50, 62, 81, 113
cognition 7, 11, 18, 145, 269, 301–2
comparison of forecasts 95, 100, 103, 105, 128, 169, 249, 257
consensus formation 101–5, 125–6, 157, 183, 261, 269–70
conservation of energy 63, 73, 79–80, 90–2, 119, 120, 122, 131, 141, 148, 161–2, 175, 185–6, 190, 195, 218, 223, 227ff, 230, 246–7, 249, 254–8, 273–4, 276, 281
corporatism 13, 102–3, 127–8, 286
costs of energy 49, 197, 200, 246
counter-modelling 242
credibility
  of models 242
  of forecasts 38, 67, 124

data selection 13–18, 120, 268, 302
decision-making 13–14, 20–5, 103, 105, 267–8, 273–4, 281, 306–7
democracy and forecasting 8, 13, 26, 200, 267–8, 305–7
demographic developments 53, 69, 140, 291
documentation of models 52–3

econometric models 3, 16–17, 19, 75, 85–7, 95–6, 103, 116, 119–20, 127–8, 142–3, 148, 161, 163, 166, 168, 173–4, 189, 216, 218, 222, 226, 229–30, 241, 257, 261, 271, 276, 303
economic growth 63, 69, 72, 78–9, 91, 113, 126, 140, 160, 183, 192–3, 196, 253–4, 280–1, 284, 286
economics/economists 15, 18–19, 25, 86, 90, 150–1, 169, 185, 243, 258, 271, 280, 285, 301
economy–energy interaction 36–7, 69, 89, 97, 121, 160, 195, 257–8
elasticities (prices, income) 135, 144–5, 190–9, 196–7, 209, 253–5, 303
electricity 44–7, 62, 69, 70, 81, 88, 113, 116ff, 137ff, 155ff, 161ff, 174, 180ff, 192ff, 212
élites 8, 22, 181–5
end-use models 121, 126, 143, 218, 222, 228ff, 270, 276, 301 *see also* backcasting
energy coefficients 15, 142, 155, 157, 160, 169–70, 172
energy developments
  in Canada 213ff
  in Denmark 157–9
  in France 181ff
  in Germany 62–3
  in the Netherlands 86–7
  in Norway 137–8
  in the UK 112ff
energy forecasts 66–7, 70–1, 78–80, 120ff, 169 *see also* low-energy forecasts, establishment and alternative forecasts
energy independence 184, 223, 224, 246, 258
energy inquiries
  administrative bodies: Canada 213ff; Denmark 163ff; Norway 140, 143–5, 149
  public procedures: Canada 213ff; France 186, 200; Germany (Enquête Commission) 62, 64, 75ff, 269, 285; Netherlands 87, 93ff, 105, 269; UK Sizewell 116; UK Windscale 93, 118, 120, 123, 127, 269, 285
energy-intensive industries 120, 137, 141, 146
energy policy
  in Canada 208–9, 229–33
  in Denmark 158
  in France 181–6, 201, 274
  in Germany 61, 73–4, 81
  in the Netherlands 87ff
  in Norway 274
  in the UK 112ff, 125
energy services 73 *see also* end-use models
engineers 181–2, 198–9, 243, 258, 285, 301
environmental movements and values 5, 16, 20,

62, 72–3, 85, 89, 93, 95–6, 98–9, 101, 103, 105, 124, 137, 141, 145–6, 150–1, 173–4, 187, 191, 196, 202, 218, 244–6, 250, 267–71, 276, 284, 301
establishment forecasts 64ff, 111ff, 116, 118ff, 124, 156, 193ff, 213ff, 220ff.
extrapolation method 87–8, 90, 127, 142, 160, 188–9, 214, 276, 295, 301

fast breeder reactors 45–7, 49–50, 53, 55, 76ff, 197, 247–8, 277
feedback in models 25, 35, 37, 51
financing of modelling and forecasting 4, 34, 92, 118, 187, 229, 246
forecast compromise 167, 171–2 *see also* consensus, negotiation
forecaster
as informant 293–5
as observer 292–3
forecasting in
Canada 28, 213ff, 268, 275–7, 285–6, 294–5
Denmark 16, 28, 157ff, 268, 271–2, 276, 284, 286, 299
France 22, 28, 186ff, 268, 272–3, 280, 284–6, 295–6, 300
Germany 7, 22, 26–8, 61–2, 63ff, 127, 268–70, 272, 276, 279–80, 284–5, 294, 300
the Netherlands 7, 22, 26–8, 85ff, 127–8, 268–70, 272, 276, 284, 286, 294, 299–300
Norway 22, 28, 137ff, 268, 271–2, 276, 279, 284–5, 299–300
the UK 7, 26–8, 110ff, 268–70, 272, 284–5, 294, 298, 300
the US 22, 28, 244ff, 268–9, 273–5, 284, 286, 294, 296, 300
forecasts as policy statements 73–4, 224
formal models 33–4, 51–2, 188–9

government and administrations 73ff, 165ff, 170ff, 186ff, 208–9, 220, 244ff, 260–1, 271–2, 273–7.
growth assumptions 5–7, 91, 195, 279–80, 303 *see also* economic growth

heating market 70, 158, 195
hydropower 114, 137

IIASA world energy model 3–6, 27, 34ff, 269, 277–8, 303
impact of models and forecasts 250ff, 259–60
industrial interests 98, 103, 146, 159, 173, 215, 225, 271
input–output models 37, 229
institutions 8, 11, 38, 145ff, 286
international power 258

judgemental adjustments 33–4, 37–8, 52, 141, 214, 218, 221, 249–50, 261

lawyers 58
legitimation 12, 14, 22–3, 27, 74, 105, 268, 270, 278, 281, 290
lifestyle changes 92, 281
linear programming 36, 49
load curve 72, 158, 195
low-energy forecasts 7, 26, 71ff, 78–9, 89–92, 98ff, 111, 118, 227–9 *see also* alternative forecasts

market-place of ideas 242, 261, 268–9, 277, 284
market shares 113–15, 157–9, 190, 222
mega-models 3, 34, 52, 302–4
methodology 13–18, 80, 89, 117, 119, 121–2, 127, 144, 157, 160–3, 165ff, 188–9, 199–201, 203, 221, 230, 253ff, 261, 268, 270–4, 290–1, 302
model complexity 38, 90–1, 269, 277, 294, 302
model consistency 51
modelling competition 298–300
modelling
the modelling process 8, 11
of self-reactive systems 290ff
of social systems 19
model monopoly 86, 128, 142, 151–2, 173, 268, 281, 298–300, 305
model paradigm 22
model publication 52, 100
multi-level modelling 9 *see also* modelling of self-reactive systems

natural gas 41–4, 88, 113ff, 208, 213ff, 247
negotiation of forecast results 13, 89, 107, 127, 163ff, 306 *see also* energy inquiries
normative forecasting 190, 221, 224, 230–3, 273, 276, 304–5
norms 11
nuclear power 75ff, 81, 87ff, 92–3, 99–102, 113, 114ff, 157–9, 180, 183–6, 188, 193–5, 197, 246, 269, 273, 296

oil 39–41, 61, 157, 183, 185, 208, 215ff, 224, 248
oil companies 67–9, 91, 99, 215
oil prices 7, 9, 105, 157
oil substitution 63, 157
organizations 18, 19–20, 145ff, 231, 268, 281–6

parameter estimation 15, 18
planning 19, 68, 70, 137–8, 157, 175, 182, 200–1, 226, 267–8, 306
centralized 22, 85–6, 91–5, 128, 142–4, 148, 154, 169, 187, 285–6, 300
policy analysis 33, 241–4, 258–62, 295–8
policy bias 6, 141, 144
policy implementation 19, 221, 259, 296–300, 305–6
policy modelling 54, 225, 273
political culture 11, 102
politics of forecasting 8, 13, 45, 102ff, 267, 271, 281–6
politics of modelling 4
power plant size 64–7, 158–9
PPBS 243
predictive forecasts 73–4, 216, 229, 230–2

prices 69, 148, 151, 157, 159, 168, 176, 203, 222–3, 258, 272, 275, 281, 303
professions 6, 14, 18–9
psychoanalysis 291

quantification 13

regulatory agencies 211ff, 229, 271–7
renewable energy 78, 91, 124, 158, 185–6, 227ff, 230
research institutes 33ff, 74–5, 173, 244ff, 260–1
robustness of forecasts 35, 48–50, 54

sales strategy 182, 184, 190, 193, 194–5, 268, 273, 296
scenario analysis 34, 39, 41ff, 67, 71–2, 78–80, 89, 91–3, 94ff, 103–5, 116, 120ff, 123–4, 128, 162, 168, 190–1, 222–3, 227ff, 231, 247–8, 255, 261, 273–4, 303–6
scientific objectivity 4, 6, 11, 34, 141, 144, 151–2, 191, 200, 262, 268, 270, 278, 281, 290, 301, 303
scientific rationality 182
sectoral models 69, 120, 122, 142, 155, 161ff, 171, 174, 191, 221–2, 257
self-fulfilling consequences 15, 217, 232, 268
self-fulfilling prophecies 19, 159, 173, 229, 272–3, 290ff
self-transformation 13
social goals 5, 258
social structure 25–6
solar energy 73, 249, 277
soft energy paths *see* alternative and low-energy forecasts
supply and demand 35–6, 69, 115, 119, 123, 125–6, 137, 141, 151, 158–9, 171, 182, 184–6, 190, 195, 197, 212, 213ff, 221ff, 227–8, 230, 253, 258, 273, 275–6, 294–6
synthetic oil 246–7, 277
system dynamics 11, 53, 71, 303

technical models 116, 121, 127, 143, 161–2, 166, 174 *see also* backcasting, end-use models
technocracy 8, 11–13, 27, 90, 127–8, 182, 188, 190, 198–9, 201, 204, 267–8, 272–3, 306
technology 34, 76
time horizon 4, 16, 64, 89, 115–16, 187, 203
trade unions 93, 118, 185, 190

uniformity of forecasts 170–1
uranium 41–4
utilities 64–7, 142–3, 146–7, 155–7, 165ff, 170ff, 184ff, 201, 271, 273–4

values 5, 7

zero-growth forecasts *see* low-energy and alternative forecasts